AF682914

Boundary Elements
and
Other Mesh Reduction Methods
XXVIII

WITPRESS

WIT Press publishes leading books in Science and Technology.
Visit our website for the current list of titles.
www.witpress.com

WITeLibrary

Home of the Transactions of the Wessex Institute.
Papers presented at BEM/MRM XXVIII are archived in the WIT elibrary in volume 42 of WIT Transactions on Modelling and Simulation (ISSN 1743-355X).
The WIT electronic-library provides the international scientific community with immediate and permanent access to individual papers presented at WIT conferences.
http://library.witpress.com

Transactions Editor

Carlos Brebbia
Wessex Institute of Technology
Ashurst Lodge, Ashurst
Southampton SO40 7AA, UK

Email: carlos@wessex.ac.uk

WIT Transactions on Modelling and Simulation

Editorial Board

C Alessandri
Universita di Ferrara
Italy

M A Atherton
South Bank University
UK

J Baish
Bucknell University
USA

C D Bertram
The University of New
South Wales
Australia

D E Beskos
University of Patras
Greece

M Bonnet
Ecole Polytechnique
France

J A Bryant
University of Exeter
UK

M B Bush
The University of Western
Australia
Australia

M A Celia
Princeton University
USA

A H-D Cheng
University of Mississippi
USA

J J Connor
Massachusetts Institute
of Technology
USA

D E Cormack
University of Toronto
Canada

D F Cutler
Royal Botanic Gardens
UK

E R de Arantes e Oliveira
Insituto Superior Tecnico
Portugal

G De Mey
Ghent State University
Belgium

J Dominguez
University of Seville
Spain

Q H Du
Tsinghua University
China

S Elghobashi
University of California
Irvine
USA

A El-Zafrany
Cranfield University
UK

P Fedelinski
Silesian Technical
University
Poland

S Finger
Carnegie Mellon University
USA

J I Frankel
University of Tennessee
USA

M J Fritzler
University of Calgary
Canada

L Gaul
Universitat Stuttgart
Germany

G S Gipson
Oklahoma State University
USA

S Grilli
University of Rhode Island
USA

K Hayami
National Institute of Informatics
Japan

J A C Humphrey
Bucknell University
USA

D B Ingham
The University of Leeds
UK

N Kamiya
Nagoya University
Japan

D L Karabalis
University of Patras
Greece

J T Katsikadelis
National Technical
University of Athens
Greece

H Lui
State Seismological Bureau Harbin
China

W J Mansur
COPPE/UFRJ
Brazil

R A Meric
Research Institute for Basic Sciences
Turkey

J Mikielewicz
Polish Academy of Sciences
Poland

K Onishi
Ibaraki University
Japan

E L Ortiz
Imperial College London
UK

M Predeleanu
University Paris VI
France

D Qinghua
Tsinghua University
China

S Rinaldi
Politecnico di Milano
Italy

T J Rudolphi
Iowa State University
USA

G Schmid
Ruhr-Universitat Bochum
Germany

A P S Selvadurai
McGill University
Canada

X Shixiong
Fudan University
China

P Skerget
University of Maribor
Slovenia

V Sladek
Slovak Academy of Sciences
Slovakia

T Speck
Albert-Ludwigs-Universitaet
Freiburg
Germany

J Stasiek
Technical University of Gdansk
Poland

S Syngellakis
University of Southampton
UK

M Tanaka
Shinshu University
Japan

N Tosaka
Nihon University
Japan

T Tran-Cong
University of Southern Queensland
Australia

W S Venturini
University of Sao Paulo
Brazil

J F V Vincent
The University of Bath
UK

J R Whiteman
Brunel University
UK

Z-Y Yan
Peking University
China

K Yoshizato
Hiroshima University
Japan

G Zharkova
Institute of Theoretical and Applied Mechanics
Russia

TWENTY-EIGHTH WORLD CONFERENCE ON
BOUNDARY ELEMENTS AND
OTHER MESH REDUCTION METHODS

BEM/MRM XXVIII

Conference Chairmen

C. A. Brebbia
Wessex Institute of Technology, UK

J. T. Katsikadelis
National Technical University of Athens, Greece

International Scientific Advisory Committee

C Alessandri
D E Beskos
M Bonnet
P Broz
M Bush
C-S Chen
A H-D Cheng
T G Davies
A J Davies
G De Mey
V G DeGiorgi
J Dominguez
A El-Zafrany
G Fasshauer
J Frankel
L Gaul
G S Gipson
K Hayami
Y C Hon
M Hribersek
M S Ingber
D B Ingham
M A Jaswon
M Kanoh
A J Kassab
E Kita
G Kuhn
V Leitao
G D Manolis
W J Mansur
J C Miranda Valenzuela
K H Muci-Kuchler
Y Ochiai
K Onishi
D Poljak
V Popov
H Power
M Predeleanu
J J Rencis
V Roje
T J Rudolphi
G Rus Carlborg
B Sarler
E Schnack
A P S Selvadurai
X Shu
L Skerget
V Sladek
S Syngellakis
A Tadeu
M Tanaka
N Tosaka
T Tran-Cong
W S Venturini
O von Estorff
T Wu
S-P Zhu

Organised by

Wessex Institute of Technology, UK.

Sponsored by

International Journal of Engineering Analysis with Boundary Elements
(EABE)

Boundary Elements and Other Mesh Reduction Methods XXVIII

Editors

C. A. Brebbia
Wessex Institute of Technology, UK

J. T. Katsikadelis
National Technical University of Athens, Greece

WIT*PRESS* Southampton, Boston

C. A. Brebbia
Wessex Institute of Technology, UK

J. T. Katsikadelis
National Technical University of Athens, Greece

WIT Press
Ashurst Lodge, Ashurst, Southampton, SO40 7AA, UK
Tel: 44 (0) 238 029 3223; Fax: 44 (0) 238 029 2853
E-Mail: witpress@witpress.com
http://www.witpress.com

For USA, Canada and Mexico

Computational Mechanics Inc
25 Bridge Street, Billerica, MA 01821, USA
Tel: 978 667 5841; Fax: 978 667 7582
E-Mail: infousa@witpress.com
http://www.witpress.com

British Library Cataloguing-in-Publication Data

A Catalogue record for this book is available
from the British Library

ISBN: 1-84564-164-7
ISSN: 1746-4064 (print)
ISSN: 1743-355X (on-line)

The texts of the papers in this volume were set individually by the authors or under their supervision. Only minor corrections to th: text may have been carried out by the publisher.

No responsibility is assumed by the Publisher, the Editors and Authors for any injury and/or damage to persons or property as a matter of products liability, negligence or otherwise, or from any use or operation of any methods, products, instructions or ideas contained in the material herein.

© WIT Press 2006.
Printed in Great Britain by Athenaeum Press Ltd., Gateshead.

All rights reserved. No part of this publication may be reproduced, stored in a retrieval system, or transmitted in any form or by any means, electronic, mechanical, photocopying, recording, or otherwise, without the prior written permission of the Publisher.

Preface

The present volume contains the edited proceedings of the 28th World Conference on Boundary Elements and other Mesh Reduction Methods (BEM/MRM28) held at the beautiful island of Skiathos, Greece, in May 2006. This is the second conference of this series organized in Greece. The first (BEM23) took place in the island Lemnos in May 2001.

This series of international conferences, organized annually since 1978, with the collaboration of distinguished engineers and scientists at various places of the world, has served as the established forum for the advancement of the Boundary Element Method. It has attracted innovative contributions in the areas of fundamental principles, theoretical, computational and algorithmic aspects of the method as well as advanced applications to small and large scale engineering problems of the real world. All leading researchers in the five continents of the world, involved in the development of the BEM, have attended more than a few of these meetings and have presented original papers, which have helped to develop the BEM into a powerful modern computational tool for solving problems of engineering praxis. It is not an exaggeration to say that the proceedings of these BEM conferences together with the International Journal of Engineering Analysis with Boundary Elements constitute the BEM Treasure Chest of the BEM community.

The development of mesh reduction methods to simplify the computational techniques has attracted the attention of the community. As the BEM is inherently a mesh reduction method, the world conference has widened its scope to encompass these methods, hence the new name BEM/MRM. This extension has added vitality to these international scientific meetings by attracting new promising researchers in the field.

The present volume captures the results of excellent BEM/MRM work carried out by researchers from various parts of the world. It covers advanced formulations, advanced structural applications, damage and fracture mechanics, dynamics and vibrations, fluid flow, heat and mass transfer, electrical and electromagnetic problems and computational techniques.

The editors would like to express their gratitude to the members of the

International Scientific Advisory Committee and to other colleagues, for their assiduous review of abstracts and follow-up papers included in this book. Their diligent work has ensured the high quality of this volume. Finally, the authors are to be commended for the excellent contributions, which advance the area of computational methods and ensure the longevity of the BEM after nearly 30 years and present the MRM as the computational tool of the 21st century

The editors
Skiathos, 2006

Contents

Section 1: Advanced formulations

Section 2: Advanced structural applications

Section 3: Heat and mass transfer

Section 4: Electrical engineering and electromagnetics

Section 5: Fluid flow

Section 6: Computational techniques

Section 7: Dynamics and vibrations

Section 8: Damage fracture and mechanics

Section 1
Advanced formulations

Explicit formulations for advanced Green's functions with built-in boundaries

G. S. Gipson[1] & B. W. Yeigh[2]
[1]School of Civil and Environmental Engineering, Oklahoma State University, USA
[2]Saint Louis University, USA

Abstract

This paper amplifies upon a previously presented BEM formulation where the two-dimensional logarithmic fundamental solution is transformed so as to automatically accommodate rectangular boundaries with fixed boundary conditions. Explicit derivations are presented using conformal mapping. Computational examples and comparisons with the standard procedure illustrate the advantages of the method.
Keywords: Green's function, fundamental solution, boundary elements, conformal mapping, explicit formulations.

1 Introduction

In 1986, Gipson et al. [1] presented boundary element results for phreatic surface and subsurface flow using an advanced Green's function that inherently accounted for certain boundary conditions common to such analyses. Due to the nature of that presentation and space limitations in the proceedings, the details of the Green's function derivation were relegated to a reference in what has since become a difficult-to-obtain technical report [2]. In the years since the publication [1], there have been numerous requests made of the original authors to provide more substantive details of how the advanced Green's function was obtained. Also during this time, the global scope of boundary element technology has been expanded to more directly embrace the meshless methodology, which was a primary theme in the original work. This paper represents an attempt to fill the gap in the archival literature left by the omission

WIT Transactions on Modelling and Simulation, Vol 42, © 2006 WIT Press
www.witpress.com, ISSN 1743-355X (on-line)
doi:10.2495/BE06001

of the derivation, and also demonstrates how the technique can be used to accommodate other similar boundary value problems.

2 Background

Fixed rectangular boundaries with no-flux, no temperature, no displacement, etc. type boundary conditions recur frequently in practical engineering analyses. The Laplace and Poisson equations describe a wide variety of steady-state physical problems including heat transfer, electrostatics, and groundwater seepage. Terminology applicable to this latter physical application will be used here since recurrent rectangular boundaries occur so frequently in geotechnical analysis of soil-structure interaction. Examples include sheetpiles, dam structures, and U-locks. Boundary elements, because of its boundary data nature, is a natural choice for analyzing soil seepage phenomena. The domains, which theoretically are infinite half-spaces, are rarely represented as such in any analysis technique. Particularly, in finite elements, finite differences, and boundary elements, the standard procedure is to extend the mesh of analysis a *goodly distance* (using the vernacular of St. Venant) away from the region of prime interest and make an assumption about the nature of the physical phenomenon *far away*. Usually, the geotechnical engineer assumes that there is a horizontal impermeable boundary at some distance below the surface of the earth as well as vertical impermeable planes sufficiently far downstream and upstream from the region of analysis. Finite elements and finite differences both suffer from the drawback that representing the *far away* approximation typically involves the input of considerably more nodes and consequently more unknowns into the region. The size and cost of the analysis grow geometrically with the improvement in these approximate boundary conditions.

This does not happen with boundary elements. This technique is inherently superior to finite elements and finite differences in that far fewer equations and discretization effort are needed to represent the problem. Only the boundaries of the region are involved in the solution.

However, the modeling efficiency of the entire process can be improved even further, and this aspect of the general problem forms the subject of this research. The definition of the situation is best described pictorially as in fig. 1. Denote by ϕ the piezometric head which is obtained as a solution of Laplace's equation in steady-state seepage. It seems that in the synthesis of the vast majority of groundwater seepage problems, the three impermeable boundaries as described above exist in the same general configuration. These are characterized geometrically by a boundary at $y = y_{\min}$ with a no-flux boundary condition ($\partial\phi/\partial y = 0$); and two boundaries, located respectively at $x = x_{\min}$ and $x = x_{\max}$, which are also impermeable ($\partial\phi/\partial x = 0$). Since these three boundary configurations occur so frequently, it would be advantageous to eliminate the need for their explicit discretization in the boundary element representation of the problem.

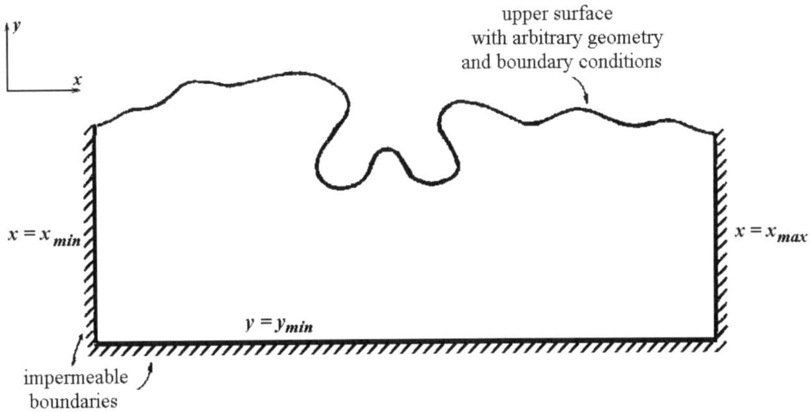

Figure 1: The basic geometrical configuration with three rectilinear, impermeable boundaries and an arbitrary upper boundary.

The way to eliminate these boundaries is to make their existence an inherent part of the mathematical formulation of the problem. In boundary elements, this is done by using an appropriate Green's function which has the no-flux conditions at certain points built-in. This is as opposed to using the free-space *fundamental solution* that is most often resorted to in boundary element methods. To the authors' knowledge, the only published Green's function for this type of problem are in references [1] and [2]. The following is a step-by-step derivation of the Green's function.

3 Derivation of the Green's function

We will assume that the homogeneous, isotropic, steady-state form of Darcy's law applies to the problem. We have

$$\frac{\partial^2 \phi}{\partial x^2} + \frac{\partial^2 \phi}{\partial y^2} = 0 \tag{1}$$

where ϕ is the total head. The Green's function ϕ^* for this problem is defined by

$$\frac{\partial^2 \phi}{\partial x^2} + \frac{\partial^2 \phi}{\partial y^2} = -\delta(x - x_0)\delta(y - y_0) \tag{2}$$

where $\delta(x)$ is the Dirac delta function. The physical interpretation of having a point source (sink) of unit intensity at point (x_0, y_0) is used in the definition of ϕ^*. The fundamental solution, or equivalently, the Green's function for unbounded space is easily obtained from eqn. (2) by switching to planar polar coordinates centered at (x_0, y_0). Denote by r the radial distance from (x_0, y_0). Eqn. (2) becomes:

$$\frac{1}{r}\frac{d}{dr}\left(r\frac{d\phi^*}{dr}\right) = -\frac{\delta(r)}{2\pi r} \tag{3}$$

Two integrations yields

$$\phi^* = -\frac{1}{2\pi}\ln(r) + C_1 \ln(r) + C_2$$

C_1 and C_2 may be used to fit special types of boundary conditions. They are usually unnecessary for the fundamental problem and are set to zero.
We take ϕ^* to be

$$\phi^* = -\frac{1}{2\pi}\ln(r) \tag{4}$$

a well-known result.

Although we could in principle manipulate eqn. (2) until the desired boundary conditions of fig. 1 are met, it is simpler to adapt eqn. (4) to the desired form if possible. This is can be done for a wide variety of problems using the method of images and the complex variable technique of conformal mapping. The basic premise is that the Dirichlet and Neumann problems can be solved for any simply-connected two-dimensional region which can be mapped conformally by an analytic function on to the unit circle or the half plane [3].

Denote by w the complex (u, v) plane as shown in fig. 2. If a point source with potential given by eqn. (4) is placed at (u_0, v_0), we can create a physical scenario equivalent to having an impermeable boundary along the u – axis. This is done by placing an image charge of the same strength at point $(u_0, -v_0)$. The new Green's function is obtained by superposing these two potentials:

$$\phi_H^* = -\frac{1}{4\pi}\ln\left[(u-u_0)^2 + (v-v_0)^2\right] - \frac{1}{4\pi}\ln\left[(u-u_0)^2 + (v+v_0)^2\right] \tag{5}$$

and the region of analysis is now reduced to a half-plane depicted in fig. 2.

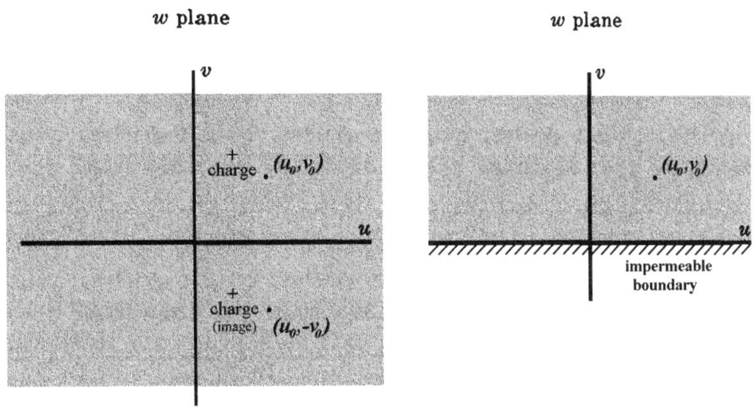

Figure 2: The entire complex plane with symmetrically placed identical point sources (left) is equivalent to a half-plane formulation (right) with an impermeable boundary at $v = 0$.

The new half-plane region may be mapped conformally into the z_1-plane (fig. 3) with

$$w = \cos\left(\frac{\pi z_1}{L}\right) \tag{6}$$

where L is an arbitrary non-zero length scale.

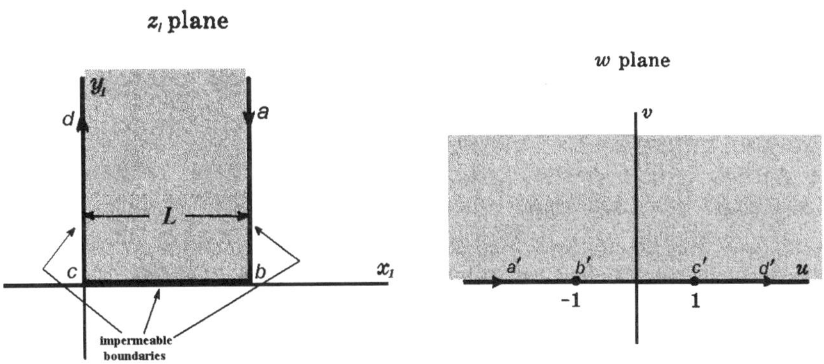

Figure 3: Mapping of the w half-plane into the z_1 plane. The boundary conditions invoked along $v = 0$ are mapped as well.

Eqn. (6) is equivalent to setting

$$\begin{aligned} u &= \sin x_1 \cosh y_1 \\ v &= \cos x_1 \sinh y_1 \end{aligned} \tag{7}$$

Note that due to the mapping, the no-flux boundary condition indicated in fig. 2 is invoked on all three finite boundaries.

The next step is to translate the entire region in the y_1 – direction such that $y = y_{min}$ is the lower boundary, and simultaneously perform a *stretching and translation* mapping in the x-direction so as to fix the left and right boundaries at $x = x_{min}$ and $x = x_{max}$, respectively.

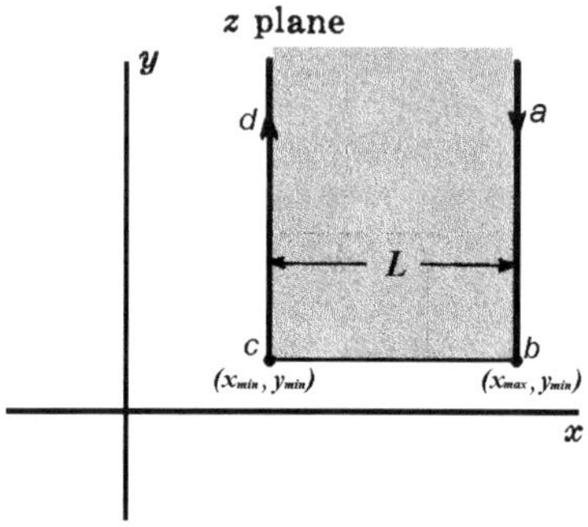

Figure 4: Mapping of the z_1 plane from fig. 3 on to the global coordinate configuration depicted in fig. 1. Note that the previously arbitrarily specified parameter L is now set equal to $x_{max} - x_{min}$.

The appropriate mapping is

$$z = x_1 + iy_1$$

with

$$x_1 = \frac{x - x_{min}}{x_{max} - x_{min}} L$$
$$y_1 = y - y_{min} \tag{8}$$

It is simplest to now set $L = x_{max} - x_{min}$. The mapped region is shown in fig. 4.

From eqns. (7) and (8), we have the final form of u and v:

$$u = \cos\left[\frac{\pi(x - x_{min})}{x_{max} - x_{min}}\right]\cosh\left[\frac{\pi(y - y_{min})}{x_{max} - x_{min}}\right]$$
$$v = \sin\left[\frac{\pi(x - x_{min})}{x_{max} - x_{min}}\right]\sinh\left[\frac{\pi(y - y_{min})}{x_{max} - x_{min}}\right] \tag{9}$$

and eqns.(9) are then substituted into eqn. (5) to obtain the desired Green's function. Note that the functional form appears differently than its original appearance in Gipson et al [1] due to the use of a different mapping (sine function instead of cosine) in eqn. (6).

Once programmed into a boundary element computer code, this Green's function will allow a user to solve a groundwater flow problem (or any equivalent problem governed by Laplace's equation) by specifying boundary elements only on the upper surface, and the parameters x_{min}, x_{max}, and y_{min}. This is a very useful result, and it has delivered perhaps the simplest possible numerical device for solving general Laplace problems of the type defined by the geometry of fig. 1.

4 Illustration

Explicit examples applicable to the seepage problem are available in reference [1]. Fig. 5, recast from that work, is an exemplary illustration to show the advantages produced by the more sophisticated Green's function. Not only does it reduce the amount of boundary element discretization considerably, but on the no-flux surfaces where no elements are required, the boundary conditions are satisfied *exactly*. Use of the advanced Green's function also reduces computational time because of the smaller number of simultaneous equations produced. In this example, 43 linear elements were used, and 44 simultaneous equations had to be solved. Equivalent accuracy with the standard fundamental solution and 77 linear elements required 78 nodes and simultaneous equations.

5 Other special Green's functions

With the conformal mapping method and other creative applications of imaging, other useful Green's functions may be produced. For instance, if the impermeable boundaries in fig. 1 are replaced with boundaries held at zero potential (the equivalent Dirichlet problem), a typical geometry that occurs in heat transfer and diffusion is the result. We may reuse all the mappings in Section 3 with the one difference that a *negative* image charge is placed at $(u_0, -v_0)$ in fig. 2 instead of positive. This renders the previous impermeable boundaries as zero-potential boundaries. The only change in the formulas occurs in eqn. (5) with a sign difference preceding the second term:

$$\phi_H^* = -\frac{1}{4\pi}\ln\left[(u-u_0)^2+(v-v_0)^2\right]+\frac{1}{4\pi}\ln\left[(u-u_0)^2+(v+v_0)^2\right] \tag{10}$$

It should be further noted that if any portion of the implicit zero-potential is nonzero, explicit boundary elements can be placed at the location with the proper potential assigned. By the principle of superposition, the zero potential is therefore overridden.

If only two of the perpendicular boundaries are required to be held at zero potential, the configuration can be obtained by skew-reflecting the pair of charges that produced eqn. (10) around the v-axis. The effect is to place positive charges at (u_0, v_0) and $(-u_0, -v_0)$, and negative charges at $(u_0, -v_0)$ and $(-u_0, v_0)$. The new Green's function follows once again from Section 3 as:

$$\phi_H^* = -\frac{1}{4\pi}\ln\left[(u-u_0)^2+(v-v_0)^2\right]+\frac{1}{4\pi}\ln\left[(u-u_0)^2+(v+v_0)^2\right]$$
$$-\frac{1}{4\pi}\ln\left[(u+u_0)^2+(v+v_0)^2\right]+\frac{1}{4\pi}\ln\left[(u+u_0)^2+(v-v_0)^2\right]$$

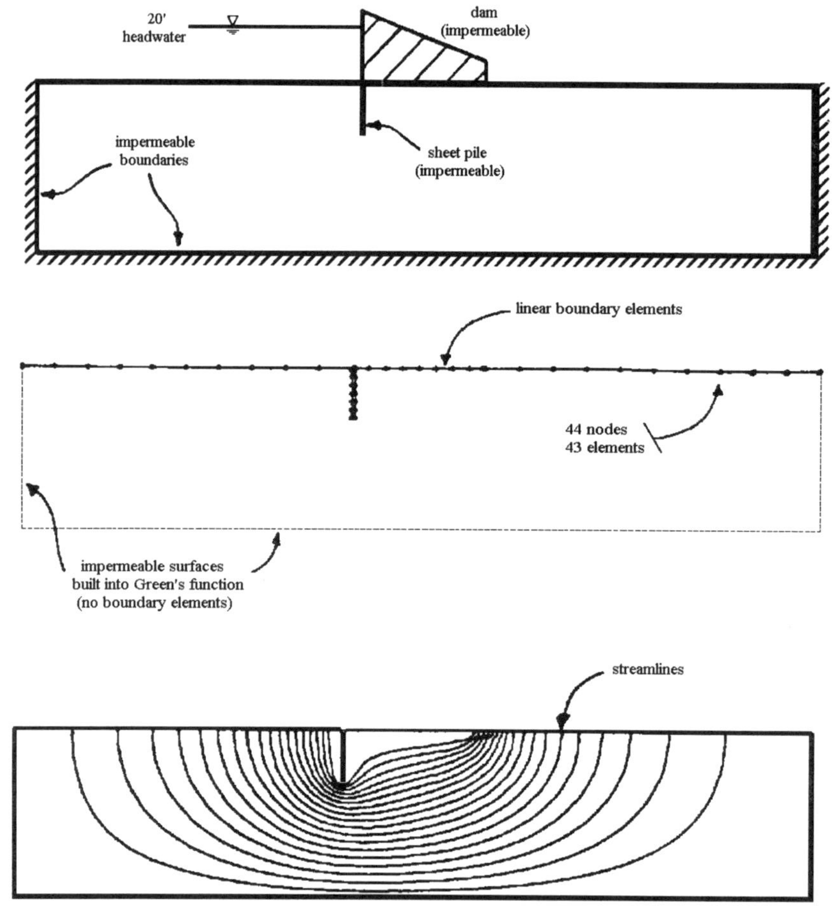

Figure 5: Boundary element analysis of a dam/sheetpile problem depicting the idealized synthesis (top), the boundary element model developed from using the advanced Green's function (middle), and the results of the analysis in the form of streamline contours (bottom).

This same idea of imaging charges and/or using conformal mapping can be continued theoretically indefinitely to produce other useful configurations for the boundary element analyst.

6 Summary and conclusions

The use of conformal mapping coupled with the method of images has been shown to produce an enhancement to the conventional boundary element method. Implicit boundaries satisfying exact boundary conditions can reduce the modeling effort and computational time while increasing the accuracy. Explicit derivations of Green's functions used in previous work have been presented. The method is a natural mesh reduction technique and can be coupled with other MRM's to further progress the efficiency of computational mechanics.

Acknowledgement

The authors would like to acknowledge the USAE Waterways Experiment Station which provided the original funding for this work.

References

[1] Gipson, G.S., Camp, C.V., Radhakrishnan, N. Phreatic surface and subsurface flow with boundary elements using an advanced Green's function. *Betech 86*, eds. J.J. Conner & C.A. Brebbia. Computational Mechanics Publications: Southampton, pp 385-94, 1986.

[2] Gipson. G.S. Applications of Boundary Elements to the Problems of the U.S. Corps of Engineers. Report to Battelle Columbus Laboratories, Contract #DAAG29-81-D-0100; 1985.

[3] Churchill. R.V., Brown, J.W., Verhey, R.F., *Complex Variables and Application 3rd ed.*, McGraw-Hill, pp. 195-238, 1974.

The meshless analog equation method: a new highly accurate truly mesh-free method for solving partial differential equations

J. T. Katsikadelis
School of Civil Engineering, National Technical University of Athens, Greece

Abstract

A new purely meshless method to solve PDEs is presented. The method is based on the concept of the *analog equation* of Katsikadelis, hence its name meshless analog equation method (MAEM), which converts the original equation into a simple solvable substitute one of the same order under a fictitious source. The fictitious source is represented by MQ-RBFS. Integration of the analog equation allows the approximation of the sought solution by new RBFs. Then inserting the solution into the PDE and BCs and collocating at the mesh-free nodal points yields a system of linear equations, which permit the evaluation of the expansion coefficients. The method exhibits key advantages of over other RBF collocation methods as it is highly accurate and the matrix of the resulting linear equations is always invertible. The accuracy is increased using optimal values of the shape parameters of the multiquadrics by minimizing the potential that produces the PDE. Without restricting its generality, the method is illustrated by applying it to the general second order elliptic PDE. The studied examples demonstrate the efficiency and high accuracy of the developed method.

1 Introduction

The interest in mesh-free methods to solve PDEs has grown noticeably in the past 20 years. This is mainly due to the fact that (i) Mesh generation over complicated 2D and especially 3D domains is a very difficult problem and may require long time, in some cases weeks or months, to create a well behaved mesh (ii) The convergence rate of the traditional methods is of second order. The mesh-free MQ-RBFs (multiquadric radial basis functions) method developed by

WIT Transactions on Modelling and Simulation, Vol 42, © 2006 WIT Press
www.witpress.com, ISSN 1743-355X (on-line)
doi:10.2495/BE06002

Kansa [1] has attracted the interest of the investigators, because it enjoys exponential convergence and is very simple to implement. The primary disadvantage of the MQ scheme is that it is global, hence, the coefficient matrices resulting from this scheme are full and suffer from ill-conditioning, particularly as the rank increases. Extended research has been performed by many investigators to overcome this drawback and several techniques have been proposed to improve the conditioning of the matrix [2], which, however, complicate the implementation of the MQ-RBFs method and render it rather problem dependent. Moreover, although the performance of the method depends on the shape parameter of MQs, there is no widely accepted recipe for choosing the optimal shape parameters. Therefore, extended research is ongoing to optimize these parameters [3]. Nevertheless, all these quantities are chosen rather arbitrarily or empirically.

In this paper a new meshless RBFs method is presented, which overcomes the drawbacks of the standard MQ-RBFs method. The method is based on the concept of the analog equation of Katsikadelis, according to which the original equation is converted into a substitute equation, the analog equation, under a fictitious source. The fictitious source is represented by radial basis functions series of multiquadric type. Integration of the analog equation yields the sought solution as series of new radial basis functions. To make this idea more concrete we consider the elliptic BVP

$$Lu = g \quad \text{in } \Omega \tag{1}$$

$$Bu = \bar{g} \quad \text{on } \Gamma \tag{2}$$

If $u = u(\mathbf{x})$ is the sought solution of eqn (1) and L, B linear operators. If $\tilde{L}$ is another linear operator of the same order as L, we obtain

$$\tilde{L}u = b \quad \text{in } \Omega \tag{3}$$

where $b = b(\mathbf{x})$ is an unknown fictitious source. Eqn (3) under the boundary condition (2) can give the solution of the problem, if the fictitious source $b(\mathbf{x})$ is first established. In this context the fictitious source is approximated by MQ RBFs series. Thus we can write

$$\tilde{L}u = \sum_{j=1}^{M+N} a_j f_j \quad \text{in } \Omega \tag{4}$$

where $f_j = \sqrt{r^2 + c^2}$, $r = \|\mathbf{x} - \mathbf{x}_j\|$ and M, N represent the number of collocation points inside Ω and on Γ, respectively.

Eqn (4) is integrated to yield the solution

$$u_p = \sum_{j=1}^{M+N} a_j \hat{u}_j \tag{5}$$

where $\hat{u}_j = \hat{u}_j(r)$ is the solution of

$$\tilde{L}\hat{u}_j = f_j \tag{6}$$

Since the $\tilde{L}$ is arbitrary, it is chosen so that the solution of eqn (6) can be established, e.g. if $\tilde{L}$ of the second order we can choose $\tilde{L} = \nabla^2$ or if L is of the fourth order we can chose $\tilde{L} = \nabla^4$. Subsequently, the solution (5) is inserted into the PDE (1) and BC (2) to yield

$$\sum_{j=1}^{M+N} L\hat{u}_j(\mathbf{x})a_j = g \quad \text{in } \Omega \tag{7}$$

$$\sum_{j=1}^{M+N} B\hat{u}_j(\mathbf{x})a_j = \overline{g} \quad \text{on } \Gamma \tag{8}$$

Collocation eqns (7) and (8) at nodal points in Ω and on Γ, Fig. 1, yields a system of linear equations, which permit the evaluation of the expansion coefficients.

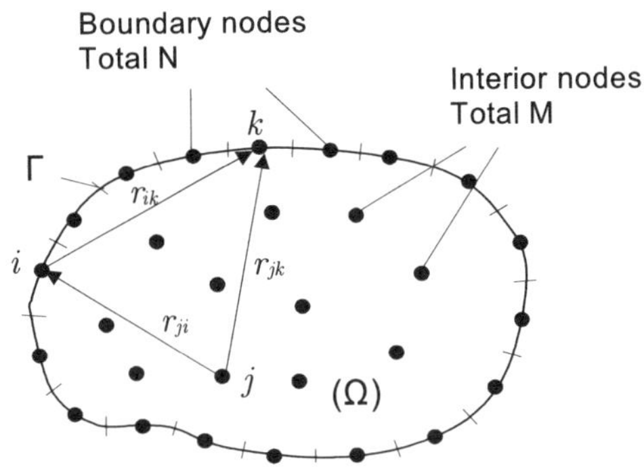

Figure 1.

The major advantage of the presented formulation is that it results in coefficient matrices, which are not ill-conditioned and thus they can be always inverted. Moreover, since the accuracy of the solution depends on a shape parameter of the MQs and the position of the collocation points, a procedure is developed to optimize these parameters by minimization of the functional that produces the PDE as *Euler-Lagrange* equation [4] under the inequality constraint that the condition number of the coefficient matrix ensures invertibility. This procedure requires the evaluation of a domain integral during the minimization process, which is facilitated by converting it to a boundary integral using DRM. The method is illustrated by applying it to the solution of the general second order elliptic PDE. Several examples are studied, which demonstrate the efficiency and accuracy of the method.

2 Problem statement

We consider the partial differential equation

$$A\,u_{,xx} + 2Bu_{,xy} + Cu_{,yy} + Du_{,x} + Eu_{,y} + Fu = g(\mathbf{x}) \quad \text{in } \mathbf{x} \in \Omega \tag{9}$$

subject to the boundary conditions

$$\begin{aligned} u &= \alpha(\mathbf{x}), \quad \mathbf{x} \in \Gamma_u \\ \kappa u + \nabla u \cdot \mathbf{m} &= \gamma(\mathbf{x}), \quad \mathbf{x} \in \Gamma_m \end{aligned} \tag{10a,b}$$

where $\Gamma = \Gamma_u \cup \Gamma_m$ is the boundary of Ω, which may be multiply-connected; $u = u(\mathbf{x})$ is the unknown field function; $A, B, \ldots, F$ position dependent coefficients satisfying the ellipticity condition $B^2 - AC < 0$ and $\mathbf{m} = (An_x + Bn_y)\mathbf{i} + (Bn_x + Cn_y)\mathbf{j}$ is a vector in the direction of the con-normal on the boundary. Finally, $\kappa(\mathbf{x})$, $\alpha(\mathbf{x})$ and $\gamma(\mathbf{x})$ are functions specified on Γ. We consider the functional [4].

$$J(u) = \int_\Omega \left[\frac{1}{2}(Au_{,x}^2 + 2Bu_{,x}\, u_{,y} + Cu_{,y}^2 - Fu^2) + gu \right] d\Omega + \int_\Gamma \left(\frac{1}{2}\kappa u^2 - \gamma u \right) ds \quad (11)$$

We can easily show that the condition $\delta J(u) = 0$ yields the boundary value (9), (10) provided that

$$A_{,x} + B_{,y} = D\,, \qquad B_{,x} + C_{,y} = E \quad (12a,b)$$

Therefore, the solution of Eqn (9) under the boundary conditions (10a,b) make $J(u) = \min$.

The boundary value problem (9), (10) under the conditions (12) for suitable meaning of the coefficients occurs in many physical problems such as thermostatic, elastostatic, electrostatic and seepage problems, where the involved media exhibit heterogeneous anisotropic properties.

3 The MAEM solution

The analog equation is obtained from eqn (3), if we take $\tilde{L} = \nabla^2$. Thus we have

$$\nabla^2 u = b(\mathbf{x}) \quad (13)$$

and (6) becomes

$$\nabla^2 \hat{u}_j = f_j \quad (14)$$

which for $f_j = \sqrt{r^2 + c^2}$ yields after integration

$$\hat{u}_j = \frac{1}{9} f^3 + \frac{1}{3} fc^2 - \frac{c^3}{3} \ln(c + f) + G\ln(r) + F \quad (15)$$

where $G = 0$ for $r = 0$, otherwise it is arbitrary. As we will see, the arbitrary constants G, F play an important role in the method, because together with the shape parameter c control the conditioning of the coefficient matrix and the accuracy of the results.

The solution is approximated by

$$u \simeq \sum_{j=1}^{M+N} a_j \hat{u}_j \quad (16)$$

and it is forced to satisfy the governing equation and the boundary conditions. Thus, it is inserted into eqns (9) and (10) to yield the system of linear equations

$$\mathbf{Aa} = \mathbf{b} \quad (17)$$

where

$$\mathbf{A} = \begin{bmatrix} L\hat{u}_{ij} & B\hat{u}_{ij} \end{bmatrix}^T, \ \mathbf{b} = \begin{Bmatrix} \mathbf{g} & \boldsymbol{\alpha} & \boldsymbol{\gamma} \end{Bmatrix}^T \quad (18)$$

in which B is the operator defined by eqns (10a,b).

Apparently, the approximation (16) with the *new radial basis functions* $\hat{u}_j$ is better than the conventional one with $f_j = \sqrt{r^2 + c^2}$, because they can accurately approximate not only the field function itself but also the first and second derivatives. This shown in Fig. 2.

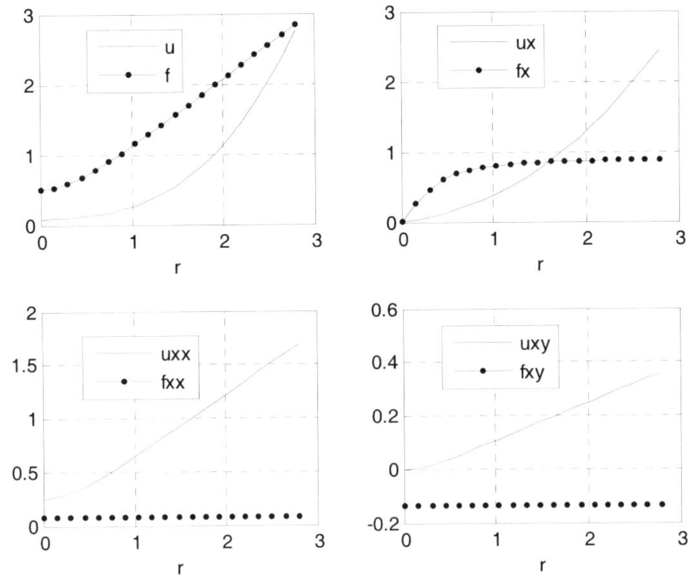

Figure 2: Variation of $f(r)$, $\hat{u}(r)$ and their derivatives along the line $y = 0.5x$ ($c = 0.5$, $G = 1.\mathrm{e}-2$, $F = 0$).

4 Optimal values of shape parameter, constants G, F and centers of EBFs

The coefficients a_j evaluated from eqn (17) can be used to obtain optimal values of the shape parameter and centers of the multiquadrics by minimizing the functional (11). The evaluation of the domain integral is facilitated, if it is converted to boundary line integral using DRM [5]. Thus denoting by

$$R(\mathbf{x}) = \frac{1}{2}\left(Au_{,x}^2 + 2Bu_{,x}\, u_{,y} + Cu_{,y}^2 - Fu^2 \right) + gu \tag{19}$$

and approximating the integrand of the domain integral by

$$R(\mathbf{x}) \simeq \sum_{j=1}^{M} \overline{a}_j \hat{u}_j(r) \tag{20}$$

we obtain

$$\int_\Omega R(\mathbf{x}) d\Omega \simeq \sum_{j=1}^{M} \overline{a}_j \int_\Omega \hat{u}_j(r) d\Omega \tag{21}$$

Application of (20) at the collocation points yields

$$\overline{\mathbf{a}} = \mathbf{U}^{-1}\mathbf{R}, \quad \mathbf{U} = [\hat{u}(r_{ji})], \; \mathbf{R} = \{R(x_i)\} \tag{22}$$

Subsequently, using the Green' reciprocal identity

$$\int_{\Omega}\left(v\nabla^2 u - u\nabla^2 v\right)d\Omega = \int_{\Gamma}\left(vu,_n - uv,_n\right)ds \tag{23}$$

for $v = 1$ and $u = \hat{w}_j$, where $\hat{w}_j$ is a particular solution $\nabla^2 \hat{w}_j = \hat{u}_j$ we obtain

$$\int_{\Omega} R(\mathbf{x})d\Omega \simeq \mathbf{1}^T\hat{\mathbf{Q}}\ \ \mathbf{U}^{-1}\mathbf{R} \tag{24}$$

where $\hat{Q}_{jk} = \int_{\Gamma_k} \hat{w}_{j,n}(r_{jk})ds$ and $\mathbf{1}^T = \{1 \;\; 1 \;\; \cdots \;\; 1\}$.

It is apparent that the functional $J(u)$ depends on the following sets of parameters.

(i). The shape parameter c and the arbitrary constants G, F.

(ii). The $2M + 2N$ coordinates x_j, y_j of the centers.

Therefore, we can search for the minimum using various levels of optimization depending on the design parameters that we wish to be involved in the optimization procedure. Although, the functional $J(u)$ is quadratic with respect to a_j, the inclusion however of c and x_j, y_j requires direct minimization methods for nonlinear objective functions.

5 Numerical results

5.1 Example 1

As a first example we consider a benchmark problem [6]. This problem is governed by Poisson's equation

$$\nabla^2 u = -\frac{10^6}{52} \quad \text{in } \Omega = (-0.3, 0.3)\times(-0.2, 0.2) \tag{25a}$$

$$u = 0 \quad \text{on } \Gamma \tag{25b}$$

The exact value of u at the center is $u(0,0) = 310.10$. This value was recovered using the described solution procedure with $N = 80$ boundary nodes, $M = 9\times 11 = 99$, $c = 0.228$, $G = 1.\mathrm{e}-11$, $F = 5$.

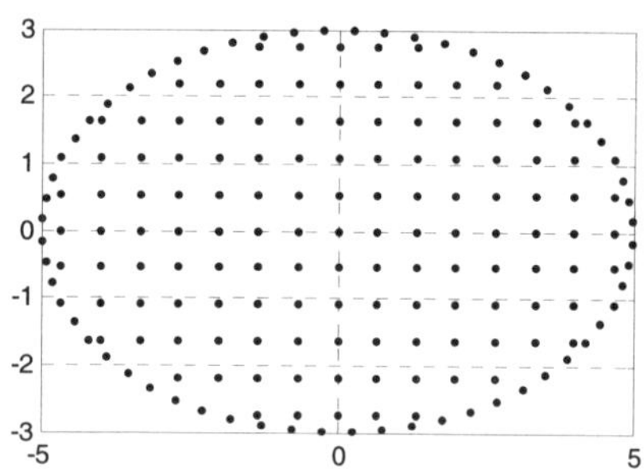

Figure 3: Elliptic domain and nodal points.

5.2 Example 2

As a second example we obtain the solution of the following boundary value problem for complete second order equation

$$(1+y^2)u_{xx}+2xyu_{xy}+(1+2x^2)u_{yy}+xu_x+yu_y+u = 7x^2-5xy+5y^2+4 \quad \text{in } \Omega \qquad (26)$$

where Ω is the ellipse with semi-axes $a=5,\ b=3$.

Three types of boundary conditions have studied

(i) $u=\alpha(\mathbf{x})$ on Γ (Dirichlet)

(ii) $\nabla u\cdot\mathbf{m}=\gamma(\mathbf{x})$ on Γ (Neumann)

(iii) $\nabla u\cdot\mathbf{m}=\gamma(\mathbf{x})$ on Γ_m, $u=\alpha(\mathbf{x})$ on Γ_u (mixed)

where $\Gamma_m=\{y=b\sqrt{1-x^2/a^2},\ 0\le x\le a\}$, $\Gamma_u=\Gamma-\Gamma_m$ and

$$\alpha(\mathbf{x})=x^2-xy+y^2$$

$$\gamma(\mathbf{x})=\left[\frac{(1+y^2)x}{a}+\frac{xy^2}{b}\right](2x-y)+\left[\frac{x^2y}{a}+\frac{(1+2x^2)y}{b}\right](-x+2y)$$

The analytical solution is $u_{exact}=x^2-xy+y^2$. The results obtained with $N=60$, $M=125$, $c=7$, $G=5\text{e}-9$, $F=0$ are shown in Fig. 5.

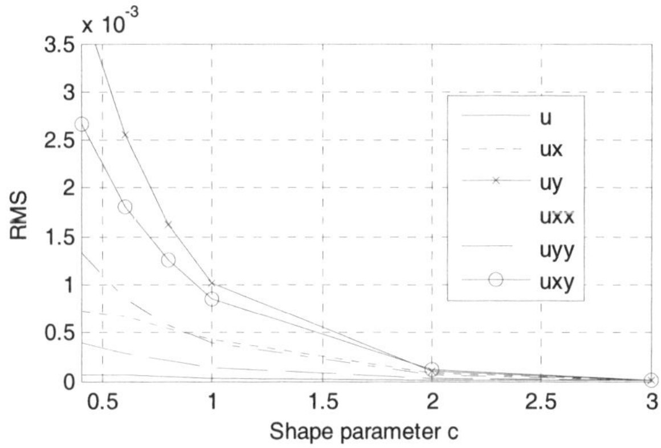

Figure 4: Dependence of RMS on c in Example 2 (case i).

Fig. 4 shows the convergence of $RMS=\frac{1}{m}\sqrt{\sum_{i=1}^{m}\{[u(i)-u_{exact}(i)]/u_{exact}(i)\}^2}$ with increasing shape parameter. In all three cases the computed results are practically identical with the exact ones.

5.3 Example 3

As third example the following boundary value problem studied in [7] has been solved

$$\nabla^2 u=(2x^2y^2+2x^2y+2xy^2-6xy)e^{(x+y)} \quad \text{in } \Omega=(0,1)\times(0,1) \qquad (27)$$

$$u=0 \quad \text{on } \Gamma$$

The analytical solution is

$$u_{exact}=x(x-1)y(y-1)e^{(x+y)}$$

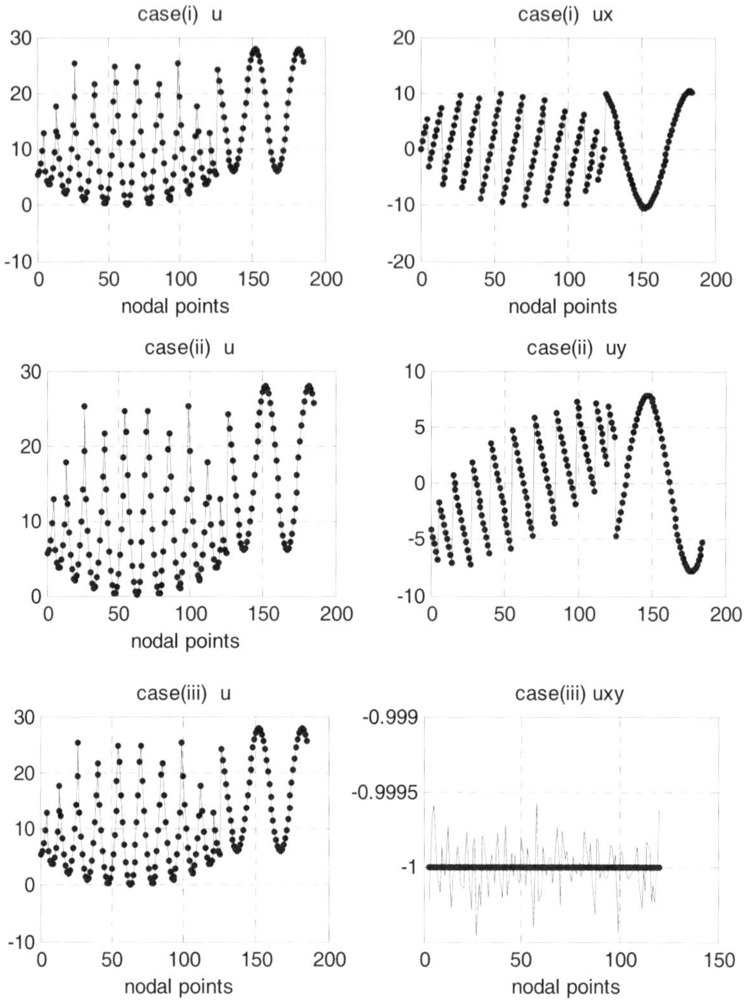

Figure 5: Nodal values of the solution and its derivatives. Solid line: computed.

The solution computed with $c = 0.56$, $G = 8.61e-11$, $F = 11.1$ is shown in Fig. 6 as compared with the exact one. Moreover, the RMS for the solution and its derivatives is shown in Table 1. The computed potential is $J(u)_{comp} = -0.1065$, while $J(u)_{exact} = -0.1067$.

6 Conclusions

A new truly meshless method, the MAEM (Meshless Analog Equation Method) is developed for solving PDEs, which describe the response of non-homogeneous anisotropic bodies of arbitrary geometry. Without restricting its generality, the method is illustrated by second order elliptic PDEs under general boundary conditions. The developed method is based on the concept of the analog equation, which converts the original equation into a Poisson's equation under a fictitious source. Using MQ-RBFs to approximate the fictitious source and integration leads to the approximation of the sought solution by new RBFs, which have key advantages over the direct MQ-RBFs collocation method, summarized as:

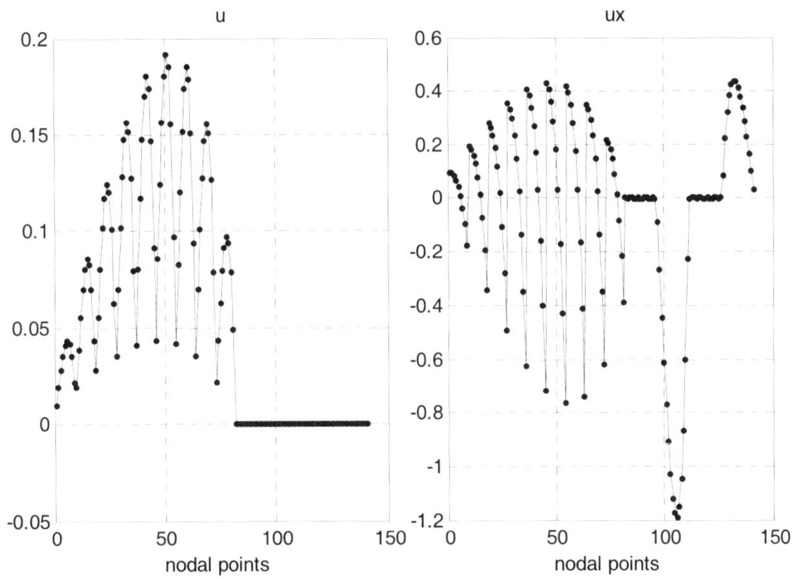

Figure 6: Nodal values of the solution and its derivative. Solid line: exact.

Table 1: RMS of the solution and its derivatives in Example 3.

$RMS \times 10^4$								
c	G	F	u	u_x	u_y	u_{xx}	u_{yy}	u_{xy}
0.56	8.61e-11	11.1	2.683	5.356	5.356	3.214	3.214	7.517

- Since the method allows the control of the condition number, an invertible coefficient matrix for the evaluation of the RBFs expansion coefficients can be always established.
- The method gives accurate results, because the *new RBFs* approximate accurately not only the solution itself but also its derivatives.
- Optimum values of the shape parameters and centers of RBFs can be established by minimizing the potential that yields the PDE. Therefore, the uncertainty of choice of shape parameter is eliminated. It was also observed from the studied examples that a regular mesh of nodal point gives good

results and the solution was not sensitive to the position of the RBFs centers.

- The method depends only on the order of the differential operator and not on the specific problem.

Moreover, as other RBFs methods:

- It is truly meshless, hence no domain (FEM) or boundary (BEM) discretization and integration is required. It also avoids establishment of fundamental solutions and evaluation of singular integrals.
- The method can be in a straightforward manner employed for the solution of problems in other dimensions, of other type (parabolic and hyperbolic) or of higher order as wells as for nonlinear ones.

References

[1] Sharan, M., Kansa, E.J. and Gupta, S., Application of the multiquadric method for the solution of elliptic partial differential equations, *Appl. Math. And Comp.* **84**, pp. 275-302, 1987.

[2] Kansa, E.J. and Hon, Y.C., Circumventing the ill-conditioning problem with multiquadric radial basis functions: Applications to elliptic partial differential equations, *Comp. and Math. App.* **39**, pp. 123-137, 2005.

[3] Kansa, E.J., Highly accurate methods for solving elliptic partial differential equations: In: Brebbia, C.A., Divo, E. and Poljak, D. (eds.), *Boundary Elements XXVII*, WIT Press, Southampton, pp. 5-15, 2005.

[4] Katsikadelis, J.T., A BEM based meshless variational method for the second order elliptic partial differential equation using optimal multiquadrics, In: Georgiou, G. et al. (eds), *Proceedings of the 5th GRACM 05 International Congress on Computational Mechanics,* Limassol, Cyprus 29 June-1 July 2005, **2**, pp. 903-910.

[5] Katsikadelis, J.T., *Boundary Elements: Theory and Applications*, Elsevier Science, London, 2002.

[6] Cameron, A. D.; Casey, J. A.; Simpson, G. B., Benchmark tests for thermal analysis. *Glasgow: NAFEMS Publications*, 1986.

[7] Fedoseyev, A.I., Friedman, M.J. and Kansa, E.J., Improved multiquadric method for elliptic partial differential equations via PDE collocation on the boundary, *Comp. Math. With Applications*, 43, pp, 439-455, 2002.

Solving Poisson's equations by the Discrete Least Square meshless method

H. Arzani[1] & M. H. Afshar[2]
[1]*Shahid Rajaee University, Lavizan, Tehran, Iran*
[2]*Department of Civil Engineering,*
Iran University of Science & Technology

Abstract

A meshless method is proposed in this paper for the solution of two-dimensional elliptic problems. The proposed method does not require any mesh so it is truly a meshless method. The approach termed generically the "Discrete Least Square meshless method" is applied to discrete the governing differential equations in inner and boundary nodes. A functional is defined as the sum of the squared residual of the governing differential equation and the boundary conditions at the nodal points. Moving least-square (MLS) interpolation is used to construct the shape function's values, which have high continuity in the problem domain. To evaluate the accuracy of the method as an alternative meshless method the development and theory of this new approach is presented in the context of the solution of 2D elliptic equations. Numerical results show that the method possesses high accuracy with low computational effort.
Keywords: meshless method, Discrete Least Square, elliptic problems.

1 Introduction

The idea of using finite difference simplicity and finite element capability of handling complex geometries are the subject of many researches. This is mainly because the mesh generation part of the solution has shown to be a very time consuming challenge especially in finite element applications. The idea of developing methods requiring no mesh has led to the emerging of a new class of the so-called 'meshless' methods. Many of the meshless methods developed so far require background mesh to carry out numerical integration. The integration cells, however, need not be compatible with nodes and thus they can be generated more easily than the FEM meshes. The existing meshless methods can

WIT Transactions on Modelling and Simulation, Vol 42, © 2006 WIT Press
www.witpress.com, ISSN 1743-355X (on-line)
doi:10.2495/BE06003

be generally divided in two main categories depending on the way the discrete equations are formed.

First are the methods based on the weak form of the given differential equation. All these methods use one form of the weighted residual such as Galerkin or Petrov-Galerkin for discretization of the governing differential equations. In this category, one can find Smooth Particle Hydrodynamics (SPH) by Monaghan [1], which is the oldest of the meshless methods, Reproducing Kernel Particle method (RKPM) [2]. These methods use finite integral for function approximation; Partition of Unity (PU) method [3]; hp cloud method [4]; Diffuse Element Method (DEM) by Nayroles et al. [5]; Element Free Galerkin (EFG) by Belytschko et al. [6]. Atluri and Zhu [7] and Zhu et al. [8] suggested the local Petrov-Galerkin and local boundary integral equation (LBIE) approach in which integration is performed locally on each subdomain. A common feature of all these methods is the need for numerical integration requiring a mesh of quadrature points in the domain. Construction of appropriate integration cells, however, is a difficult job and can make meshfree methods less effective. For these reasons, Beissel and Belytschko [9] suggested a nodal integration procedure instead of using Gaussian quadrature in establishing the coefficients of the system of equations. J.X. Zhou et al [10] proposed a nodal integration procedure based on Voronoi diagram for general Galerkin meshless methods. Some of the meshless methods use finite series for function approximation which include Polynomial Point Interpolation Method (PPIM), Radial Point Interpolation Method or Radial Basis Function (RBF) by Chen et al. [11] and well-known Moving Least Square Method (MLS) described by Lancaster and Salkauskas [12] and used by Nayroles et al. [5].

Second are the methods starting directly from the governing equation such as finite point method by Onate et al. [13]. These methods are often arrived at using a point collocation weighted residual formulation of the problem. The collocation method, however, can suffer from the stability problem as that encountered in the nodal integration. In addition, it requires higher-order derivatives of shape functions and results in non-symmetric stiffness matrices.

In this paper we use a fully least square approach in both of the governing differential equation discretization and function approximation which are the main components of every meshless method. The outline of this paper is organized as follows. The moving least square approximation for establishing shape functions is briefly described in section 2. The Discrete Least Square method for discretizing the governing differential equation is presented in section 3. Two elliptic problems solved and the results are presented in section 4. And we close with some concluding remarks in section 5.

2 Moving Least Square (MLS) method

The method of Moving Least Squares (MLS) has been widely used for function approximation by meshless community. The advantages of MLS are three folds: first, there is no need for explicit meshes in the construction of MLS shape functions. Second, high order continuity of shape functions so constructed eliminates the necessity of using weak form of governing equations as required in finite element method (FEM) using standard shape functions. In addition, higher order continuity, if required, is not introduced at the expense of increasing the unknown parameters as usually practiced in FEM. Third; the availability of smooth derivatives eliminates the need for costly procedure of gradient recovery, which is usually required by standard FEM.

In MLS, the function to be approximated is represented by:

$$u^h(\mathbf{x}) = \sum_{i=1}^{m} p_i(\mathbf{x})\, a_i(\mathbf{x}) \equiv \mathbf{p}^T(\mathbf{x})\,\mathbf{a}(\mathbf{x}) \tag{1}$$

Here $\mathbf{p}^T(\mathbf{x})$ is a set of linearly independent polynomial basis and $\mathbf{a}(\mathbf{x})$ represents the unknown coefficients to be determined by the fitting algorithm. The polynomial bases of order m in one and two dimension are given by:

$$\mathbf{p}^T(x) = [1, x, x^2, \ldots, x^m] \tag{2}$$

$$\mathbf{p}^T(\mathbf{x}) = \mathbf{p}^T(x, y) = [1, x, y, x^2, xy, y^2, \ldots, x^m, \ldots, xy^{m-1}, y^m] \tag{3}$$

In the MLS approximation, at each point ($\mathbf{x}$), $\mathbf{a}(\mathbf{x})$ is chosen to minimize the sum of weighted squared residuals defined by:

$$J = \frac{1}{2}\sum_{I=1}^{n} w(|\mathbf{x}\text{-}\mathbf{x}_I|)\left[\mathbf{p}^T(\mathbf{x}_I)\,\mathbf{a}(\mathbf{x}) - u_I\right]^2 \tag{4}$$

Where u_I is nodal value of the function to be approximated, n is the number of nodes and $w(|\mathbf{x}\text{-}\mathbf{x}_I|)$ is the weight function defined to have compact support. The weight functions are chosen to have the following properties:

1) $w(|\mathbf{x}\text{-}\mathbf{x}_I|) > 0$ On a subdomain

2) $w(|\mathbf{x}\text{-}\mathbf{x}_I| = 0$ Outside the subdomain (5)

3) $\int_\Omega w(|\mathbf{x}\text{-}\mathbf{x}_I|)d\Omega = 1$ A normality property

4) $w(|\mathbf{x}\text{-}\mathbf{x}_I|)$ A monotonically decreasing function

5) $w(|\mathbf{x}\text{-}\mathbf{x}_I|) \rightarrow \delta(s)$ *as* $|\mathbf{x}\text{-}\mathbf{x}_I| = \mathrm{h} \rightarrow 0$ where $\delta(\mathrm{s})$, is the Dirac delta function.

Many weight functions are established and used by different researchers. In this paper, we use a cubic spline weight function defined as:

$$w(r) = \begin{cases} \dfrac{2}{3} - 4r^2 + 4r^3 & \text{for} \quad r \le \dfrac{1}{2} \\ \dfrac{4}{3} - 4r + 4r^2 - \dfrac{4}{3}\mathrm{r}^3 & \text{for} \quad \dfrac{1}{2} < r \le 1 \\ 0 & \text{for} \quad r > 1 \end{cases} \tag{6}$$

In which, $r = s / s_{\max}$, $s = \left\| \mathbf{x} - \mathbf{x}_I \right\|$ and $s_{\max}$ is the radius of the support.
Eqn (4) can be written in matrix form as

$$\mathbf{J} = (\mathbf{Pa} - \mathbf{u})^T \mathbf{W} (\mathbf{Pa} - \mathbf{u}) \tag{7}$$

Where

$$\mathbf{u}^T = (u_1, u_2, \ldots, u_n) \tag{8}$$

$$\mathbf{P} = \begin{bmatrix} p_1(\mathbf{x}_1) & p_2(\mathbf{x}_1) & \cdots & p_m(\mathbf{x}_1) \\ p_1(\mathbf{x}_2) & p_2(\mathbf{x}_2) & \cdots & p_m(\mathbf{x}_2) \\ \vdots & \vdots & \vdots & \vdots \\ p_1(\mathbf{x}_n) & p_2(\mathbf{x}_n) & \cdots & p_m(\mathbf{x}_n) \end{bmatrix} \tag{9}$$

and

$$\mathbf{W}(\mathbf{x}) = \begin{bmatrix} w(|\mathbf{x} - \mathbf{x}_1|) & 0 & \ldots & 0 \\ 0 & w(|\mathbf{x} - \mathbf{x}_2|) & \ldots & 0 \\ \vdots & \vdots & \vdots & \vdots \\ 0 & 0 & \ldots & w(|\mathbf{x} - \mathbf{x}_n|) \end{bmatrix} \tag{10}$$

The coefficients **a** are found by minimizing $\mathbf{J}$ with respect to these coefficients. Carrying out the differentiation:

$$\frac{1}{2}\frac{\partial \mathbf{J}}{\partial \mathbf{a}} = \mathbf{A}(\mathbf{x})\mathbf{a}(\mathbf{x}) - \mathbf{B}(\mathbf{x})\mathbf{u} = 0 \tag{11}$$

Where

$$\mathbf{A} = \mathbf{P}^T \mathbf{W}(\mathbf{x})\mathbf{P} \tag{12}$$

$$\mathbf{B} = \mathbf{P}^T \mathbf{W}(\mathbf{x}) \tag{13}$$

Solving the above equation for the unknown parameters.

$$\mathbf{a}(\mathbf{x}) = \mathbf{A}^{-1}(\mathbf{x})\mathbf{B}(\mathbf{x})\mathbf{u} \tag{14}$$

The approximation of the unknown function can now be written as

$$\mathbf{u}^h(\mathbf{x}) = \sum_{I=1}^{n} \mathbf{N}_I(\mathbf{x})\mathbf{u}_I \tag{15}$$

where the shape functions are defined as:

$$\mathbf{N} = \mathbf{p}^T(\mathbf{x})\mathbf{A}^{-1}(\mathbf{x})\mathbf{B}(\mathbf{x}) \tag{16}$$

In this case, $\mathbf{u}_I \neq \mathbf{u}^h(x_I)$, so the parameters $\mathbf{u}_I$ cannot be treated like nodal values of the unknown function. The shape functions are not strict interpolates since they do not pass through the data. The shape functions do not satisfy the Kronecker delta condition:

$$N_i(\mathbf{x}_j) \neq \delta_{ij} = \begin{cases} 1 & if\ i = j \\ 0 & if\ i \neq j \end{cases} \tag{17}$$

where $N_i(\mathbf{x}_j)$ is the shape function of node *i* evaluated at node *j* and δ_{ij} is the Kronecker delta.

3 Discrete Least Square (DLS) method

Consider the following differential equation

$$L(\mathbf{u}) = \mathbf{f} \qquad \text{In } \Omega \tag{18}$$

$$B(\mathbf{u}) = \mathbf{g} \qquad \text{On } \Gamma \tag{19}$$

Where L and B are the differential operator defined on the problem domain Ω and its boundary (Γ), respectively. The philosophy of least square is to find an approximate solution that results in minimum residual error when substituted into equations (18) and (19). The first step is to assume the form of approximate solution ($\mathbf{u}^h$), including a total of m parameters which can be adjusted to minimize the error. This is sometimes called a trial solution, and can be represented by:

$$\mathbf{u}(\mathbf{x}) \cong \mathbf{u}^h(\mathbf{a}, \mathbf{x}) \tag{20}$$

Where $\mathbf{a}$ is the vector of unknown parameters and $\mathbf{x}$ represents the independent variables of the domain. The error is measured by the residuals that result when $\mathbf{u}^h$ is substituted into eqns (19) and (20).

$$\mathbf{R}_\Omega(\mathbf{a}, \mathbf{x}) = L(\mathbf{u}^h) - \mathbf{f} \qquad \text{for } \mathbf{x} \text{ in } \Omega \tag{21}$$

$$\mathbf{R}_\Gamma(\mathbf{a}, \mathbf{x}) = B(\mathbf{u}^h) - \mathbf{g} \qquad \text{for } \mathbf{x} \text{ on } \Gamma \tag{22}$$

Where $\mathbf{R}_\Omega$ and $\mathbf{R}_\Gamma$ are called interior and boundary residuals, respectively. Finally, a weighted sum of squared residuals is minimized over the domain (Ω), establishing the best values of the parameters $\mathbf{a}$. In the Discrete Least Square formulation, the squared residuals are evaluated and summed at the set of points $\mathbf{x}_i$ chosen to represent the problem domain Ω and its boundary (Γ).

$$\mathbf{I}_d(\mathbf{a}) = \frac{1}{2}\sum_{i=1}^{ne}\left[\mathbf{R}_\Omega(\mathbf{a}, \mathbf{x}_i)\right]^2 + \frac{\alpha}{2}\sum_{i=1}^{nb}\left[\mathbf{R}_\Gamma(\mathbf{a}, \mathbf{x}_i)\right]^2 \tag{23}$$

where *ne* and *nb* are the number of points chosen on the domain Ω and the boundary Γ, respectively. The factor α in above equation is the relative weight of the boundary residuals with respect to the interior residuals. It is the same of weight coefficient in general penalty method for boundary condition imposing and is equal unit in this paper.

Minimization of the eqn (23) leads to:

$$\frac{\partial \mathbf{I}}{\partial \mathbf{a}} = \sum_{i=1}^{ne}\left[\frac{\partial \mathbf{R}_{\Omega}(\mathbf{a},\mathbf{x}_i)}{\partial \mathbf{a}}\right]\left[\mathbf{R}_{\Omega}(\mathbf{a},\mathbf{x}_i)\right] + \alpha\sum_{i=1}^{nb}\left[\frac{\partial \mathbf{R}_{\Gamma}(\mathbf{a},\mathbf{x}_i)}{\partial \mathbf{a}}\right]\left[\mathbf{R}_{\Gamma}(\mathbf{a},\mathbf{x}_i)\right] = 0 \quad (24)$$

Substitution of $\mathbf{u}^h = \sum_{i=1}^{nn} N_i \mathbf{u}_i = \boldsymbol{Nu}$ in eqns (21), (22) and (24) yields the final system of equations.

$$\boldsymbol{KU=F}$$

System of algebraic equations should be solved for the vector of unknown parameters $\boldsymbol{U}$. Here $nn = ne + nb$ denotes the total number of nodes used to represent the problem domain of it body.

4 Numerical Investigation

In this section, two numerical examples in the area of elliptic problems are solved and results are presented to illustrate the performance of the proposed discrete least square meshless method. We consider two dimensional steady state heat conduction or seepage equation in a homogeneous orthotropic body.

$$\frac{\partial}{\partial x}\left(D\frac{\partial \phi}{\partial x}\right) + \frac{\partial}{\partial y}\left(D\frac{\partial \phi}{\partial y}\right) = S \quad (25)$$

Subject to appropriate Dirichlete and Newmann boundary conditions

$$\phi = \bar{\phi} \quad \text{on } \Gamma_{\mathrm{u}}$$
$$D\frac{\partial \phi}{\partial \mathrm{n}} = \bar{\mathrm{q}} \quad \text{on } \Gamma_{\mathrm{q}} \quad (26)$$

The residuals on interior and boundary nodes are defined by

$$\mathbf{R}_{\Omega} = D\frac{\partial^2 \phi}{\partial x_{\mathrm{j}}^2} - S \neq 0 \qquad \mathrm{j} = 1,2$$
$$\mathbf{R}_{\Gamma_{\mathrm{u}}} = \phi - \bar{\phi} \quad (27)$$
$$\mathbf{R}_{\Gamma_{\mathrm{q}}} = D\frac{\partial \phi}{\partial x_{\mathrm{j}}} n_{\mathrm{j}} - \bar{\mathrm{q}}$$

n_{j} is the jth component of the outward unit normal vector to the boundary Γ_{q}.

General differential operators in eqns (18), (19) are defined as.

$$L(\cdot) = \left[D\frac{\partial^2(\cdot)}{\partial x_{\mathrm{j}}^2}\right] \qquad \text{on } \Omega \text{ and } \mathrm{j} = 1,2 \quad , \quad \mathbf{f} = S \quad (28)$$

$$\mathbf{B}(.) = 1.0 \quad , \quad \mathbf{g} = \bar{\phi} \quad \text{on } \Gamma_u$$
$$\mathbf{B}(.) = D\frac{\partial(.)}{\partial x_j} n_j \quad , \quad \mathbf{g} = q \quad \text{on } \Gamma_q \tag{29}$$

Application of DLS method leads to the following system of equations

$$\mathbf{k} \ \varphi = \mathbf{f} \tag{30}$$

Where φ is the vector of unknown parameters $\left[\phi_1, \phi_2, ..., \phi_n\right]^T$.

$$\mathbf{k}_{lm} = \sum_{i=1}^{ne} \left[D \frac{\partial^2 N_l}{\partial x_j^2} \right]_i^T \left[D \frac{\partial^2 N_m}{\partial x_j^2} \right]_i + \sum_{i=1}^{nb} \left[BN_l \right]_i^T \left[BN_l \right]_i \tag{31}$$

$$\mathbf{f}_l = \sum_{i=1}^{ne} \left[D \frac{\partial^2 N_l}{\partial x_j^2} \right]_i^T S_i + \sum_{i=1}^{nb} \left[BN_l \right]_i^T g_i \tag{32}$$

4.1 Poisson equation

Consider the solution of the Poisson's equation.

$$\nabla^2 u(x, y) = \mathbf{sin}\, \pi x\, \mathbf{cos}\, \pi y \qquad \Omega(x, y): \{0 \le x \le 1, \quad 0 \le y \le 1\}$$

Boundary conditions given as

$u = 0$ $\quad x = 0$

$u = 0$ $\quad x = 1$

$u = 0$ $\quad y = 0$

$u = 0$ $\quad y = 1$

The exact solution of the governing equation is given by

$$u = \frac{1}{2\pi^2} \mathbf{sin}\, \pi x\, \mathbf{cos}\, \pi y$$

Numerical solutions are obtained on two sets of nodal spacing. First with 121 nodes (11×11) and second with 676 nodes (26×26). Polynomial order is chosen zero order $\mathbf{p} = [x^0 y^0] = [1]$ and subdomain of every node includes one nearest node on both side and both direction (np=1.0, ns=3.0). Figures 1 and 2 show numerical and exact solution of two sections (0.2, 0.5meter) of the problem domain for two nodal distributions.

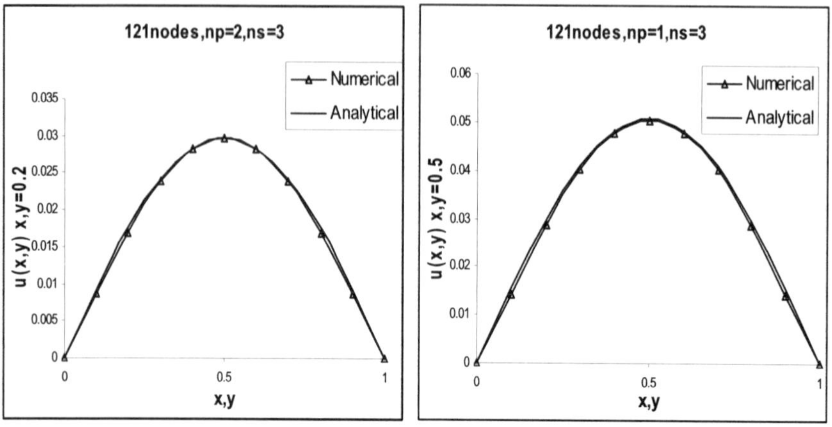

Figure 1: Section plot of Laplace solution with 121nodes.

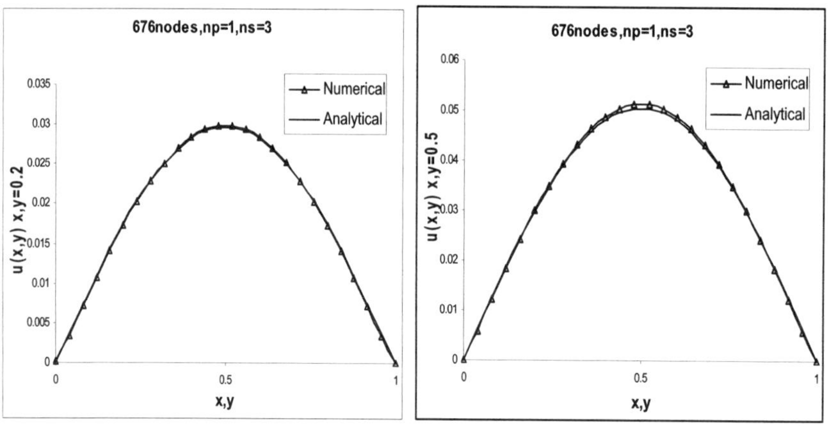

Figure 2: Section plot of Laplace solution with 676 nodes.

4.2 Seepage problem

We consider Seepage problem with this governing equation and Dirichlete and Newmann boundary conditions.

$$\frac{\partial^2 \phi}{\partial x^2} + \frac{\partial^2 \phi}{\partial y^2} = 0 \qquad \Omega(x, y) : \begin{Bmatrix} 0 \le x \le 2 \\ 0 \le y \le 1 \end{Bmatrix}$$

Subject to

$\phi = 35.0 \qquad 0 \le x \le 1 \quad , \quad y = 1$

$\phi = 0.0 \qquad 1 \le x \le 2 \quad , \quad y = 1$

$\frac{\partial \phi}{\partial n} = 0.0 \qquad$ on other boundaries

Problem domain discretizes 3321(41×81) nodes $(\Delta x = \Delta y = 0.025)$. A polynomial of zero order $p = [x^0 y^0] = [1]$ is used (np=1). Same as previous example every subdomain includes two nearest neighbor nodes on both side and both direction (ns=3). Figure 3 shows a countor plot of ϕ results in Problem domain. As shown in figure the distribution of potential are smooth inparticular near the Newmann boundaries.

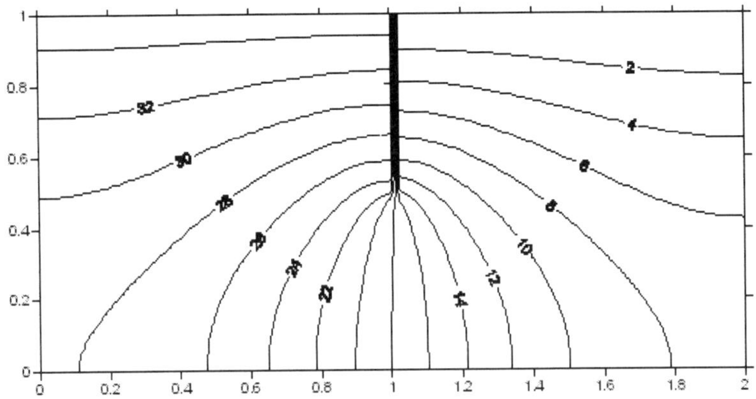

Figure 3: Seepage problem solution (np=1, ns=3).

5 Concluding remarks

In this paper, we present Discrete Least Squares (DLS) meshless method for the solution of elliptic problems. A fully Least Squares method is used in both function approximation and the discretization of the governing differential equations. The meshless shape functions are derived using the Moving Least Squares (MLS) method of function approximation. The discretized equations obtained via a discrete least squares method in which the sums of the squared residuals minimized with respect to unknown nodal parameters. The proposed method has the additional advantages of the producing symmetric, positive definite matrices even for non-self adjoint operators. The method is tested against two elliptic examples in two dimensional steady state forms.

References

[1] Monaghan, J.J., An introduction to SPH, Comput. Phys. Comm. **48** (1988) 89-96.

[2] Liu, W.K., Li, S., Adee, J. & Belytschko, T., Reproducing Kernel Particle methods, Int. Journal For Numerical Methods fluids, **20**, (1995) 1081-1106.

[3] Melenk, J.M. & Babuska, I., The partition of unity finite element method: basic theory and applications, Comput. Methods Appl. Mech. Engng.1999, **139**, 289-314.

[4] Durate, C.A. & Oden, J.T., HP clouds –An h-p meshless method, Numer. Meth. Partial Diff. Eqns. 1996; **12**, 673-705.

[5] Nayroles, B., Touzot, G. & Villon, P., Generalizing the finite element method: diffuse approximation and diffuseelements , Comput. Mech. **10** (1992), 307-318

[6] Belytschko, T., Liu, Y. & Gu, L., Element Free Galerkin methods, Int Journal For Numer Methods Engng, **37**, (1994) 229-256.

[7] Atluri, S.N, Zhu, T., A New meshless local Petrov-Galerkin (MLPG) approach in computational mechanics. Computational Mechanics 1998; **22**, 117- 127.

[8] Zhu, T., Zhang, J. &. Atluri, S.N., A local boundary integral equation (LBIE) method in computational mechanics, and meshless discretization approach. Computational Mechanics 1998; **21**, 223-235.

[9] Beissel, S. & Belytschko, T., Nodal integration of the Element Free Galerkin method. Comput. Methods. Appl. Mech. Engng **139**, (1996) 49-74.

[10] Zhou, J.X., Wen, J.B., Zhang, H.Y. & Zhang, L., A nodal integration and post- processing technique based on Voroni diagram for Galerkin meshless methods. Comput. Methods Appl. Mech. Engng. 2003; **192**, 3831-3843.

[11] Chen, J.K. & Beraun, J.E., A generalized smoothed particle hydrodynamics method for nonlinear dynamic problems, Comput. Methods. Appl. Mech. Engng, **190** (2000) 225- 239.

[12] Lancaster, P. & Salkauskas, K., Surfaces generated by moving least square methods Math. Comput. **37** (1981).

[13] Onate, E., Idelsohn, S., Zienkiewicz, O.C., Taylor, R.L. & Sacco, C., A stabilized finite point method for analysis of fluid mechanics problems, Comput. Methods. Appl. Mech. Engng. **139** (1996), 315-346.

Evaluation of Asian option by using RBF approximation

E. Kita, Y. Goto, F. Zhai & K. Shen
Graduate School of Information Sciences, Nagoya University, Nagoya, Japan

Abstract

This paper describes the evaluation of the price of the Asian option by using the radial bases function (RBF) approximation. In the previous study, we described the evaluation method of the European and the American options. In this paper, the Asian option is considered. A governing differential equation is discretized with the Crank-Nicholson scheme and the RBF approximation. The system of equations is solved for the option price. The numerical results are compared with the FDM solutions in order to confirm the validity of the formulation.

1 Introduction

Recently, financial derivatives have been widely used and their importance has increased. The importance of the derivative transaction is increasing for the adequate sharing of the financial risk. The option transaction is one of the most important financial derivatives and therefore, several schemes have been presented by many researchers for their pricing [1, 2].

Several financial options have been developed; the European option, American option, Look-Back option, Exotic option and so on. In previous studies, the authors described the pricing of the European and American options [3, 4]. The Asian option will be considered in this paper. In the Asian option, the payoff is performed according to the time-average value of the asset price. The Asian option can be classified into the average rate option, the average strike option and so on. While, in the former, the payoff depends on the difference between the time-average value of the asset of the asset price and the expiration price, in the latter, the payoff depends on the difference between the average value and the asset price on an expiration date. In this paper, we focused on the average strike option.

WIT Transactions on Modelling and Simulation, Vol 42, © 2006 WIT Press
www.witpress.com, ISSN 1743-355X (on-line)
doi:10.2495/BE06004

The price of the average strike option can be evaluated as the solution of the Black-Scholes differential equation by taking the payoff condition on an expiration date. The Black-Scholes equation is discretized according to the Crank-Nicolson scheme on the time axis and the option price is approximated with Radial Bases Function with unknown parameters at each time step. The initial value of the parameter on the expiration date is determined from the payoff condition. Then, the parameters on the pricing day are evaluated according to the backward algorithm from the expiration date to the pricing date. The numerical solutions are compared with the finite difference solutions.

The remaining of the paper is organized as follows. In section 2, the evaluation of the average strike option is formulated. The numerical examples are shown in section 3. Finally, the obtained results are summarized in section 4.

2 Formulation

The Asian option is also known as the average option. In the American and European options, the payoff depends on the difference between the asset price $S(t)$ and the expiration price. On the other hand, in the Asian option, the time-average value of the asset price $S(t)$ is estimated first and then, the payoff is exercised according to the difference between the time-average value and the expiration price or the asset price on the expiration date. If the payoff depends on the expiration price, the Asian option is called as the average rate option. If the payoff depends on the asset price on the expiration date, the Asian option is called as the average strike option.

In this study, we will consider the European-type average strike option.

2.1 Governing equation and boundary condition

First, we will define the time-average value of the asset price S as the function:

$$I = \int_0^t S(\tau) d\tau. \tag{1}$$

In the European-type average strike option, the payoff depends on the difference between the time-average value and the asset price on the expiration date. The governing differential equation of the option is given as:

$$\frac{\partial V}{\partial t} + S\frac{\partial V}{\partial I} + \frac{1}{2}\sigma^2 S^2 \frac{\partial^2 V}{\partial S^2} + rS\frac{\partial V}{\partial S} - rV = 0 \tag{2}$$

If the function R is defined from the asset price S as

$$\begin{aligned} R &= \frac{1}{S}\int_0^t S(\tau) d\tau \\ &= \frac{I}{S}, \end{aligned} \tag{3}$$

the price V is given as

$$V(S, R, t) = SH(R, t). \tag{4}$$

Substituting equations (3) and (4) to equation (2), we have

$$\frac{\partial H}{\partial t} + FH = 0, \tag{5}$$

where the operator F is defined as

$$F = \frac{1}{2}\sigma^2 R^2 \frac{\partial^2}{\partial R^2} + (1-r)R\frac{\partial}{\partial R}. \tag{6}$$

The payoff condition of the average strike option on the expiration date $t = T$ is defined as follows, in the case of European-call type,

$$\max\left(S - \frac{1}{T}\int_0^t S(\tau)d\tau, 0\right) \tag{7}$$

and, in the case of European-put type,

$$\max\left(\frac{1}{T}\int_0^t S(\tau)d\tau - S, 0\right). \tag{8}$$

where $\max(a_1, a_2)$ means the bigger one among a_1 and a_2.

Now, we consider the pricing of the average strike option in the call-type.

Substituting equations (3) and (4) to (7), we have the payoff condition on the expiration date $t = T$;

$$SH(R,T) = S\max\left(1 - \frac{R}{T}, 0\right),$$

and therefore,

$$H(R,T) = \max\left(1 - \frac{R}{T}, 0\right). \tag{9}$$

Finally, the governing equation and the boundary condition of the average strike option are given by equation (5) and (9), respectively.

2.2 Solution using RBF

Discretizing the equation (5) with Crank-Nicolson Scheme, we have

$$\frac{H(t+\Delta t) - H(t)}{\Delta t} + (1-\theta)FH(t+\Delta t) + \theta FH(t) = 0 \tag{10}$$

where the parameter θ is taken in the range of $0 \leq \theta \leq 1$.

Defining the parameters $H(t) = H^m$ and $H(t+\Delta t) = H^{m+1}$, we have

$$AH^{m+1} = BH^m \tag{11}$$

where

$$A = 1 + (1 - \theta)\Delta t F$$
$$B = 1 - \theta \Delta t F.$$

The derivative price H governed with the equation (5) is approximated with the RBF function as

$$H = \sum_{n=1}^{N} \lambda_n \phi_n \tag{12}$$

where N and λ_j denote the total number of data points and the unknown parameters, respectively.

Substituting equation (12) to equation (11), we have

$$A \sum_{n=1}^{N} \lambda_n^{m+1} \phi_n = B \sum_{n=1}^{N} \lambda_n^{m} \phi_n$$
$$\sum_{n=1}^{N} A\phi_n \lambda_n^{m+1} = \sum_{n=1}^{N} B\phi_n \lambda_n^{m} \tag{13}$$

2.3 Algorithm

The algorithm of the solution procedure is defined as

1. Distribute N data points on $0 \leq R \leq R_{max}$ and discretize $0 \leq t \leq T$ with T/M.
2. Solve equation (12) to evaluate H^T on the expiration date $t = T$.
3. Approximate H^T by equation (12) to evaluate λ_n^T on the expiration date $t = T$.
4. $t \leftarrow T - \Delta t$.
5. Solve equation (13) to estimate λ_n^t.
6. $t \leftarrow t - \Delta t$.
7. IF $t \neq 0$, go to step 5.
8. Evaluate H^0 from equation (12) and λ_n^0 on the date $t = 0$.

3 Numerical example

First we will adopt the radial bases function:

$$\phi(R, R_j) = \sqrt{c^2 + \|R - R_j\|^2} \tag{14}$$

where $r_j^2 = \|S - S_j\|$.

The parameters are specified in Table 1. The total number of the data points are 101. They are distributed uniformly in the range of $0 \leq R \leq 1.0$. For comparison with the finite difference solutions, the time-step size is taken as $\Delta t = 0.0005$; the number of the time-step is $M = 1000$.

Table 1: Parameters for numerical result.

Expiration date	$T = 0.5$ [year]
Risk free interest rate	$r = 0.1$
Volatility	$\sigma = 0.4$
Crank-Nicolson method	$\theta = 0.5$
Maximum R	$R_{max} = 1.0$
Time step size	$\Delta t = 0.0005$
Number of time step	$M = 1000$
Number of stock data points	$N = 101$

Table 2: The condition number to each c.

RBF parameter c	Condition number
0.02	1.32×10^6
0.025	5.65×10^6
0.03	2.47×10^7
0.035	1.09×10^8
0.04	4.86×10^8
0.045	2.18×10^9
0.05	9.87×10^9

For determining the parameter c in the equation (13), we will estimate the condition number of the coefficient matrix $B\phi_j$ in equation (12). The results are shown in Table 2. In the case of the parameter $c = 0.04$, the numerical results are shown in Fig. 1.

In Fig. 1, the abscissa and the ordinate denote R and H, respectively. Figure 1 indicates that the price H dose not well converge to 0.

For improving the computational accuracy, instead of the above RBF (14), we will take the another RBF:

$$\phi(R, R_j) = \frac{1}{\sqrt{c^2 + \|R - R_j\|^2}} \tag{15}$$

The parameter c in equation (15) is taken as $c = 0.04$ and the other parameters are specified in Table 1. The numerical results are shown in Fig. 2. The finite difference solutions are shown in Fig. 3. We notice from Figs. 1 and 2 that the use of the equation (15) obtain good convergence to improve the computational accuracy.

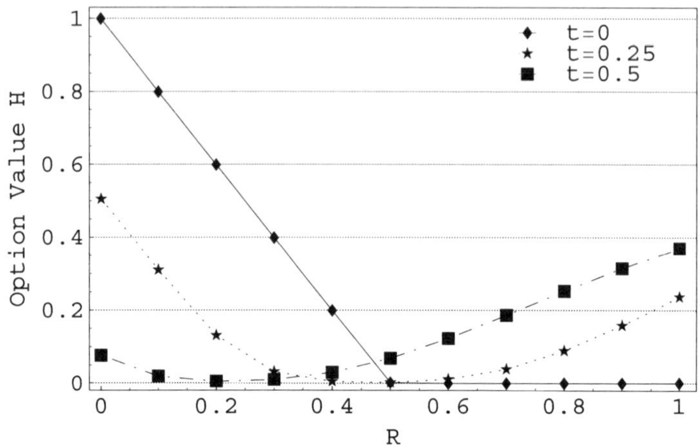

Figure 1: Values of European average strike call option, RBF: Multiquadrics, $c = 0.04$.

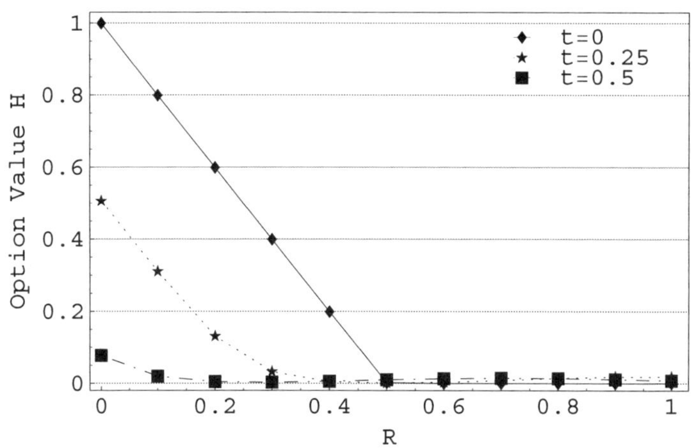

Figure 2: Values of European average strike call option, RBF: Reciprocal Multiquadrics, $c = 0.04$.

4 Conclusions

This paper described the evaluation of the Asian option by using the RBF approximation. The Asian option can be classified into the average rate option and the average strike options. While, in the former, the payoff depends on the difference between the time-average value of the asset price and the expiration price, in the

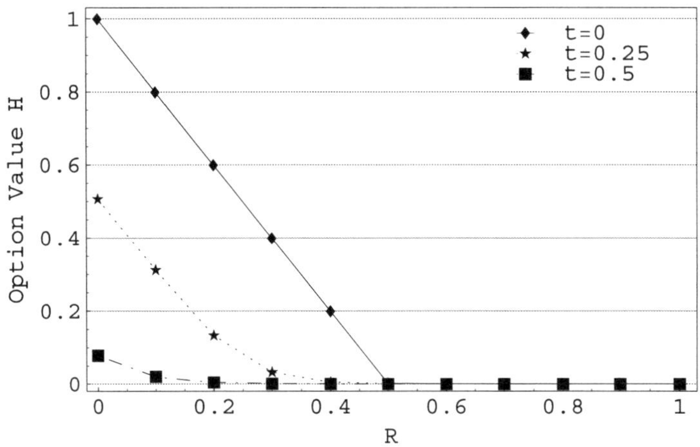

Figure 3: Values of European average strike call option, FDM.

latter, the payoff depends on the average value and the asset price on the expiration date. In this paper, we focused on the average strike option.

In the average strike option, the introduction of the new function leads to the different governing equation as the European and the American options. The use of the Crank-Nicholson scheme and the RBP approximation transforms the governing differential equation to the system of equations. The system of equations are solved in the backward algorithm from the expiration time $t = T$ to the time $t = 0$.

First, the multi quadratic RBF was adopted for the analysis. The results show that the convergence of the solution is not good. Next, the reciprocal multi quadratic RBF was applied. The results converged well and agreed well with the finite difference solutions.

In the future plan, we are going to apply the formulation to the other options.

References

[1] G. Courtadon. A more accurate finite difference approximation for valuation of options. *Journal of Financial and Quantitative Analysis*, Vol. 17, pp. 697–703, 1982.

[2] P. Wilmott, J. Dewynne, and S. Howison. *Option Pricing: Mathematical Models and Computation*. Oxford Financial Press, 1993.

[3] E. Kita and Y. Goto. Evaluation of the european stock option by using the rbf approximation. In A. Kassab, C. A. Brebbia, E. Divo, and D. Poljak, editors, *Boundary Elements XXVII (Orlando, USA, 2005)*, pp. 91–98, 2005.

[4] Y. Goto and E. Kita. Estimation of american option using radial bases function approximation. In V. M. A. Leitao, C. J. S. Alves, and C. A. Duarte, editors, *Proceedings of the ECCOMAS Thematic Conference on Meshless Methods (Meshless 2005)*, pp. D41.1–4, 2005.

Bumps modeling using the principal shear stresses

P. P. Prochazka & A. Yiakoumi
Association of Czech Concrete Engineers, CTU in Prague, Civil Engineering, Prague, Czech Republic

Abstract

In several previous papers by the authors the problem of bumps occurrence has been solved by distinct element methods, namely by the free hexagon method. The latter method proved its significance in the description of the most probable nucleation of debonding of rock mass, when cracks or flaws occurred, which at the moment of bumps create the contact surface of the moving part of the coal seam. The movement of particles after bumps is described. In this paper another approach is used. Starting with the aim of describing whether the bumps occur or not, or under which condition they appear, the continuum of both the rock seam and the overburden (rock) is considered. The decision whether the bumps are triggered can then be derived from conditions at different points of the coal seam. For the solution, the boundary element method is used and elastic behavior (more precisely brittle behavior of the material of coal) is assumed.
Keywords: bumps in deep mines, the most probable disconnecting curve, slip condition, constitutive behavior.

1 Introduction

Bumps or rock bursts are phenomenon, which occurs in deep mines and mostly it is qualified as a sudden release of internal energy concentrated at the face of the the shaft in longwall mining. This extreme concentration can be caused by material changes in the overburden rock massif, of distant local disturbances, like creation of cracks, emission of gas, human activities in the neighborhood of the site of mining, etc. In any case, it is necessary not to forget the fact that along the upper part of the shaft where the ceiling and the rock are in contact, a tip of a crack (notch) causes a natural accumulation of energy. From classical theories

WIT Transactions on Modelling and Simulation, Vol 42, © 2006 WIT Press
www.witpress.com, ISSN 1743-355X (on-line)
doi:10.2495/BE06005

based on elastic solution such a point is classified as a mass point of singularities, and even involving plasticity this point is the most dangerous from the standpoint of failure of the structure. The latter fact is very important for our next considerations.

It is well known that fracture mechanics problems can be substituted by contact problems with introducing proper new parameters, which are connected with certain law along the contact. This law has to be selected very attentively. In most cases generalized Mohr-Coulomb hypothesis is adopted and the material parameters can be obtained from straightforward experiments. Such experiments are difficult to carry out in the case of assessment of cracks. This is why we consider contact problem instead the problem from the field of fracture mechanics.

Both parts, the overburden and the coal seam (layer), are mechanically described by boundary elements. As the mechanical behavior of a part of the overburden is assumed to be in plastic state, particularly in the vicinity of the tip of the crack, a subregion, or subdomain, of this part is treated separately (not disconnected from the continuum). A very powerful trick is used: the polarization tensor is introduced and the problem is solved in other quantities. This enables us to solve plasticity in a small part of the overburden domain. The plastic strains are involved in eigenstrains, which in this case serve the plastic strains. It will be seen that in the case of the boundary element method it is very useful. For the sake of clarity the structure under consideration is depicted in Fig. 1.

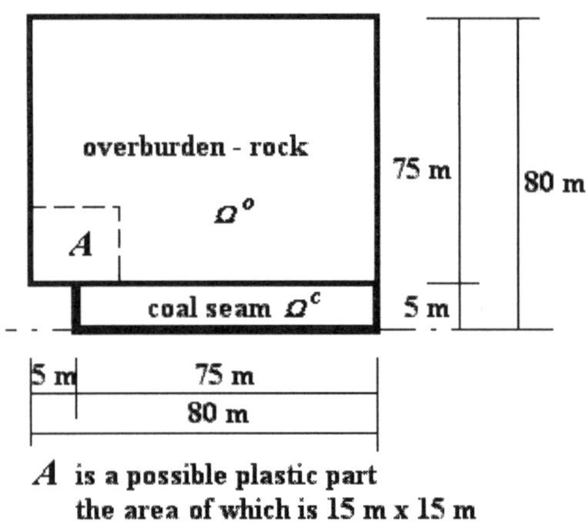

Figure 1: Geometry of the structure under consideration.

Cundall in [1] established basic ideas on distinct elements, which can describe large displacements in structures. Dynamic equilibrium is considered and ball elements simulate the 3D movement of particles. A generalization is given in [2] concerning granular material. Onck et al. came with hexagonal elements, which, in opposite to Cundall [1], enable one to describe also stresses in the structure. A disadvantage of this approach is the fact that FEM is used for a description of internal behavior in each element and there is a numerical discrepancy in the formulation. This disadvantage is removed in [4], where the boundary element method is used for description of material behavior inside the elements. Free hexagons are applied in both last cases. Contact problems are analyzed in the classical work by Duvant and Lions, [5]. In [6], basic ideas of damage in material are introduced. For the purpose of the creation of numerical models the introduction of coupled modeling (scale modeling in stands and numerical models) appears to be very promising. This was published in [7] and [8], for example. This paper intends to find out the principles on which a suggested prediction of bumps can be based. The large displacements are not considered, as they are not important from the point of view of the safety of miners. The decision appears decisive, whether or not the bumps occur.

2 Generalized Hooke's law

Our aim in this paper is to overcome the problems appearing in discrete formulations with partial or total cavities created in the vicinity of the tip of the crack at the upper part of the shaft, which consists in the unique decision, when the bumps occur. These problems are due to crack formulations, or equivalent contact formulations, involving the influence of the tip of the crack. In our formulation the debonding zones will be stated and the disconnecting of fibers from matrix will be described by additional quantities - eigenparameters. The eigenparameters will also describe the inelastic behavior of the matrix.

First, let us introduce the eigenparameters in Hooke's law:

$$\begin{aligned} \boldsymbol{\sigma} &= \mathbf{L}(\boldsymbol{\varepsilon} - \boldsymbol{\mu}) = \mathbf{L}\boldsymbol{\varepsilon} + \boldsymbol{\lambda} \\ \boldsymbol{\varepsilon} &= \mathbf{M}\boldsymbol{\sigma} + \boldsymbol{\mu} = \mathbf{M}(\boldsymbol{\sigma} - \boldsymbol{\lambda}) \end{aligned} \tag{1}$$

where $\boldsymbol{\sigma}$ is the stress field, $\boldsymbol{\varepsilon}$ is the strain tensor, $\mathbf{L}$ is the elastic material stiffness matrix of linear elasticity, $\mathbf{M}$ is its compliance, $\boldsymbol{\mu}$ is the eigenstrain tensor and $\boldsymbol{\lambda}$ is the eigenstress tensor. In the matrix the eigenstrain may stand for plastic strain and the eigenstress for relaxation stress. The above formulation may serve for calculation of plastic states in material. If the eigenparamters are zero, the stiffness matrix turns to elastoplastic, in the case of elastoplastic states, for example. If eigenparameters vary, the matrix $\mathbf{L}$ can be considered as constant. The last description of plastic states leads us to the most general case, when both eigenparameters are non-zero and the stiffness matrix changes. A similar assertion holds for the second law (1).

Let us now denote the domain of overburden as Ω^o and the coal Ω^c. All quantities belonging to the overburden will bear the superscript o and that belonging to the coal seam will have the superscript c. These subdomains are not overlapping. Two-dimensional case is studied.

3 Integral formulation

The aim of this paragraph is to show some possible relations involving stresses (or strains) and eigenstresses (or eigenstrains) in the structures. Consider the domain $\Omega^o \in R^2$ with the boundary Γ^o, describing the overburden and $\Omega_c \in R^2$ with the boundary Γ^c, which describes the coal seam. The common part is denoted by $\Gamma^i \equiv \Gamma^o \cap \Gamma^c$ and the outer boundary $\Gamma \equiv \Gamma^o \cup \Gamma^c - \Gamma^i$. The boundary Γ is split as follows: $\Gamma \equiv \Gamma_u \cup \Gamma_p, \Gamma_u \cap \Gamma_p = 0$.

On Γ_u the displacements $u_i = \overline{u}_i, i = 1,2,3$ are prescribed, while on Γ_P the tractions $p_i = \overline{p}_i$, $i = 1,2,3$ are given. The primed quantities are prescribed.

Let us split the process into two steps. In the *first step* consider elastic behavior of both domains Ω^o and Ω^c, so that linear homogeneous and isotropic Hooke's law with material stiff nesses $\mathbf{L}_o^0$ being valid on Ω^o and $\mathbf{L}_c^0$ on Ω^c holds:

$$(\sigma_o^0)_{ij} = (L_o^0)_{ijkl}(\varepsilon_o^0)_{kl} \quad \text{in} \quad \Omega^o \quad \text{and} \quad (\sigma_c^0)_{ij} = (L_c^0)_{ijkl}(\varepsilon_c^0)_{kl} \quad \text{in} \quad \Omega^c,$$
$$u_i^0 = \overline{u}_i \ \text{ on } \Gamma_u, \ p_i^0 = \overline{p}_i \ \text{ on } \Gamma_p. \tag{2}$$

and the boundary quantities in this step are denoted by superscript 0. The BEM solution makes no difficulties, since for the homogeneous and isotropic elasticity the fundamental solution exists. We get the boundary displacements $\mathbf{u}^0$, the tractions $\mathbf{p}^0$, the "small" strain tensor ε^0, and the distribution of stresses σ^0. These quantities are considered to be known in what follows. Note that compliance matrices $\mathbf{M}_o^0$ and $\mathbf{M}_c^0$ can be formulated as the inverse to stiffness matrices $\mathbf{L}_o^0$ and $\mathbf{L}_c^0$, respectively.

In the *second step* the shape of the original body (the same as is in the first case) is held together with the boundary displacements $\mathbf{u} = \overline{\mathbf{u}} \in \Gamma_u$ and tractions $\mathbf{p} = \overline{\mathbf{p}} \in \Gamma_p$. The body under study in the second step is the real one; it means that it involves plastic behavior in Ω^o but elastic behavior in Ω^c. The real displacements $\mathbf{u}$, strain ε, and stress σ are unknown and the Hooke's law holds, involving also the eigenparameters:

$$\sigma_{ij} = L_{ijkl}\varepsilon_{kl} + \lambda_{ij}, \ \lambda_{ij} = -L_{ijkl}\mu_{kl} \ \text{in } \Omega \equiv \Omega^o \cup \Omega^c. \tag{3}$$

The symmetric polarization stress tensor τ is introduced as ($\mathbf{L}^0 = \mathbf{L}_o^0$ on Ω^o and $\mathbf{L}^0 = \mathbf{L}_c^0$ on Ω^c):

$$\sigma_{ij} = L^0_{ijkl}\varepsilon_{kl} + \tau_{ij} \tag{4}$$

Then we introduce new variables:

$$u'_i = u_i - u^0_i \text{ in } \Omega \text{ , } u'_i = 0 \text{ on } \Gamma_u \tag{5}$$

and also

$$\varepsilon'_{ij} = \varepsilon_{ij} - \varepsilon^0_{ij} \text{ , } \sigma'_{ij} = \sigma_{ij} - \sigma^0_{ij} \text{ in } \Omega \text{ , } p'_i = 0 \text{ on } \Gamma_p \tag{6}$$

From (4) and (6_2) one gets:

$$\sigma'_{ij} = L^0_{ijkl}\varepsilon'_{kl} + \tau_{ij} \text{ in } \Omega \tag{7}$$

Since both σ and σ^0 are statically admissible, it holds:

$$\frac{\partial \sigma'_{ij}}{\partial x_j} = \frac{\partial (L^0_{ijkl}\varepsilon'_{kl} + \tau_{ij})}{\partial x_j} = 0 \quad \text{in } \Omega \tag{8}$$

$$\tau_{ij} - [L]_{ijkl}\varepsilon_{kl} - \lambda_{ij} = 0 \quad \text{in } \Omega \tag{9}$$

and

$$[L]_{ijkl} = L_{ijkl} - L^0_{ijkl} .$$

Recall that the boundary conditions in the primed system are written as:

$$\mathbf{u}'_i = 0 \text{, on } \Gamma_u \text{ , } \sigma_{ij} n_j = 0 \text{ on } \Gamma_p \tag{10}$$

Let us concentrate our attention on basic properties of the primed system. In the domain Ω the static equations (8) hold. The kinematical equations read:

$$\varepsilon'_{ij}(\mathbf{x}) = \frac{1}{2}\left(\frac{\partial u'_i(\mathbf{x})}{\partial x_j} + \frac{\partial u'_j(\mathbf{x})}{\partial x_i} \right). \tag{11}$$

The relation (7) can be written in more details as:

$$\sigma'_{ij} = 2G^0\varepsilon'_{ij} + \delta_{ij}\frac{2G^0\nu^0}{1-2\nu^0}\varepsilon'_{kk} + \tau_{ij} = 2G^0\varepsilon'_{ij} + \lambda^0\delta_{ij}\varepsilon'_{kk} + \tau_{ij} \text{ ,} \tag{12}$$

and Kronecker's delta $\delta_{ij} = 1$ for $i = j$ and in the contrary it is equal to zero. Material constants G^0 and λ^0 are Lame's constants (G^0 is also referred to as the shear modulus). All these constants are valid in the elastic medium and all these material constants are uniform in the Ω, as they create $\mathbf{L}^0$.

We briefly mention the integral formulation of nonlinear behavior of materials. In standard way the integral formulation of elasticity with involving the eigenparameters reads:

$$\begin{aligned} u'_m(\xi) = \int_{\Gamma_p} p'_i(\mathbf{x}) u^*_{im}(\mathbf{x};\xi) \mathrm{d}\Gamma(\mathbf{x}) - \int_{\Gamma_u} u'_i(\mathbf{x}) p^*_{im}(\mathbf{x};\xi) \mathrm{d}\Gamma(\mathbf{x}) + \\ + \int_{\Omega} \varepsilon^*_{ijm}(\mathbf{x};\xi) \tau_{ij}(\mathbf{x}) \mathrm{d}\Gamma(\mathbf{x}), \quad \xi \in \Omega, \quad k = 1,2,3, \end{aligned} \tag{13}$$

Note that the equation (13) is derived for points of observer belonging to the domain. Since the kernels with asterisks are used for the elastic medium, the primed system possesses a solution provided that the values of primes quantities and distribution of polarization tensor are known. This is obviously not true and we generalize (13) in the well-known way:

$$\begin{aligned} c_{m\alpha}(\xi) u'_\alpha(\xi) = \int_{\Gamma_u} p'_i(\mathbf{x}) u^*_{im}(\mathbf{x};\xi) \mathrm{d}\Gamma(\mathbf{x}) - \int_{\Gamma_p} u'_i(\mathbf{x}) p^*_{im}(\mathbf{x};\xi) \mathrm{d}\Gamma(\mathbf{x}) + \\ + \int_{\Omega} \varepsilon^*_{ijm}(\mathbf{x};\xi) \tau_{ij}(\mathbf{x}) \mathrm{d}\Omega(\mathbf{x}), \qquad k = 1,2,3. \end{aligned} \tag{14}$$

In (14) we introduced the matrix $\mathbf{c}$, which arises from positioning the point ξ on the boundary Γ, due to the singular nature of the kernel $\mathbf{p}^*$. The matrix $\mathbf{c}$ possesses the known properties: at point $\xi \in \Omega$, the matrix $\mathbf{c}$ is the unit matrix; at point $\xi \in \Gamma$ and in the vicinity of this point on the boundary is smooth, the matrix $\mathbf{c} = \frac{1}{2}\mathbf{I}$, where $\mathbf{I}$ is the unit matrix (the matrix $\mathbf{c}$ is the zero matrix in the case $\xi \notin \Omega$); at point ξ, positioned at a vertex on the boundary, the values of $\mathbf{c}$ are dependent on the angle of the vertex.

Suppose $\xi \in \Omega$. Differentiating (13) by ξ_n and using kinematical equations one obtains:

$$\begin{aligned} \varepsilon'_{mn}(\xi) = \int_{\Gamma_u} p'_i(\mathbf{x}) h^*_{imn}(\mathbf{x};\xi) \mathrm{d}\Gamma(\mathbf{x}) - \int_{\Gamma_p} u'_i(\mathbf{x}) j^*_{imn}(\mathbf{x};\xi) \mathrm{d}\Gamma(\mathbf{x}) + \\ + \int_{\Omega} \tau_{ij}(\mathbf{x}) \varepsilon^*_{ijmn}(\mathbf{x};\xi) \mathrm{d}\Omega(\mathbf{x}) + \mathrm{C}[\tau_{mn}(\xi)], \qquad k, l = 1,2,3. \end{aligned} \tag{15}$$

where C is the convicted term, which arises at the internal point $\xi \in \Omega$ when exchanging integration and differentiation. Substitution for the stress polarization tensor (9), (15) leads to the following expression:

$$\begin{aligned}\varepsilon_{mn}(\xi) = \varepsilon^0_{mn}(\xi) + \int_{\Gamma_u} p'_i(\mathbf{x}) h^*_{imn}(\mathbf{x};\xi)\mathrm{d}\Gamma(\mathbf{x}) - \int_{\Gamma_p} u'_i(\mathbf{x}) i^*_{jimn}(\mathbf{x};\xi)\mathrm{d}\Gamma(\mathbf{x}) + \\ + \int_{\Omega} \bar{\varepsilon}^*_{ijmn}(\mathbf{x};\xi)[(L_{ij\alpha\beta}(\mathbf{x}) - L^0_{ij\alpha\beta})\varepsilon_{\alpha\beta}(\mathbf{x}) + \lambda_{ij}(\mathbf{x})]\mathrm{d}\Omega(\mathbf{x}) + \\ + \mathrm{C}[(L_{mn\alpha\beta}(\xi) - L^0_{mn\alpha\beta})\varepsilon_{\alpha\beta}(\xi) + \lambda_{mn}(\xi)].\end{aligned} \tag{16}$$

Using Hooke's law, primed stress field can be obtained. Let us identify the internal cells in the domain by subdomains Ω_k, k = 1,..., N. On these subdomains the stiffness matrix $\mathbf{L}(\mathbf{x})$ turns to $\mathbf{L}^k$, and the compliance $\mathbf{M}(\mathbf{x})$ turns to $\mathbf{M}^k$, no longer depending on a position. Moreover, suppose the eigenstrain $\mu_{ij}(\mathbf{x})$ to be introduced on each subdomain is uniform, i.e. equal to μ^k_{ij}. Then:

$$\begin{aligned}\sigma_{mn}(\xi) = \sigma^0_{mn}(\xi) + \int_{\Gamma_u} p'_i(\mathbf{x}) d^*_{imn}(\mathbf{x};\xi)\mathrm{d}\Gamma(\mathbf{x}) - \int_{\Gamma_p} u'_i(\mathbf{x}) i^*_{imn}(\mathbf{x};\xi)\mathrm{d}\Gamma(\mathbf{x}) + \\ - \sum_{k=1}^{N} M^0_{ij\alpha\beta} \int_{\Omega_k} \bar{\sigma}^*_{ijmn}(\mathbf{x};\xi)\sigma_{\alpha\beta}(\mathbf{x})\mathrm{d}\Omega(\mathbf{x}) + \sum_{k=1}^{N} \mu^k_{ij} \int_{\Omega_k} \bar{\sigma}^*_{ijmn}(\mathbf{x};\xi)\mathrm{d}\Omega(\mathbf{x}) + \\ + \{\text{convected term}\}\end{aligned} \tag{17}$$

Similarly, (13) turns to:

$$\begin{aligned}c_{m\alpha}(\xi) u'_\alpha(\xi) = -M^0_{ij\alpha\beta} \int_{\Omega} \sigma^*_{ijm}(\mathbf{x};\xi)\sigma_{\alpha\beta}(\mathbf{x})\mathrm{d}\Omega(\mathbf{x}) + \\ + \int_{\Gamma_u} p'_i(\mathbf{x}) u^*_{im}(\mathbf{x};\xi)\mathrm{d}\Gamma(\mathbf{x}) - \int_{\Gamma_p} u'_i(\mathbf{x}) p^*_{im}(\mathbf{x};\xi)\mathrm{d}\Gamma(\mathbf{x}) + \\ + \sum_{k=1}^{N} M^k_{ij\alpha\beta} \int_{\Omega_k} \sigma^*_{ijmn}(\mathbf{x};\xi)\sigma_{\alpha\beta}(\mathbf{x})\mathrm{d}\Omega(\mathbf{x}) + \sum_{k=1}^{N} \mu^k_{ij} \int_{\Omega_k} \sigma^*_{ijmn}(\mathbf{x};\xi)\mathrm{d}\Omega(\mathbf{x}).\end{aligned} \tag{18}$$

Using standard procedure for boundary element method and eliminating both $\mathbf{u}'$ and $\mathbf{p}'$ (what is possible because of a special nature of the problem), we get relation between stresses and eigenstrains. In the end we identify the elastic medium with the elastic part of Ω^o, i.e. with $\Omega^o - A$, see Fig. 1, and no eigenparameters are considered in this domain. The same is valid for the coal seam, as the brittle behavior is assumed there.

Now we can decide whether the nonlinear behavior will occur in the neighborhood of the tip of the crack (being positioned at the upper right corner of the shaft). In what follows, von Mises-Huber-Hencky plasticity will be adopted in the overburden, while the coal seam remains linear (brittle) till some peak point (shear strength). It will be seen that the development of plasticity plays decisive role in bumps occurrence. The plastic development is strictly dependent on the speed of mining.

4 Examples

The coal seam is assumed to be about 1 000 m deep. The overburden and the coal seam are created from material with the following properties:
Rock: $E_{\text{elastic}} = 21$ GPa, $E_{\text{yeild}} = 17$ GPa, $E_{\text{residual}} = 17$ GPa, $\sigma_{+} = 0.1\,\text{GPa}$
$\nu_{\text{elastic}} = 0.33$, $\nu_{\text{yield}} = 0.33$
Coal seam: $E_{\text{elastic}} = 17$ GPa, $\nu_{\text{elastic}} = 0.28$, strength = peak stress = 16 GPa
$\tau_b = 2.7$ GPa, $\tan\varphi = 0.28$.

In Fig. 2 hypsography of principal stresses are drawn to show us that the stress is concentrated in the close vicinity of the crack tip, being positioned at the upper part of the face of the shaft. Concentrations are very high, and the division into internal cells influences the accuracy of the results. In elastic case infinite peaks can be expected.

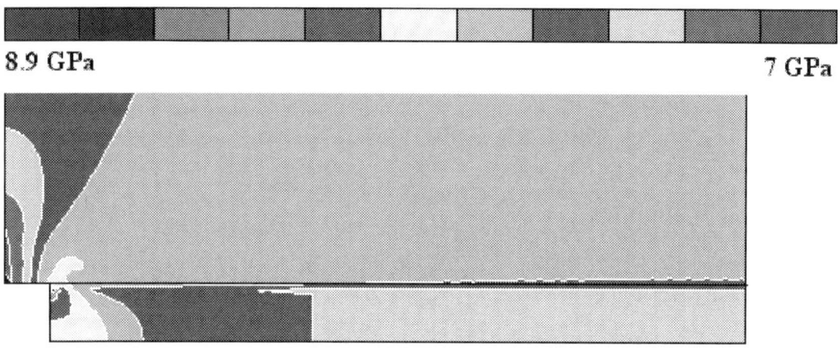

Figure 2: Hypsography of principal shear stresses – elastic case.

A different situation is described in Fig. 3 where plasticity is reached and a part of the region A is fully plasticized. At the most dangerous point the stresses principally drop and the danger of bumps occurrence is basically lower than before. Such a situation can occur when very slow mining takes place.

For the sake of completeness, distribution of horizontal displacements are shown in Fig. 4. Because the coal seam is very deep, the lower boundary of the body can be considered as the axis of symmetry. From these three pictures it is seen that the most probable shape of the contact zone between the steady and moving part of the coal after the bumps approaches a parabola. This is also in agreement with the experiments either on site or in laboratory.

5 Conclusions

In our study a numerical model for prediction of bumps occurrence is described. It starts with formulation on a continuum, contrary to the previous discrete models describing this very dangerous phenomenon. The numerical model presented cannot describe the movement of particles after bumps occurrence, but can decide after some additional information, whether the miners are endangered or when the situation is safe. Note that a similar situation occurs when assessing the stability of slopes. The designer is mostly interested about whether the failure of such a slope is not probable. If the threat of failure of the slope is high, then the designer will not be concerned of how the movement of the soil mass continues.

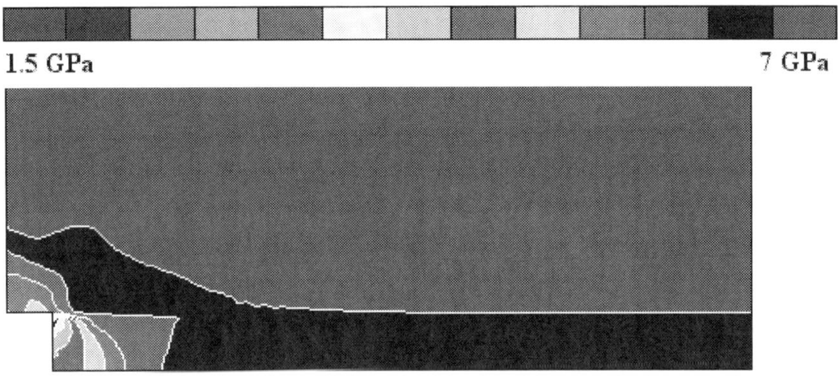

Figure 3: Hypsography of principal shear stresses – fully plastic case.

Figure 4: Hypsography of horizontal displacements – fully plastic state.

The safety margin of the bumps can be determined from the situation on possible slip surfaces, given from the contour lines of principal shear stress obtained from our computation.

A very important problem still remains: How to connect the velocity of mining together with the material properties with the velocity of development of plastic regions in the neighborhood of the crack tip. This will require long ranged study with different materials in the laboratory. On the other hand, the rock and coal in deep mines behaves quite differently to the material, which is not loaded, i.e. is nearby the terrain. This is due to a principal difference between aeromechanics and mining engineering.

Acknowledgement

Financial support of this research was provided by the Grant agency of the Czech Republic, No. 103/05/0334.

References

[1] Cundall, P.A. 1971. A computer model for simulation progressive large-scale movements of blocky rock systems. Symposium of the international society of rock mechanics: 132-150.

[2] Moreau, J.J. 1994. Some numerical methods in multibody dynamics: Application to granular materials. Eur. J. Mech. Solids, 13, (4): 93-114.

[3] Onck, P. & van der Giessen, E. 1999. Growth of an initially sharp crack by grain boundary cavitation. Jour. Mech. And Physics of Solids 28: 328-352.

[4] Procházka, P. 2004. Application of Discrete Element Methods to Rock Bumps. Engineering Fracture Mechanics, 45: 254-267.

[5] Duvant, J. & Lions, J.P. 1972. Variational Inequalities in Mechanics. DUNOD, Paris.

[6] Kachanov, L.M. 1992, Introduction to Continuum Damage Mechanics. Martinus Nijhoff Publishers, Dordrecht, Netherlands.

[7] Procházka, P. & Trčková, J. 2000. Coupled modeling of Concrete Tunnel Lining. Our World in Concrete and Structures, Singapore: 125-132.

[8] Trčková, J. & Procházka, P. 2001. Coupled modeling of tunnel face stability. Proc. ISRM 2001 – 2nd ARMS, A.A. Balkema Publishers: 283-286.

Integral equations for elastic problems posed in principal directions: application for adjacent domains

A. N. Galybin[1] & Sh. A. Mukhamediev[2]
[1] *Wessex Institute of Technology, Southampton, UK*
[2] *Institute of Physics of the Earth, Moscow, Russia*

Abstract

This article addresses a new type of boundary condition in plane elastic boundary value problems. Principal directions are given on a contour separating interior and exterior domains; the stress vector is continuous across the contour. Solvability of this problem is investigated and the number of linearly independent solutions is determined. Some special cases in which the problem is underspecified have been reported.
Keywords: plane elasticity, boundary value problems, principal directions, complex potentials.

1 Introduction

Classical boundary value problems, BVP, of the plane elasticity require one of the following surface conditions to be known on the entire boundary of a domain (see Muskhelishvili [1]): (i) stress vector; (ii) displacement vector; or (iii) certain combinations of stress and displacement components (mixed problems). In these cases the BVP is well posed and possesses a unique solution.

Galybin and Mukhamediev [2] and Galybin [3] considered different types of BVPs in which magnitudes of stresses, displacements or forces are not specified on the boundary. It has been shown that the BVPs of this type may have a finite number of solutions or be unsolvable. Solvability depends on the, so-called, index of singular integral equations (see, e.g., Gakhov [4]). It can be determined in every particular case from the analysis of principal directions (of the stress tensor), orientation of displacement or forces on the entire boundary of a considered domain.

WIT Transactions on Modelling and Simulation, Vol 42, © 2006 WIT Press
www.witpress.com, ISSN 1743-355X (on-line)
doi:10.2495/BE06006

This article investigates solvability of a new plane elastic BVP with a certain combination of boundary conditions, which has not been addressed before (sections 2 and 3). Namely, principal directions are given on a contour separating interior and exterior domains; the stress vector is continuous across the contour. The problem has direct applications in geodynamics for identification of stresses in adjacent tectonic plates. The article also presents some special (degenerated) cases in which the problem is underspecified (section 4).

2 Singular integral equation of the problem

2.1 Problem formulation in terms of stress functions

Let Γ be a closed contour separating the complex plane into interior Ω^+ and exterior Ω^- domains. Stress states in both domains can be expressed through sectionally holomorphic functions (complex potentials) $\Phi(z)$ and $\Psi(z)$ of complex variable $z=x+iy$ by the Kolosov–Muskhelishvili solution (no body forces)

$$P(z,\bar{z})=\Phi(z)+\overline{\Phi(z)},\quad D(z,\bar{z})=\bar{z}\Phi'(z)+\Psi(z) \tag{1}$$

Here the harmonic function P (mean stress) and complex-valued function D (stress deviator) represent the following combinations of stress components σ_{xx}, σ_{yy} and σ_{xy}.

$$P(z,\bar{z})=\frac{\sigma_{xx}(z,\bar{z})+\sigma_{yy}(z,\bar{z})}{2},\quad D(z,\bar{z})=\frac{\sigma_{yy}(z,\bar{z})-\sigma_{xx}(z,\bar{z})}{2}+i\sigma_{xy}(z,\bar{z}) \tag{2}$$

The stress vector on Γ has the following complex form

$$N^{\pm}(\zeta)+iT^{\pm}(\zeta)=P^{\pm}(\zeta)+\frac{\overline{d\zeta}}{d\zeta}\overline{D^{\pm}(\zeta)},\quad \zeta\in\Gamma \tag{3}$$

Hereafter a function of a single variable stands for the boundary value of this function, "±" denote the boundary values obtained by approaching Γ from domains $\Omega^{\pm}$ respectively; $N^{\pm}$ and $T^{\pm}$ are normal and shear components of the stress tensor on Γ.

Principal directions of the stress tensor, angles $\varphi(z,\bar{z})$, are connected to the argument of the stress deviator, $\alpha(z,\bar{z})$, as follows

$$\varphi(z,\bar{z})=-\frac{1}{2}\alpha(z,\bar{z}),\quad \alpha(z,\bar{z})=\arg D(z,\bar{z}) \tag{4}$$

The following boundary value problem, BVP, of holomorphic functions is considered further on: *find stress potentials $\Phi(z)$ and $\Psi(z)$ by the following boundary conditions*

$$N^{+}(\zeta)+iT^{+}(\zeta)=N^{-}(\zeta)+iT^{-}(\zeta),\quad \zeta\in\Gamma \tag{5}$$

$$\arg D^{\pm}(\zeta)=\alpha^{\pm}(\zeta),\quad \zeta\in\Gamma \tag{6}$$

where two real valued functions $\alpha^{\pm}(\zeta)$ are known provided that boundary values of principal directions are given on the contour.

Boundary conditions (6) can also be expressed as follows

$$\mathrm{Im}\left[e^{-i\alpha^{\pm}(\zeta)}D^{\pm}(\zeta)\right]=0, \quad \zeta\in\Gamma \tag{7}$$

As soon as potentials are found, the stress fields (i.e., stress functions and stress components) in both the exterior and interior domains can be determined by formulas (1) and (2).

2.2 Reduction to singular integral equations

Complex potentials satisfying (5) can be obtained from the representation provided by Savruk [5] in the following form

$$\Phi(z)=\frac{1}{2\pi i}\int_{\Gamma}\frac{g'(t)}{t-z}dt, \quad \Psi(z)=\frac{-1}{2\pi i}\int_{\Gamma}\frac{\overline{g'(t)}}{t-z}d\bar{t}-\frac{1}{2\pi i}\int_{\Gamma}\frac{\bar{t}g'(t)}{(t-z)^2}dt \tag{8}$$

Here function $g'(t)$ is the derivative of a complex-valued function $g(t)$ proportional to the jump of the displacement vector across the contour with the coefficient $2G(1+\kappa)^{-1}$ (G is the shear modulus, $\kappa=3-4\nu$ for plane strain and $\kappa=(3-\nu)/(1+\nu)$ for plain stress, ν is Poisson's ratio). It should be noted that the considered BVP does not require specification of elastic constants if one has no intention of analysing displacements.

Single valuedness of displacements imposes the following condition on $g(t)$

$$\int_{\Gamma}g'(t)dt=0 \tag{9}$$

After simple transformations expressions of the stress functions take the form

$$P(z,\bar{z})=\mathrm{Re}\left(\frac{1}{\pi i}\int_{\Gamma}\frac{g'(t)}{t-z}dt\right)$$

$$D(z,\bar{z})=\frac{-1}{2\pi i}\int_{\Gamma}\frac{\overline{g'(t)}}{t-z}d\bar{t}-\frac{1}{2\pi i}\int_{\Gamma}\frac{(\bar{t}-\bar{z})g'(t)}{(t-z)^2}dt \tag{10}$$

These representations are valid for any point of the interior as well as exterior domains of the entire complex plane separated by for the contour Γ.

Boundary values of holomorphic functions are found by the Sokhotski–Plemelj formulae $2\varphi^{\pm}=\pm g'+\mathbf{I}(g')$ (e.g., Gakhov [4]). Then for boundary values of the stress functions one obtains (see also [5] for the boundary values of the second integral in the expression for D)

$$
\begin{aligned}
P^{\pm}(\zeta) &= \mathrm{Re}\left(\pm g'(\zeta) + \frac{1}{\pi i} \int_{\Gamma} \frac{g'(t)}{t-\zeta} dt \right) \\
D^{\pm}(\zeta) &= \mp \frac{d\bar{\zeta}}{d\zeta} \mathrm{Re}(g'(\zeta)) - \frac{1}{2\pi i} \int_{\Gamma} \left(\frac{\overline{g'(t)}}{t-\zeta} \frac{d\bar{t}}{dt} + \frac{(\bar{t}-\bar{\zeta}) g'(t)}{(t-\zeta)^2} \right) dt
\end{aligned} \tag{11}
$$

Substitution of the second formula in (11) into boundary conditions (7) results in the following system of singular integral equations

$$
\begin{cases}
\sin\left(2\theta + \alpha^{+}\right) \mathrm{Re}\, g' - \mathrm{Im}\left(e^{-i\alpha^{+}} \mathbf{J}\left(g', \bar{g}'\right)\right) = 0 \\
\sin\left(2\theta + \alpha^{-}\right) \mathrm{Re}\, g' + \mathrm{Im}\left(e^{-i\alpha^{-}} \mathbf{J}\left(g', \bar{g}'\right)\right) = 0
\end{cases} \tag{12}
$$

where operator **J**(…) denotes the integral term

$$
\mathbf{J}\left(g', \bar{g}'\right) = \frac{1}{2\pi i} \int_{\Gamma} \left(\frac{\overline{g'(t)}}{t-\zeta} e^{-2i\theta(t)} + \frac{(\bar{t}-\bar{\zeta}) g'(t)}{(t-\zeta)^2} \right) dt \tag{13}
$$

One can form a linear combination of both equations in (12) by excluding the complex conjugated operator from the system. This results in the following single complex equation

$$
\begin{aligned}
&\left(e^{i\alpha^{-}} \sin(\alpha^{+} + 2\theta) + e^{i\alpha^{+}} \sin(\alpha^{-} + 2\theta)\right) \ \mathrm{Re}\, g' \\
&+ \sin(\alpha^{+} - \alpha^{-}) \ \mathbf{J}\left(g', \bar{g}'\right) = 0
\end{aligned} \tag{14}
$$

Equation (14) is equivalent to the system of equation (12) if $\alpha^{+} \pm \alpha^{-} \neq 0, \pm\pi$ (this condition provides complex valuedness of (14)). In some cases of simple geometry it is not satisfied (see section 4).

Singular integral equation (14) is homogeneous therefore, according to Noether's theorems [4], the number of its independent solutions is determined by the index of the problem (next section).

3 Solvability of integral equations

3.1 Reduction to the Riemann problem for holomorphic functions

Let us extract the dominant part of SIE (13) by separating singular and regular terms in the integral operator in (12), which leads to

$$
\mathbf{J}\left(g', \bar{g}'\right) = e^{-2i\theta}\, \mathbf{I}\left(\mathrm{Re}\, g'\right) + \mathbf{R}\left(g', \bar{g}'\right) \tag{15}
$$

Here **I**(..) and **R**(…) are singular and regular operator respectively. They are expressed as follows ($\zeta \in \Gamma$)

$$\mathbf{I}(g')=\frac{1}{\pi i}\int_{\Gamma}\frac{g'(t)}{t-\zeta}dt,\quad \mathbf{I}^2(g')=g' \tag{16}$$

$$\mathbf{R}(g',\overline{g}')=\frac{1}{2\pi i}\int_{\Gamma}\frac{2\left(e^{-2i\theta(t)}-e^{-2i\theta(\zeta)}\right)\operatorname{Re}g'(t)+\left(e^{-2i\arg(t-\zeta)}-e^{-2i\theta(t)}\right)\overline{g}'(t)}{t-\zeta}dt \tag{17}$$

In the latter formula one can notice that the argument is bounded and continuous on an arbitrary smooth contour. It also satisfies the Hölder condition, and its value at the origin coincides with the angle of the tangent inclination to the x-axis, thus

$$\lim_{t\to\zeta}e^{-2i\arg(t-\zeta)}=\lim_{t\to\zeta}\frac{\bar{t}-\bar{\zeta}}{t-\zeta}=\frac{d\bar{\zeta}}{d\zeta}=e^{-2i\theta(\zeta)} \tag{18}$$

By substituting (15) into (14) and neglecting the regular integrals one obtains the dominant equation in the form

$$\begin{aligned}&\left(e^{i\alpha^-}\sin(\alpha^+ +2\theta)+e^{i\alpha^+}\sin(\alpha^- +2\theta)\right)\ \operatorname{Re}g'\\&+e^{-2i\theta}\sin(\alpha^+-\alpha^-)\ \mathbf{I}\left(\operatorname{Re}g'\right)=0\end{aligned} \tag{19}$$

Solvability of (19) is determined by the coefficient of the correspondent Riemann problem that is found as follows (Gakhov [4])

$$G(\zeta)=\frac{e^{i\alpha^+(\zeta)}\sin\left(\alpha^-(\zeta)+2\theta(\zeta)\right)}{e^{i\alpha^-(\zeta)}\sin\left(\alpha^+(\zeta)+2\theta(\zeta)\right)} \tag{20}$$

The index of a function is determined as its increment after the complete traverse of the contour in positive direction (counter clockwise) divided by 2π. It is evident that the index of G depends only on the difference of principal directions (because the ratio of the sines does not contribute into the increment of G), therefore

$$\operatorname{Ind}G=\left.\frac{\alpha^+(\zeta)-\alpha^-(\zeta)}{2\pi}\right|_{\Gamma}=-\frac{1}{\pi}\left.\left(\varphi^+(\zeta)-\varphi^-(\zeta)\right)\right|_{\Gamma}=2\mathrm{K} \tag{21}$$

where $|_{\Gamma}$ denotes the increment.

Thus, for an arbitrary, simply connected domain, bounded by a smooth closed contour and for any non-negative index, 2K, the solution of the dominant equation will in general include a polynomial of 2K order (or 2K-1 order if stresses vanish at infinity). This means that up to 2K+1 complex constants or (if each complex constant is counted as 2 real constants) 4K+2 real constants are included into the solution. For any negative index no bounded solutions exist.

This analysis has to be acknowledged in numerical implementation. Thus, after discretisation of (19) followed, for instance, by the collocation technique, the system for the determination of unknowns should have less rank then the number of unknowns (provided that 2K≥0), which means that 4K+2 real parameters cannot be determined.

4 Degenerated cases

In some cases the operator **J**(…) is radically simple and the analysis of solvability should be revised. Such cases include simple geometries and/or special cases of load. Several special cases are considered in this section. All of them are special ones because normal stresses on the interface do not violate boundary conditions (7) and the potential Φ(z) can be determined with certain arbitrariness. Thus, the complete problem of stress tensor determination is underspecified and has an infinite number of solutions.

4.1 Joined half-planes

Let Γ be the real axis, then $\overline{\zeta}=\zeta=x$ and $e^{i\theta}=1$, which immediately results in

$$\mathbf{R}(g',\overline{g}')=0,\quad \mathbf{J}(g',\overline{g}')=\mathbf{I}(\operatorname{Re}g'),\quad \overline{\mathbf{I}(\operatorname{Re}g')}=-\mathbf{I}(\operatorname{Re}g') \tag{22}$$

Therefore the system of SIE takes the form

$$\begin{cases}\sin\alpha^{+}\operatorname{Re}g'-i\cos\alpha^{+}\,\mathbf{I}(\operatorname{Re}g')=0\\ \sin\alpha^{-}\operatorname{Re}g'+i\cos\alpha^{-}\,\mathbf{I}(\operatorname{Re}g')=0\end{cases} \tag{23}$$

Both these two equations are of the dominant form. Their solvability depends upon indexes, $2K^{+}$ and $2K^{-}$, of the correspondent Riemann problems; these are determined as follows

$$2K^{+}=\operatorname{Ind}\left(e^{-2i\alpha^{+}}\right)=\left.\frac{\varphi^{+}(x)}{\pi}\right|_{-\infty}^{\infty},\quad 2K^{-}=\operatorname{Ind}\left(e^{-2i\alpha^{-}}\right)=-\left.\frac{\varphi^{-}(x)}{\pi}\right|_{-\infty}^{\infty} \tag{24}$$

The unknown function $\operatorname{Re}g'$ should satisfy both equations simultaneously, which is only possible if

$$\sin(\alpha^{+}+\alpha^{-})=0 \tag{25}$$

If boundary values of principal directions do not satisfy (25) only the trivial solution of the system exists, $\operatorname{Re}g'=0$.

If (25) is satisfied, it follows from (24) that for solvability $2K^{+}=2K^{-}=2K$ and solution of (23), in accordance with Gakhov [4], takes the form

$$\operatorname{Re}g'(x)=\cos\alpha(x)\,e^{\Lambda(x)}x^{-K}P_{2K-1}(x) \tag{26}$$

$$\Lambda(x)=\frac{1}{2\pi i}\int_{-\infty}^{\infty}\frac{\ln\left(-t^{-2K}e^{-2i\alpha(t)}\right)}{t-x}dt \tag{27}$$

where $P_{2K-1}(z)$ is a polynomial of order 2K-1 with arbitrary complex coefficients (for the case when stresses vanish at infinity). Taking into account the fact that the imaginary part of Λ(x) is constant due to the following integral

$$\int_{-\infty}^{\infty}\frac{\ln|t|}{t-x}dt=2x\int_{0}^{\infty}\frac{\ln(t)}{t^{2}-x^{2}}dt=\frac{\pi^{2}}{2}$$

(Prudnikov et al, [6]) and one can notice that the coefficients of the polynomial may be considered as real provided that the imaginary part of $\Lambda(x)$ in (26) is omitted.

It is evident that neither trivial solution not solution (29) allows one to identify $\Phi(z)$, because Img′ (i.e., the density of crack opening displacements) can be chosen arbitrary.

4.2 Crack in the plane

Let a crack of length $2l$ be situated on the interval $(-l,l)$ in the complex plane. Stresses at infinity are assumed to be zero.

It is evident that the system derived for the case of half-planes remains. The analysis is also similar to that described above, however some corrections in the determination of the indices have to be introduced to account for the open contour.

The problem is reduced to the case of half-planes by putting $G(x)=1$ on $|x|>l$. Then the ends of the interval represent the points of discontinuity of $G(x)$. This becomes evident if the asymptotic behaviour of D at the crack tips is considered. Independent of the load it can be written in the form

$$\sqrt{2\pi r}\,D = \left(K_I^{\pm} + 3iK_{II}^{\pm}\right)e^{-i\vartheta/2} - \left(K_I^{\pm} - iK_{II}^{\pm}\right)e^{-5i\vartheta/2} \tag{28}$$

where K_I and K_{II} are stress intensity factors, the indices "±" refer to the right and left crack tips correspondingly and angle ϑ is the polar angle in local coordinate system (r,ϑ) with the origin at the crack tip. Now the argument of $\alpha=\alpha(x)$ can be calculated. In particular, for points lying near the tips of the crack, the argument of D does not depend on K_I and can be determined as follows

$$\alpha(\pm l \mp 0) = \arg K_{II}^{\pm} = \frac{\pi}{2}\left(1 - \operatorname{sgn} K_{II}^{\pm}\right) \quad \Rightarrow \quad e^{2i\alpha(\pm l \mp 0)} = 1 \tag{29}$$

It is also seen that the argument α would gain the increment of $\pi/2$ if the point passed the crack end.

Thus, the coefficient of the Riemann problem $G(x)$ for infinite contour $(-\infty,\infty)$ has discontinuities at points $x=\pm l$ and satisfies the Hölder condition everywhere except these points. The index is further calculated by summing the index due to rotations of the principal directions on the crack surfaces and the index due to discontinuity of G at $x=\pm l$ (which adds unity).

Therefore the solution of the system should have the form similar to (26) but include a polynomial of 2K degree. It is evident that this case is also a special one.

4.3 Unit circle

Let Γ be the boundary of the unit circle and $\zeta=e^{i\theta}$ be point on it $(-\pi<\theta\leq\pi)$. Then the regular operator in (17) with account for single-valuedness (9) becomes

$$\mathbf{R}(g',\overline{g}')=\frac{2a}{\zeta^2},\quad 2a=\frac{1}{\pi i}\int_{\Gamma}\frac{\operatorname{Re}(g'(t))}{t}dt=\frac{1}{\pi}\int_{-\pi}^{\pi}\operatorname{Re}\left(g'(e^{i\vartheta})\right)d\vartheta. \tag{30}$$

The constant $2a$ is real, owing to the fact that $\operatorname{Re}g'$ is real (the coefficient 2 is introduced for convenience).

The SIE takes the form

$$\begin{gathered}\left(e^{i\alpha^-}\sin(\alpha^+ +2\theta)+e^{i\alpha^+}\sin(\alpha^- +2\theta)\right)\ \operatorname{Re}g' \\ +e^{-2i\theta}\sin(\alpha^+ -\alpha^-)\ \left(\mathbf{I}\ \left(\operatorname{Re}g'\right)+2a\right)=0\end{gathered} \tag{31}$$

This equation is slightly different from the dominant equation because of the presence of the terms with constant a. However it can be solved in the same way as the dominant equation. By introducing piecewise holomorphic functions $T^{\pm}(z)-a$ that are the boundary values of the Cauchy type integral of the function $\operatorname{Re}g'$, i.e. $\operatorname{Re}g'=T^+-T^-$, $\mathbf{I}(\operatorname{Re}g')=T^++T^--2a$ one transforms (31) to the following homogeneous Riemann boundary value problem

$$T^+(\zeta)=G(\zeta)T^-(\zeta),\quad \zeta\in\Gamma \tag{32}$$

where the coefficient G is specified by eqn (20).

This problem has non-trivial solutions if and only if the index of the problem, 2K, is non-negative. The number of homogeneous solutions of the Riemann problem is equal to (2K+1). However the number of independent solutions of the dominant equation is 2K because of the condition that the Cauchy-type integral vanishes at infinity.

Homogeneous solutions for the Riemann problem can be written in the form (Gakhov [4])

$$T^+(z)=a+e^{\Lambda^+(z)}P_{2\mathrm{K}-1}(z),\quad T^-(z)=a+e^{\Lambda^-(z)}z^{-2\mathrm{K}}P_{2\mathrm{K}-1}(z) \tag{33}$$

where $P_{2K-1}(z)$ is a polynomial of order 2K-1 with arbitrary complex coefficients, $\Lambda^{\pm}(z)$ are holomorphic functions determined by the following contour integral

$$\Lambda^{\pm}(z)=\frac{1}{2\pi i}\int_{\Gamma}\frac{\ln\left(-t^{-2\mathrm{K}}G(t)\right)}{t-z}dt \tag{34}$$

By applying the Sokhotski-Plemelj formulae one finds the boundary values of the piecewise holomorphic function $\Lambda^{\pm}(\zeta)$ and further the formal solution of SIE (31) is obtained as the difference $\operatorname{Re}(g')$,$=T^+-T^-$. This solution has to satisfy the real valuedness of the function sought, which leads to certain restrictions imposed on the coefficients of the arbitrary polynomials. Details of these calculations are found in [2], where the complete analysis for the interior domain is presented for the case when the second boundary conditions is formulated in terms of normal derivative of principal directions. Here we emphasise that the solution of (31) can only determine $\operatorname{Re}g'$ while $\operatorname{Im}g'$ can be chosen arbitrary, which again shows that the complete problem is underspecified.

4.4 Absence of shear stresses

Let us assume that shear stresses on the interface are absent. This means that the following expression is valid

$$\mathrm{Im}\left(e^{2i\theta(\zeta)} D^{\pm}(\zeta)\right)=0, \quad \zeta \in \Gamma \tag{35}$$

It is evident that (35) replaces boundary conditions (7) because the interface coincides with a stress trajectory of one orthogonal family. Therefore, the condition $\alpha^{+}\pm\alpha^{-}\neq 0,\pm\pi$ is not valid and one should consider the system in (12) that is further reduced to a single equation (because non-integral terms vanish while the integral operators are the same in both equations of the system). This results in the following integral equation

$$\mathrm{Im}\left(\mathbf{I}(\mathrm{Re}\, g')+e^{2i\theta}\,\mathbf{R}(g',\overline{g}')\right)=0 \tag{36}$$

It can be shown that this equation is a Fredholm-type equation by taking into account that

$$\overline{\mathbf{I}(\mathrm{Re}\, g')}=-\mathbf{I}(\mathrm{Re}\, g')+\mathbf{Reg}(\mathrm{Re}\, g') \tag{37}$$

where **Reg**(…) is a regular operator.

Therefore, one arrives at a single integral equation for the determination of two real valued functions. One can separate all terms containing Img' in the right-hand side and all terms containing Reg' in the left-hand side of the equation. The right-hand side can be temporary considered as a known function. Then, in accordance with the Fredholm theorems, this non-homogeneous equation is solvable if and only if the homogeneous onc is non-solvable and vice versa. This eventually means that equation (36) is either unsolvable or has an infinite number of solutions because Img' can be chosen arbitrary. Therefore this case also belongs to the degenerate cases.

5 Conclusions

Solvability of the BVP formulated in terms of principal directions given on a contour separating interior and exterior domains has been investigated for the case when the stress vector is continuous across the contour. The index of the corresponding singular integral equation, 2K, has to be non-negative to provide existence of solutions. For any negative index no bounded solutions exist. It has been shown that the solution includes a polynomial of 2K order, which means that up to 4K+2 real constants remain to be free parameters of the solution.

Special cases in which the problem is underspecified have been identified; these include adjacent half-planes, a crack in the plane, the circumference of a circle and the case when the interface is free of shear stresses.

References

[1] Muskhelishvili, N.I. *Some basic problems of the mathematical theory of elasticity*, P. Noordhoff, Groningen: the Netherlands, 1963.

[2] Galybin, A.N., Mukhamediev, Sh.A. Plane elastic boundary value problem posed on orientation of principal stresses. Journal of the Mechanics and Physics of Solids. **47** (11), 2381-2409, 1999.

[3] Galybin, A.N. 2002. Plane boundary value problem posed by displacement and force orientations on a closed contour. *Journal of Elasticity*. Vol. **65**, 169-184.

[4] Gakhov, F.D. Boundary value problems. Dover Publications, Inc., New York. 1990.

[5] Savruk, M.P. Two-dimensional problems of elasticity for body with cracks, Naukova Dumka, Kiev, 1981.

[6] Prudnikov, A.P., Brychkov, Ju.A., Marichev, O.I. Integrals and series. Nauka, Moscow, 1981.

Matrix decomposition MFS algorithms

A. Karageorghis & Y.-S. Smyrlis
Department of Mathematics and Statistics, University of Cyprus, Nicosia, Cyprus

Abstract

We describe the application of the Method of Fundamental Solutions (MFS) to elliptic boundary value problems in rotationally symmetric problems. In particular, we show how efficient matrix decomposition MFS algorithms can be developed for such problems. The efficiency of these algorithms is optimized by using Fast Fourier Transforms (FFTs).

1 Introduction

The Method of Fundamental Solutions (MFS) is a meshless boundary method which has become popular in recent years primarily because of its simplicity. Implementational details as well as a wide range of applications of the MFS can be found in the survey papers [1–3]. The MFS has recently been used for the solution of several elliptic boundary value problems in rotationally symmetric problems. At first, the MFS was used to solve the axisymmetric version of the governing equations [4–6]. However, the fundamental solutions of these equations involve the potentially troublesome evaluation of complete elliptic integrals. Further, when the boundary conditions of the problem under consideration are not axisymmetric, this approach requires the solution of a sequence of problems in order to approximate a finite Fourier sum. More recently, a different approach has been suggested, in which the three dimensional version of the governing equations is considered. In this work, matrix decomposition algorithms are developed for the efficient solution of the resulting systems. (An overview of matrix decomposition algorithms can be found in [7]). The algorithms proposed in this approach make use of Fast Fourier Transforms (FFTs). The basic ideas for the solution of two-dimensional harmonic problems in a disk subject to Dirichlet boundary conditions can be found in [8] while the numerical analysis of these problems is carried out in [9]. The implemen-

WIT Transactions on Modelling and Simulation, Vol 42, © 2006 WIT Press
www.witpress.com, ISSN 1743-355X (on-line)
doi:10.2495/BE06007

tation of the MFS for two-dimensional biharmonic problems in a disk is described in [10], and the corresponding algorithms for annular domains are developed in [11] and [12], respectively. The solution of three-dimensional harmonic problems in axisymmetric domains is considered in [13] and the corresponding biharmonic case in [14]. The solution of these problems in hollow axisymmetric domains is described in [15]. In this paper, we first describe this approach for two-dimensional harmonic problems and then describe it for the corresponding three-dimensional case.

2 MFS formulation for two-dimensional problems

2.1 The Laplace equation

We consider the Dirichlet boundary value problem in $\mathbb{R}^2$

$$\begin{cases} \Delta u = 0 & \text{in} \quad \Omega, \\ u = f & \text{on} \quad \partial\Omega, \end{cases} \tag{2.1}$$

where Δ denotes the Laplace operator and f is a given function. The domain Ω is the disk of radius ϱ, i.e. $\Omega = \{\boldsymbol{x} \in \mathbb{R}^2 \,:\, |\boldsymbol{x}| < \varrho\}$. In the MFS, the solution u is approximated by

$$u_N(\boldsymbol{c}, \boldsymbol{Q}; P) = \sum_{j=1}^{N} c_j\, K_2(P, Q_j), \qquad P \in \overline{\Omega}, \tag{2.2}$$

where $\boldsymbol{c} = (c_1, c_2, \ldots, c_{2N})^T \in \mathbb{C}^N$ and $\boldsymbol{Q}$ is a $2N$-vector containing the coordinates of the sources $Q_j, j = 1, \ldots, N$, which lie outside $\overline{\Omega}$. The function $K_2(P, Q)$ is a fundamental solution of the Laplace equation given by

$$K_2(P, Q) = -\frac{1}{2\pi} \log |P - Q|,$$

where $|P - Q|$ denotes the distance between the points P and Q. The singularities Q_j are fixed on the boundary $\partial\tilde{\Omega}$ of a circle $\tilde{\Omega}$ concentric to Ω and defined by $\tilde{\Omega} = \{\boldsymbol{x} \in \mathbb{R}^2 : |\boldsymbol{x}| < R\}$, where $R > \varrho$. A set of collocation points $\{P_i\}_{i=1}^N$ is placed on $\partial\Omega$. If $P_i = (x_{P_i}, y_{P_i})$, then we take

$$x_{P_i} = \varrho \cos \frac{2(i-1)\pi}{N}, \qquad y_{P_i} = \varrho \sin \frac{2(i-1)\pi}{N},$$

$i = 1, \ldots, N$. If $Q_j = (x_{Q_j}, y_{Q_j})$, then

$$x_{Q_j} = R \cos 2\frac{(j-1+\alpha)\pi}{N}, \qquad y_{Q_j} = R \sin \frac{2(j-1+\alpha)\pi}{N}, \quad j = 1, \ldots, N.$$

The parameter α $(0 \leq \alpha < 1)$ indicates that the sources are rotated by an angle $2\pi\alpha/N$ with respect to the boundary points. The coefficients $\boldsymbol{c}$ are determined so that the boundary condition is satisfied at the boundary points $\{P_i\}_{i=1}^N$, that is, $u_N(\boldsymbol{c}, \boldsymbol{Q}; P_i) = f(P_i), \;\; i = 1, \ldots, N$. This yields a linear system of the form

$$G\boldsymbol{c} = \boldsymbol{f}, \tag{2.3}$$

where $\boldsymbol{f} = [f(P_1), f(P_2), \ldots, f(P_N)]^T$ and the elements of matrix G are given by $G_{i,j} = -\dfrac{1}{2\pi} \log|P_i - Q_j|)$. The matrix G is clearly *circulant*. For an extensive account of the properties of circulant matrices, see [16].

Let $G = \text{circ}(g_1, \ldots, g_N)$, then $G = U^*DU$, where $D = \text{diag}(d_1, \ldots, d_N)$, $d_j = \sum_{k=1}^N g_k \omega^{(k-1)(j-1)}$, $\omega = e^{2\pi i/N}$, and U is the unitary $N \times N$ Fourier matrix which is the conjugate of the matrix

$$U^* = \frac{1}{N^{1/2}} \begin{pmatrix} 1 & 1 & 1 & \cdots & 1 \\ 1 & \omega & \omega^2 & \cdots & \omega^{N-1} \\ 1 & \omega^2 & \omega^4 & \cdots & \omega^{2(N-1)} \\ \vdots & \vdots & \vdots & & \vdots \\ 1 & \omega^{N-1} & \omega^{2(N-1)} & \cdots & \omega^{(N-1)(N-1)} \end{pmatrix}.$$

System (2.3) can therefore be written as $UGU^*U\boldsymbol{c} = U\boldsymbol{f}$ or $D\hat{\boldsymbol{c}} = \hat{\boldsymbol{f}}$ where $\hat{\boldsymbol{c}} = U\boldsymbol{c}$, and $\hat{\boldsymbol{f}} = U\boldsymbol{f}$. The solution of this system is $\hat{c}_i = \hat{f}_i/d_i, \quad i = 1, \cdots, N$. Having obtained $\hat{\boldsymbol{c}}$, we can find $\boldsymbol{c}$ from $\boldsymbol{c} = U^*\hat{\boldsymbol{c}}$. We thus have the following matrix decomposition algorithm for solving (2.3):

Step 1. `Compute` $\hat{\boldsymbol{f}} = U\boldsymbol{f}$.

Step 2. `Construct the diagonal matrix` D.

Step 3. `Evaluate` $\hat{\boldsymbol{c}}$.

Step 4. `Compute` $\boldsymbol{c} = U^*\hat{\boldsymbol{c}}$.

In *Step* **1** and *Step* **4**, because of the form of the matrices U and U^*, the operations can be carried out via FFTs at a cost of order $O(N \log N)$ operations. FFTs can also be used for the evaluation of the diagonal matrices in *Step* **2**.

2.2 The Cauchy–Navier equations

We consider the boundary value problem in $\mathbb{R}^2$ governed by the Cauchy–Navier equations of elasticity [17]

$$(\lambda + \mu)\, u_{k,ki} + \mu\, u_{i,kk} = 0, \;\; i = 1,2 \quad \text{in} \quad \Omega, \tag{2.4}$$

where Ω is a bounded domain, subject to the Dirichlet boundary conditions

$$u_i = f_i, \;\; i = 1,2 \quad \text{on} \quad \partial\Omega. \tag{2.5}$$

A fundamental solution of system (2.4) is given by

$$G(P,Q) = \begin{pmatrix} G_{11}(P,Q) & G_{12}(P,Q) \\ G_{21}(P,Q) & G_{22}(P,Q) \end{pmatrix}$$
$$= \zeta \log r_{PQ}\, I + \frac{\eta}{r_{PQ}^2} \begin{pmatrix} (x_P - x_Q)^2 & (x_P - x_Q)(y_P - y_Q) \\ (x_P - x_Q)(y_P - y_Q) & (y_P - y_Q)^2 \end{pmatrix} \tag{2.6}$$

where

$$\zeta = -\frac{1}{2\pi} \cdot \frac{\lambda + 3\mu}{2\mu(\lambda + 2\mu)}, \qquad \eta = \frac{1}{2\pi} \cdot \frac{\lambda + \mu}{2\mu(\lambda + 2\mu)},$$

and $r_{PQ} = \left((x_P - x_Q)^2 + (y_P - y_Q)^2\right)^{1/2}$, and (x_Q, y_Q), (x_P, y_P) are the coordinates of the points Q and P, respectively. Note that for simplicity we have changed our coordinate notation from (x_1, x_2) to (x, y). Expressions (2.6) were first derived by Lord Kelvin (see [18]). For further details and the derivation of (2.6) see Kythe [19].

The displacements u_1 and u_2 are approximated by ([20, 21])

$$u_1^N(\boldsymbol{a}, \boldsymbol{b}, \boldsymbol{Q}; P) = \sum_{j=1}^{N} a_j\, G_{11}(P, Q_j) + \sum_{j=1}^{N} b_j\, G_{12}(P, Q_j), \tag{2.7a}$$

$$u_2^N(\boldsymbol{a}, \boldsymbol{b}, \boldsymbol{Q}; P) = \sum_{j=1}^{N} a_j\, G_{21}(P, Q_j) + \sum_{j=1}^{N} b_j\, G_{22}(P, Q_j), \tag{2.7b}$$

$P \in \Omega$. Here, $\boldsymbol{a} = (a_1, a_2, \ldots, a_N)^T \in \mathbb{R}^N$ and $\boldsymbol{b} = (b_1, b_2, \ldots, b_N)^T \in \mathbb{R}^N$ are the vectors of the unknown coefficients, $\boldsymbol{Q}$ is a $N-$vector containing the coordinates of the singularities (sources) Q_j, $j = 1, \ldots, N$, which lie on the pseudo–boundary.

The satisfaction of the boundary conditions leads to a system of the form

$$\left(\begin{array}{c|c} D_{11} & D_{12} \\ \hline D_{21} & D_{22} \end{array}\right) \left(\begin{array}{c} \tilde{\boldsymbol{d}}_1 \\ \hline \tilde{\boldsymbol{d}}_2 \end{array}\right) = \left(\begin{array}{c} \tilde{\boldsymbol{g}}_1 \\ \hline \tilde{\boldsymbol{g}}_2 \end{array}\right) \tag{2.8}$$

where

$$D_{ij} = \text{diag}(\lambda_1^{ij}, \ldots, \lambda_N^{ij}), \qquad i, j = 1, 2,$$

are diagonal matrices whose diagonal elements are the eigenvalues of known circular matrices, and can thus be calculated easily.

The solution of system (2.8) is equivalent to solving N independent 4×4 systems. The solution $\tilde{\boldsymbol{d}}_1 = \left(\tilde{d}_1^1, \tilde{d}_2^1, \ldots, \tilde{d}_N^1\right)$, $\tilde{\boldsymbol{d}}_2 = \left(\tilde{d}_1^2, \tilde{d}_2^2, \ldots, \tilde{d}_N^2\right)$ is given by

$$\tilde{d}_k^1 = \frac{\lambda_k^{22}\tilde{g}_k^1 - \lambda_k^{12}\tilde{g}_k^2}{\lambda_k^{11}\lambda_k^{22} - \lambda_k^{12}\lambda_k^{21}}, \quad \tilde{d}_k^2 = -\frac{\lambda_k^{21}\tilde{g}_k^1 - \lambda_k^{11}\tilde{g}_k^2}{\lambda_k^{11}\lambda_k^{22} - \lambda_k^{12}\lambda_k^{21}}, \quad k = 1, \cdots, N, \tag{2.9}$$

where $\tilde{\boldsymbol{g}}_1 = \left(\tilde{g}_1^1, \tilde{g}_2^1, \ldots, \tilde{g}_N^1\right)$, $\tilde{\boldsymbol{g}}_2 = \left(\tilde{g}_1^2, \tilde{g}_2^2, \ldots, \tilde{g}_N^2\right)$.

3 MFS formulation for three-dimensional problems

3.1 The Laplace equation

We now consider the three-dimensional boundary value problem

$$\begin{cases} \Delta u = 0 & \text{in} \quad \Omega, \\ u = f & \text{on} \quad \partial\Omega, \end{cases} \tag{3.10}$$

where, as before, Δ denotes the Laplace operator and f is a given function. The region $\Omega \subset \mathbb{R}^3$ is axisymmetric, which means that it is formed by rotating a region $\Omega' \in \mathbb{R}^2$ about the z-axis. The boundaries of Ω and Ω' are denoted by $\partial\Omega$ and $\partial\Omega'$, respectively.

The solution u is approximated by

$$u_{MN}(\boldsymbol{c}, \boldsymbol{Q}; P) = \sum_{m=1}^{M} \sum_{n=1}^{N} c_{m,n} K_3(P, Q_{m,n}), \qquad P \in \overline{\Omega}, \tag{3.11}$$

where $\boldsymbol{c} = (c_{11}, c_{12}, \ldots, c_{1N}, \ldots, c_{M1}, \ldots, c_{MN})^T \in \mathbb{C}^{MN}$ and $\boldsymbol{Q}$ is a $3MN$-vector containing the coordinates of the sources $Q_{m,n}$, $m = 1, \ldots, M$, $n = 1, \ldots, N$, which lie outside $\overline{\Omega}$. The function $K_3(P, Q)$ is a fundamental solution of the Laplace equation in $\mathbb{R}^3$ given by $K_3(P, Q) = \frac{1}{4\pi|P-Q|}$, with $|P - Q|$ denoting the distance between the points P and Q. The singularities $Q_{m,n}$ are fixed on the boundary $\partial\tilde{\Omega}$ of a solid $\tilde{\Omega}$ surrounding Ω. The solid $\tilde{\Omega}$ is generated by the rotation of the planar domain $\tilde{\Omega}'$ which is similar to Ω'. A set of MN collocation points $\{P_{i,j}\}_{i=1,j=1}^{M,N}$ is chosen on $\partial\Omega$ in the following way. We first choose N points on the boundary $\partial\Omega'$ of Ω'. These can be described by their polar coordinates $(r_{P_j}, z_{P_j}), j = 1, \cdots, N$, where r_{P_j} denotes the vertical distance of the point P_j from the z-axis and z_{P_j} denotes the z- coordinate of the point P_j. The points on $\partial\Omega$ are taken to be $x_{P_{i,j}} = r_{P_j} \cos\varphi_i$, $y_{P_{i,j}} = r_{P_j} \sin\varphi_i$, $z_{P_{i,j}} = z_{P_j}$, where $\varphi_i = 2(i-1)\pi/M$, $i = 1, \ldots, M$. Similarly, we choose a set of MN singularities $\{Q_{m,n}\}_{i=m,n=1}^{M,N}$ on $\partial\tilde{\Omega}$ by taking $Q_{m,n} = (x_{Q_{m,n}}, y_{Q_{m,n}}, z_{Q_{m,n}})$, and $x_{Q_{i,j}} = r_{Q_j} \cos\theta_i$, $y_{Q_{i,j}} = r_{Q_j} \sin\theta_i$, $z_{Q_{i,j}} = z_{Q_j}$, where $\theta_i = 2(\alpha+i-1)\pi/M$, $i = 1, \ldots, M$. As in the two-dimensional case, the angular parameter α ($0 \leq \alpha < 1$) indicates that the sources are rotated by an angle $2\pi\alpha/M$ in the angular direction. The coefficients $\boldsymbol{c}$ are determined so that the boundary condition is satisfied at the boundary points $u_{MN}(\boldsymbol{c}, \boldsymbol{Q}; P_{i,j}) = f(P_{i,j})$, $i = 1, \ldots, M$, $j = 1, \ldots, N$. This yields an $MN \times MN$ linear system of the form

$$G\boldsymbol{c} = \boldsymbol{f}, \tag{3.12}$$

for the coefficients $\boldsymbol{c}$, where the elements of the matrix G are given by

$$G_{(i-1)\,N+j,(m-1)\,N+n} = \frac{1}{4\pi|P_{i,j} - Q_{m,n}|}, \tag{3.13}$$

$i, m = 1, \ldots, M,\ \ j, n = 1, \ldots, N$. The global matrix G has the *block circulant* structure

$$G = \begin{pmatrix} A_1 & A_2 & \cdots & A_M \\ A_M & A_1 & \cdots & A_{M-1} \\ \vdots & \vdots & & \vdots \\ A_2 & A_3 & \cdots & A_1 \end{pmatrix}, \tag{3.14}$$

where the matrices $A_\ell,\ \ell = 1, \cdots, M$, are $N \times N$ matrices defined by

$$(A_\ell)_{j,n} = \frac{1}{4\pi|P_{1,j} - Q_{\ell,n}|}, \quad \ell = 1, \ldots, M \quad j,\ n = 1, \ldots, N. \tag{3.15}$$

System (3.12) can then be written as

$$G\boldsymbol{c} = \left(I_M \otimes A_1 + P \otimes A_2 + P^2 \otimes A_3 + \cdots + P^{M-1} \otimes A_M\right) \boldsymbol{c} = \boldsymbol{f}, \tag{3.16}$$

where the matrix P is the $M \times M$ permutation matrix $P = circ\,(0, 1, 0, \cdots, 0)$ and $\otimes$ denotes the matrix tensor product.

In this case, an MDA involves the reduction of the $MN \times MN$ global system to M decoupled $N \times N$ systems. This is achieved by exploiting the block circulant structure of G and uses the unitary $M \times M$ Fourier matrix U defined in Section 2. Circulant matrices are diagonalized as described in Section 2. In particular, the permutation matrix $P = \mathrm{circ}(0, 1, 0, \ldots, 0)$ is diagonalized as $P = U^* D U$, where $D = \mathrm{diag}(d_1, \ldots, d_M),\ d_j = \omega^{j-1}$. Further, from the properties of the tensor product,

$$(U \otimes I_N)\left(P^{k-1} \otimes A_k\right)(U^* \otimes I_N) = (UP^{k-1}U^*) \otimes A_k = D^{k-1} \otimes A_k,$$

$k = 1, \ldots, M$.. Premultiplication of system (3.16) by $U \otimes I_N$ yields

$$(U \otimes I_N)\left(\sum_{k=1}^{M} P^{k-1} \otimes A_k\right)(U^* \otimes I_N)(U \otimes I_N)\,\boldsymbol{c} = (U \otimes I_N)\,\boldsymbol{f}, \tag{3.17}$$

since $(U^* \otimes I_N)(U \otimes I_N) = I_{MN}$.

Therefore, $\left(I_M \otimes A_1 + D \otimes A_2 + D^2 \otimes A_3 + \cdots + D^{M-1} \otimes A_M\right)\tilde{\boldsymbol{c}} = \tilde{\boldsymbol{f}}$, where $\tilde{\boldsymbol{c}} = (U \otimes I_N)\boldsymbol{c}$, $\tilde{\boldsymbol{f}} = (U \otimes I_N)\boldsymbol{f}$. The solution of this system can therefore be decomposed into the solution of the M *independent* $N \times N$ systems

$$B_m \tilde{\boldsymbol{c}}_m = \left(A_1 + d_m\, A_2 + d_m^2\, A_3 + \cdots + d_m^{M-1}\, A_M\right)\tilde{\boldsymbol{c}}_m = \tilde{\boldsymbol{f}}_m, \tag{3.18}$$

$m = 1, 2, \ldots M$. The (r, s) entry of the matrix B_m is

$$(B_m)_{rs} = (A_1)_{rs} + \omega^{m-1}\,(A_2)_{rs} + \cdots + \omega^{(m-1)(M-1)}\,(A_M)_{rs}\,,$$

where $r, s = 1, \ldots, N$ and $m = 1, \ldots, M$. Thus we have

$$\left((B_1)_{rs}, \ldots, (B_M)_{rs}\right)^T = M^{1/2} U^* \left((A_1)_{rs}, \ldots, (A_M)_{rs}\right)^T.$$

This observation enables us to reduce the cost of constructing the matrices $B_1, \ldots, B_M$ from $O(M^2N^2)$ operations to $O(N^2 M \log M)$ operations, using FFTs. We have the following matrix decomposition algorithm for solving (3.16):

 © 2006 WIT Press
www.witpress.com, ISSN 1743-355X (on-line)

Step 1. Compute $\tilde{\boldsymbol{f}} = (U \otimes I_N)\,\boldsymbol{f}$.
Step 2. Construct the matrices $B_m = A_1 + d_m\, A_2 + \cdots + d_m^{M-1}\, A_M$, $m = 1, \ldots, M$.
Step 3. Solve $B_m \tilde{\boldsymbol{c}}_m = \tilde{\boldsymbol{f}}_m$, $m = 1, \ldots, M$.
Step 4. Compute $\boldsymbol{c} = (U^* \otimes I_N)\,\tilde{\boldsymbol{c}}$.

In *Step* **1**, because of the form of the matrix U, the operation is equivalent to performing N FFTs of dimension M. This can done at a cost of $O(NM \log M)$ operations via an appropriate FFT algorithm. Similarly, in *Step* **4**, because of the form of the matrix U^*, the operation can be carried out via FFTs at a cost of order $O(NM \log M)$ operations. In *Step* **2**, for each $r, s = 1, \ldots, N$, we need to perform an M-dimensional FFT, in order to compute the entries $(B_1)_{rs}, \ldots, (B_M)_{rs}$. This can be done at a cost of $O(N^2 M \log M)$ operations. In *Step* **3**, we need to solve M complex linear systems of order N. This is done using an LU-factorization with partial pivoting at a cost of $O(M\, N^3)$ operations.

3.2 The Cauchy–Navier equations

We consider the boundary value problem in $\mathbb{R}^3$ governed by the Cauchy-Navier equations of elasticity

$$\begin{cases} (\lambda + \mu)\, u_{k,ki} + \mu\, u_{i,kk} = 0 & \text{in} \quad \Omega, \\ u_i = f_i & \text{on} \quad \partial\Omega. \end{cases} \tag{3.19}$$

A similar MFS discretization, which exploits the block circulant structure of the coefficient matrix, leads to the solution of M independent $3N \times 3N$ linear systems, instead of the solution of a $3MN \times 3MN$ linear system.

References

[1] Fairweather, G. & Karageorghis, A., The method of fundamental solutions for elliptic boundary value problems. Numerical treatment of boundary integral equations. *Adv. Comput. Math.*, **9(1-2)**, pp. 69–95, 1998.

[2] Fairweather, G., Karageorghis, A. & Martin, P.A., The method of fundamental solutions for scattering and radiation problems. *Engng. Analysis with Boundary Elements*, **27**, pp. 759–769, 2003.

[3] Golberg, M.A. & Chen, C.S., The method of fundamental solutions for potential, Helmholtz and diffusion problems. Boundary integral methods: numerical and mathematical aspects, WIT Press/Comput. Mech. Publ., Boston, MA, volume 1 of *Comput. Eng.*, pp. 103–176, 1999.

[4] Karageorghis, A. & Fairweather, G., The method of fundamental solutions for the numerical solution of the biharmonic equation. *J. Comput. Phys.*, **69(2)**, pp. 434–459, 1987.

[5] Karageorghis, A. & Fairweather, G., The Almansi method of fundamental solutions for solving biharmonic problems. *Int. J. Numer. Meth. Engng.*, **26(7)**, pp. 1665–1682, 1988.
[6] Karageorghis, A. & Fairweather, G., The simple layer potential method of fundamental solutions for certain biharmonic problems. Internat. *J. Numer. Methods Fluids*, **9(10)**, pp. 1221–1234, 1989.
[7] Bialecki, B. & Fairweather, G., Matrix decomposition algorithms for separable elliptic boundary value problems in two space dimensions. *J. Comput. Appl. Math.*, **46(3)**, pp. 369–386, 1993.
[8] Smyrlis, Y.S. & Karageorghis, A., Some aspects of the method of fundamental solutions for certain harmonic problems. *J. Sci. Comput.*, **16(3)**, pp. 341–371, 2001.
[9] Smyrlis, Y.S. & Karageorghis, A., Some aspects of the method of fundamental solutions for certain biharmonic problems. *CMES Comput. Model Eng. Sci.*, **4(5)**, pp. 535–550, 2003.
[10] Smyrlis, Y.S. & Karageorghis, A., A linear least–squares MFS for certain elliptic problems. *Numer. Algorithms*, **35(1)**, pp. 29–44, 2004.
[11] Tsangaris, T., Smyrlis, Y.S. & Karageorghis, A., A Matrix Decomposition MFS Algorithm for Biharmonic Problems in Annular Domains. *Computers Materials and Continua*, **1(3)**, pp. 245–258, 2004.
[12] Tsangaris, T., Smyrlis, Y.S. & Karageorghis, A., Numerical analysis of the MFS for harmonic problems in annular domains. *Numer. Methods Partial Differential Equations*. To appear.
[13] Smyrlis, Y.S. & Karageorghis, A., A matrix decomposition MFS algorithm for axisymmetric potential problems. *Engng Analysis with Boundary Elements*, **28**, pp. 463–474, 2004.
[14] Fairweather, G., Karageorghis, A. & Smyrlis, Y.S., A matrix decomposition MFS algorithm for axisymmetric biharmonic problems. *Adv. Comput. Math.*, **23(1–2)**, pp. 55–71, 2005.
[15] Tsangaris, T., Smyrlis, Y.S. & Karageorghis, A., A matrix decomposition MFS algorithm for problems in hollow axisymmetric domains. *J. Sc. Comput.* To appear.
[16] Davis, P.J., Circulant matrices. John Wiley & Sons, New York-Chichester-Brisbane, pp. xv+250, 1979. A Wiley-Interscience Publication, Pure and Applied Mathematics.
[17] Kane, J.H., *Boundary Element Analysis in Engineering Continuum Mechanics*. Prentice Hall: Engelwood Cliffs, NJ, p. 676, 1994.
[18] Love, A.E.H., *A treatise on the Mathematical Theory of Elasticity*. Dover Publications, New York, pp. xviii+643, 1944. Fourth Ed.
[19] Kythe, P.K., *Fundamental solutions for differential operators and applications*. Birkhäuser Boston Inc.: Boston, MA, pp. xxiv+414, 1996.
[20] Berger, J.R. & Karageorghis, A., The method of fundamental solutions for layered elastic materials. *Engng. Anal. Bound. Elem.*, 25, pp. 877–886,2001.
[21] Kupradze, V.D., On a method of solving approximately the limiting problems of mathematical physics. *Ž Vyčisl Mat i Mat Fiz*, **4**, pp. 1118–1121, 1964.

A meshfree minimum length method

G. R. Liu[1], K. Y. Dai[1] & X. Han[2]
[1]*Centre for ACES, Department of Mechanical Engineering, National University of Singapore, Singapore, 119260*
[2]*College of Mechanical & Automotive Engineering, Hunan University, Changsha, China*

Abstract

A meshfree minimum length method (MLM) has been proposed for solids mechanics and heat conduction problems. In this method, both polynomial terms as well as modified radial basis functions (RBFs) are used to construct shape functions using arbitrarily distributed nodes based on the minimum length procedure. The shape functions constructed possess delta function property. The numerical examples show that the method achieves better accuracy than the finite element method especially for problems with steep gradients. Some numerical implementation issues for MLM are also discussed in detail.
Keywords: meshfree method, meshless method, minimum length method, radial basis function (RBF), interpolation function.

1 Introduction

Meshfree method has achieved remarkable progress in recent year to avoid the problems related to the creation and application of predefined meshes in the traditional numerical methods, such as the finite element method (FEM), the finite difference method (FDM). In general meshfree methods developed so far can be categorized into three main groups, i.e., meshfree methods based on strong-form formulation [1, 2], on weak-form formulation [3-6] and on the combination of the above two [7, 8]. Meshfree weak-form methods can usually achieve higher accuracy than strong-form methods especially in dealing with problems in solids and structures. In addition, they can treat Neumann boundary condition more easily and the results are more stable. Hence weak-form formulation will be used in this work. One of the most important issues in meshfree method is the construction of shape functions. There are two widely

WIT Transactions on Modelling and Simulation, Vol 42, © 2006 WIT Press
www.witpress.com, ISSN 1743-355X (on-line)
doi:10.2495/BE06008

used methods: moving least-squares (MLS) and point interpolation method using radial basis functions (RBFs), or RPIM (see, e.g., [5]). The MLS method uses excessive nodes which lead to the shape functions being lack of delta function property. RPIM shape functions are constructed using exactly the same of number of nodes as the number of terms of RBFs, and hence they possess the delta function property.

In this work, we present an alternative method to construct shape functions based on arbitrarily distributed nodes, which uses more bases than the number of nodes. The shape functions constructed will still have delta function property. The minimum length procedure is used to decide the participation of terms of polynomials and RBFs. The new shape functions are then used to formulate a meshfree method based on weak-form formulation (termed as MLM for short). Numerical examples show that the method has desirable accuracy as well as convergence rate, and is preferably very easy to implement.

2 Meshfree minimum length (ML) method

Consider a field variable $u(\mathbf{x})$ which is represented by a group of arbitrarily distributed nodes $\mathbf{x}_i (i = 1, 2, \cdots, N)$ in the domain Ω bounded by Γ. At any point $\mathbf{x}_Q$, the approximated value u^h is expressed as

$$u^h(\mathbf{x}, \mathbf{x}_Q) = \mathbf{b}(\mathbf{x})\mathbf{a} \tag{1}$$

where the unknown coefficient vector $\mathbf{a} = [a_1 \;\; a_2 \;\; \cdots a_m]^T$. The basis vector is given as

$$\mathbf{b} = [\mathbf{p} \quad \mathbf{r}] \tag{2}$$

where $\mathbf{p}_{(1\times n_p)} = [1 \;\; x \;\; y \;\; x^2 \;\; xy \;\; y^2 \cdots]$, n_p is the number of the polynomial terms including one constant. In the vector $\mathbf{r}_{(1\times n)} = [r_1 \;\; r_2 \;\; \cdots \;\; r_n]$, r_i is the modified MQ-RBF at node $\mathbf{x}_i$, as given by $r_i = r(\mathbf{x}_i) = [(x - x_i)^2 + (y - y_i)^2]^q$ for 2-D cases. Note that $m = n + n_p > n$. The n is the number of nodes in support domain.

Letting Eq. (1) pass through the n field nodes in support domain, we have

$$\mathbf{u}^e = \mathbf{B}_0\mathbf{a} \tag{3}$$

where $\mathbf{u}^e = [u_1 \;\; u_2 \;\; \cdots \;\; u_n]^T$.

To uniquely determine the unknown coefficients in Eq. (3), a functional adopted from the ML procedure (see, e.g., [9]) can be established as

$$\Pi = \mathbf{a}^T\mathbf{a} + \boldsymbol{\lambda}^T(\mathbf{u}_e - \mathbf{B}_0\mathbf{a}) \tag{4}$$

The derivatives of Π with respect to vectors $\mathbf{a}$ and $\boldsymbol{\lambda}$ lead to

$$\frac{\partial \Pi}{\partial \mathbf{a}} = \mathbf{0} \;\Rightarrow\; 2\mathbf{a} - \mathbf{B}_0^T\boldsymbol{\lambda} = \mathbf{0} \tag{5}$$

$$\frac{\partial \Pi}{\partial \boldsymbol{\lambda}} = \mathbf{0} \Rightarrow \mathbf{u}_e - \mathbf{B}_0\mathbf{a} = \mathbf{0} \tag{6}$$

Solving the above two equations yields $\boldsymbol{\lambda}$ and **a**, then substitution into Eq. (1) yields

$$u = \mathbf{b}\mathbf{B}_0^T[\mathbf{B}_0\mathbf{B}_0^T]^{-1}\mathbf{u}_e = \mathbf{\Phi}\mathbf{u}_e = \sum_{i=1}^{n}\phi_i u_i \tag{7}$$

where $\mathbf{\Phi} = [\phi_1(\mathbf{x}) \quad \phi_2(\mathbf{x}) \quad \phi_3(\mathbf{x}) \quad \cdots \quad \phi_n(\mathbf{x})]$ and $\phi_i(\mathbf{x})$ is an ML shape function.

In practice the moment matrix $\mathbf{B}_0$ is generated using linear or quadratic polynomials together with *n* modified MQ-RBFs. As $n < m$, the matrix $[\mathbf{B}_0\mathbf{B}_0^T]$ is invertible for arbitrarily nodal distributions.

Because the ML procedure has "output" reproducibility [9], the shape functions possess delta function property. This can be proven as follows. Using Eq. (7), it is easy to see that

$$\mathbf{B}_0\mathbf{a} = \mathbf{B}_0(\mathbf{x})\mathbf{B}_0^T[\mathbf{B}_0\mathbf{B}_0^T]^{-1}\mathbf{u}_e = \mathbf{I}\mathbf{u}_e = \mathbf{u}_e$$

Hence

$$\phi_i(\mathbf{x}_j) = \mathbf{b}^T(\mathbf{x}_j)\mathbf{B}_0^T[\mathbf{B}_0\mathbf{B}_0^T]^{-1} = \begin{cases} 1 & (i = j) \\ 0 & (i \neq j) \end{cases} \tag{8}$$

As a consequence the essential boundary conditions can be easily enforced as in the conventional FEM.

3 Discrete form

A 2-D problem in solid mechanics can be described by equilibrium equation in the domain Ω bounded by Γ and $\Gamma = \Gamma_u + \Gamma_t$.

$$\sigma_{ij,j} + b_i = 0 \quad \text{in } \Omega \tag{9}$$

where σ_{ij} is the component of stress tensor and b_i is the body force component. Boundary conditions are given as follows.

$$\sigma_{ij}n_j = \bar{t}_i \quad \text{on } \Gamma_t \tag{10}$$

$$u_i = \bar{u}_i \quad \text{on } \Gamma_u \tag{11}$$

where the superposed bar denotes the prescribed boundary values and n_i is the component of unit outward normal to the domain. Its variational weak form is expressed as

$$\int_{\Omega}\delta(\nabla_s\mathbf{u}):\boldsymbol{\sigma}d\Omega - \int_{\Omega}\delta\mathbf{u}\cdot\mathbf{b}d\Omega - \int_{\Gamma_t}\delta\mathbf{u}\cdot\bar{\mathbf{t}}d\Gamma = 0 \tag{12}$$

With the application of the derived MLM shape functions, the discretization of Eq. (12) yields (in matrix form)

$$\mathbf{Ku} = \mathbf{F} \tag{13}$$

where

$$\mathbf{K}_{IJ} = \int_{\Omega} \mathbf{B}_I^T \mathbf{D} \mathbf{B}_J d\Omega \tag{14}$$

$$\mathbf{f}_I = \int_{\Omega} \mathbf{\Phi}_I^T \mathbf{b} d\Omega + \int_{\Gamma_t} \mathbf{\Phi}_I^T \bar{\mathbf{t}} d\Gamma \tag{15}$$

In order to evaluate the integrals in Eqs. (14) and (15), a background cell structure is required, which is independent of the filed nodes. The cell can be quadrilateral or triangular. For simplicity, quadrilateral cells are used in the paper. Within each cell Gaussian quadrature is applied.

4 Numerical examples

4.1 Cantilever beam

A cantilever beam with length L and height D is studied here. It is subjected to a parabolic traction at the free end as shown in Fig. 1. The beam is assumed to have a unit thickness so that plane stress theory is valid. The analytical solution can be found in a textbook by Timoshenko and Goodier [10]. The related parameters are taken as $E = 3.0 \times 10^7 \, kPa$, $\nu = 0.3$, D = 12 m, L = 48 m and P = 1000 N.

In order to study the convergence rates of the present method, the energy norm is defined as

$$e_e = \frac{1}{2LD}\left[\int_{\Omega} (\boldsymbol{\varepsilon}^{Numer} - \boldsymbol{\varepsilon}^{Exact})^T \mathbf{D} (\boldsymbol{\varepsilon}^{Numer} - \mathbf{e}^{Exact})\right]^{1/2} \tag{16}$$

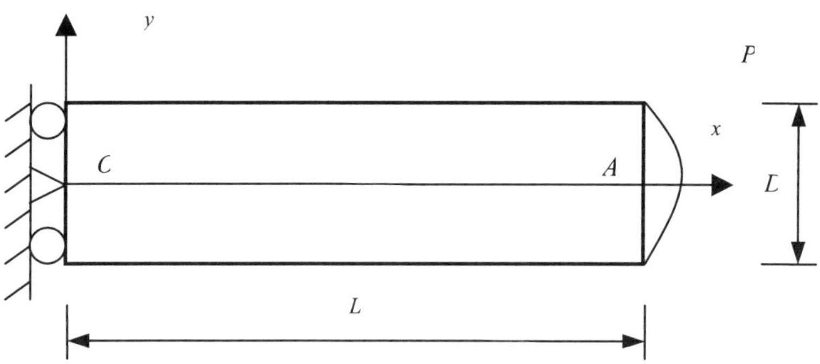

Figure 1: Cantilever beam.

As discussed in RPIM, the shape parameter q in the modified MQ-RBF has great effect on the accuracy of final results [11]. Through numerical test, it is found that, when q is close to 1.0, more accurate results are obtained. Hence if not stated otherwise, $q = 1.01$ is used in the following analysis.

The size of support domain, or the number of nodes selected in support domain, has also influence on accuracy of final results. Support domain can be a rectangle, a circle or an ellipse. An ellipse is used in this study as the nodal spacings may not be equal in two directions. The relationship is defined as

$$\frac{(x_q - x_i)^2}{a^2} + \frac{(y_q - y_i)^2}{b^2} < R^2 \tag{17}$$

where (a, b) are nodal spacings in two directions. (x_q, y_q) and (x_i, y_i) correspond to quadrature point and filed node, respectively. (17×5) regularly spaced nodes are taken for instance to examine the size of support domain. 16×4 rectangular background cells are used for integration and 3×3 Gauss quadrature order is applied in each cell. The centreline deflection of the beam is plotted in Fig. 2 when parameter r increases from 1.1 to 2.5. It can be seen that when included nodes are less than 6, the results are not accurate. Generally, the larger the number of nodes covered in support domain, the more accurate the deflection. Normally, 6-20 nodes are sufficient to give good solutions in a support domain. Accordingly, $R = 1.5$-3.0 is often used in the study.

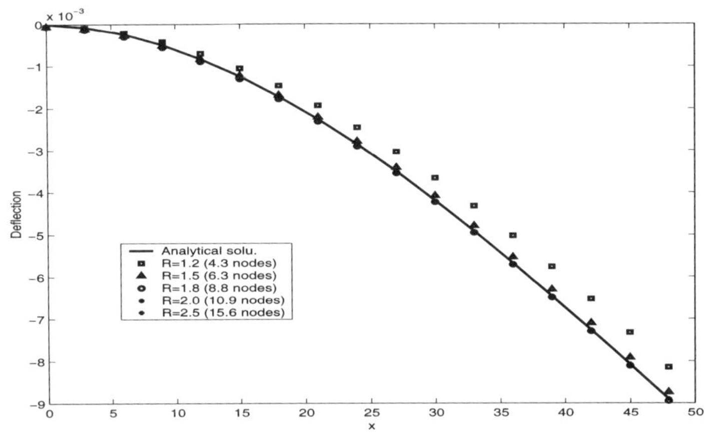

Figure 2: Centreline deflection using different sizes of support domain.

Four regular nodal patters are employed to examine the convergence rate of the present method, i.e., 11×5, 21×6, 33×9, 41×11 evenly spaced nodes. Linear and quadratic polynomials ($n_p = 3$, 6) are included respectively in the interpolation bases. For comparison, 4-node finite elements with equivalent node densities are also used for the same analysis. The convergence rates in energy norm are shown in Fig. 3. It is observed that the present method achieves a slightly higher convergence rates in energy norm when compared to the FEM. Quadratic polynomials give better results than linear ones while both of them are more accurate than those obtained by FEM. Shear stress distribution is displayed in Fig. 4 as an instance.

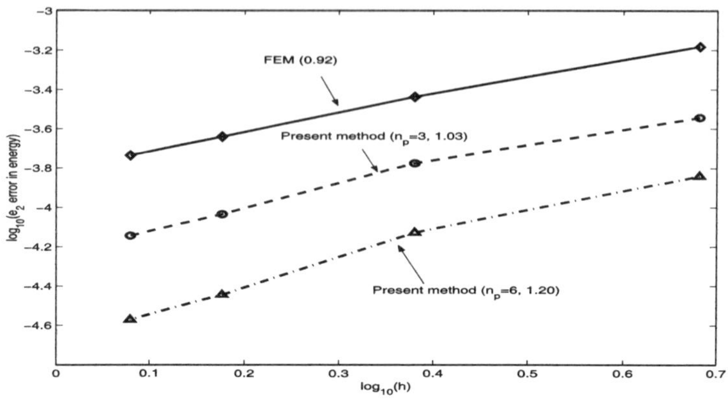

Figure 3: Convergence rates in error of error norm.

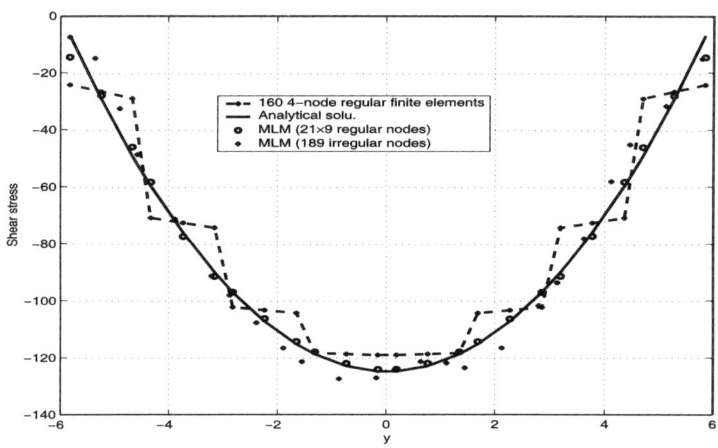

Figure 4: Shear stress of the beam at the section of $x = L/2$.

4.2 High-gradient heat conduction problem

A heat conduction problem considered here is a rectangular plate (0.5×6 in^2) with heat source

$$b(x, y) = 2s^2 \sec h^2[s(y-3)]\tanh[s(y-3)] \tag{18}$$

The boundary conditions are given by

$$\begin{gathered} T = \text{-}\tanh(3s) \ \text{ at } \ y = 0 \\ T = \tanh(3s) \ \text{ at } \ y = 6 \\ \frac{\partial T}{\partial x} = 0 \ \text{ at } x = \text{-}0.25 \text{ and } x = 0.25 \end{gathered} \tag{19}$$

The exact solution of this problem is

$$T = \tanh[s(y-3)] \tag{20}$$

As shown in the study by Belytschko et al. [3] this problem has a very high gradient near $y = 3.0$. In Eq. (18), the quantity s is a free parameter. The bigger the value of s, the higher the gradient of field T. For comparison, the 4-node finite elements with the same nodal distribution are again applied to analyze this problem. Note that $s = 40$ is used in the analysis.

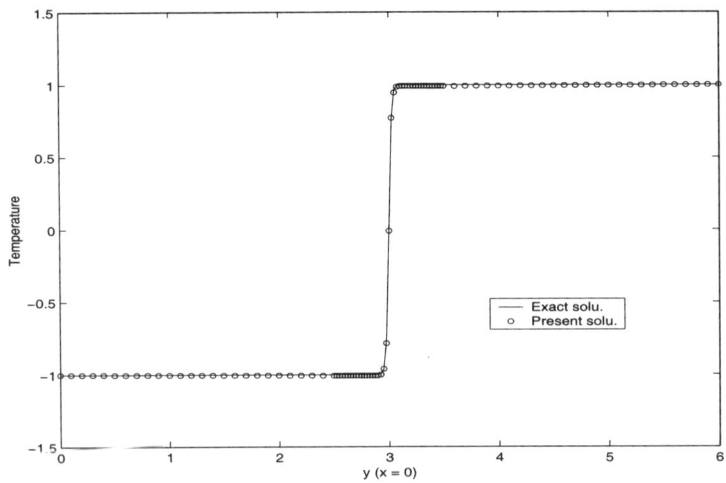

Figure 5: Comparison between the exact solution and the present solution at $x = 0$.

Figure 5 illustrates the comparison between the exact solution and the numerical solution obtained by the present method. It is observed that very good agreement is achieved. Figure 6 shows that the results for the gradient $T,_y$ by the present method are much better than those by FEM. It should be mentioned that, as only the gradient values at the quadrature points are plotted for simplicity, this is the cause that the tip value is smaller than the exact one.

5 Conclusions

In this work a meshfree minimum-length method (MLM) is proposed for 2-D solids and heat conduction problems. This method employs polynomial terms as well as modified radial basis functions as bases to interpolate filed variables. The number of bases is bigger than that of interpolated nodes and the ML scheme can select suitable basis functions automatically. Weak-form formulation is formed for 2-D elastic problems. Some numerical examples are studied. From the research the following conclusions can be drawn.

(1) Due to the delta function property of the constructed shape functions, the essential boundary conditions can be enforced conveniently as in conventional FEM.
(2) The method shows higher accuracy than the 4-node finite elements especially for problems with localized steep gradients while its convergence rate is also comparable with that of FEM.
(3) Quadratic polynomials can improve the accuracy by one order in error of energy than linear polynomials. The shape parameter q around 1.0 ($0.98 < q < 1.03;\ q \neq 1.0$) is recommended for good resolution of final results.
(4) Irregularly distributed nodal distribution performs well in the method and does not degrade prominently the accuracy of final results.

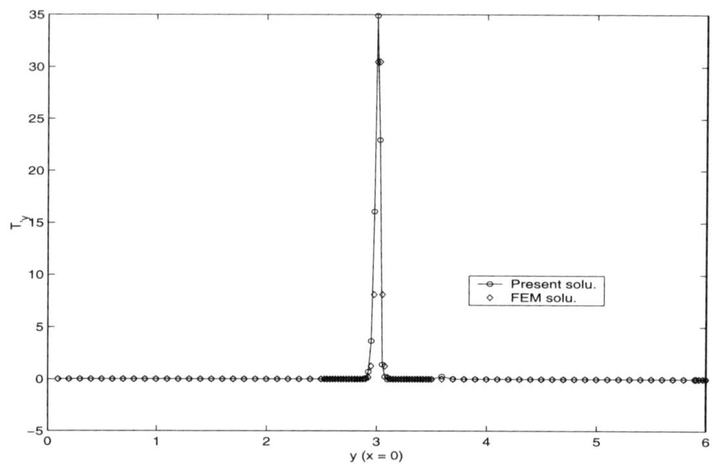

Figure 6: Comparison between the FEM solution and the present solution at $x = 0$.

References

[1] Gingold, R. A. & Moraghan J. J. U., Smooth particle hydrodynamics: theory and applications to non-spherical stars. *Man. Not. Roy. Astron. Soc.*, **181**, pp. 375–289, 1977.

[2] Liszka, T. & Orkisz J., The finite difference methods at arbitrary irregular grids and its applications in applied mechanics. *Comput. & Struct.* **11**, pp. 83–95, 1980.

[3] Belytschko, T., Lu, Y. Y. & Gu L (1994) Element-free Galerkin methods. *Int. J. Numer. Meth. Engrg.* **37**, pp. 229–256, 1994.

[4] Liu, W. K., Jun, S. & Zhang, Y. F., Reproducing Kernel Particle Methods *Inter. J. Numer. Methods Fluids.* **20**, pp. 1081–1106, 1995.

[5] Liu, G. R., *Meshfree Methods: Moving Beyond the Finite Element Method*, CRC Press: Boca Raton, FL, 2002.

[6] Liu, G. R. & Gu, Y. T., A point interpolation method for two-dimensional solids. *Int. J. Muner. Meth. Engrg*, **50**, pp. 937–951, 2001.

[7] Liu, G. R. & Gu, Y. T., A meshfree method: meshfree weak-strong (MWS) form method for 2-D solids. *Comput. Mech.* **33**, pp. 2–14, 2003.

[8] Liu, G. R., Wu, Y. L. & Ding, H., Meshfree weak-strong (MWS) form method and its application to incompressible flow problems. *Int. J. Numer. Meth. Fluids*, **46**, pp.1025–1047, 2004.

[9] Liu, G. R. & Han, X., *Computational Inverse Techniques in nondestructive evaluation*, CRC Press: Boca Raton, FL, 2003.

[10] Timoshenko, S. P. & Goodier, J. N., *Theory of Elasticity*, 3rd Edition, McGraw-Hill: New York, NY, 1970.

[11] Wang, J. G. & Liu, G. R., On the optimal shape parameters of radial basis functions used for 2D meshless methods. *Comput. Methods Appl. Mech. Eng.* **191**, pp.2611–2630, 2002.

DRM formulation for axisymmetric laser-material interactions

R. Gospavić[1], V. Popov[2], M. Srecković[3] & G. Todorović[1]
[1]*Faculty of Civil Engineering, Serbia and Montenegro*
[2]*Wessex Institute of Technology, UK*
[3]*Faculty of Electrical Engineering, Serbia and Montenegro*

Abstract

The modeling of laser-material interaction using the boundary element dual reciprocity method (BE-DRM) is presented. Thermal effects in the case of cylindrical geometry for mono as well as multi layer structures were considered. The different aspects of interaction up to the melting point of considered materials are presented. The effect of temperature dependence of the absorption coefficients on the process of laser heating was considered. The BEM formulation is based on the fundamental solution for the Laplace equation. The numerical results for spatial as well as temporal temperature distribution inside the material bulk are presented. Two cases were considered: a mono-layer and a multi layer case. In the case of a mono-layer structure DRM and DRM-MD approaches were used, and the numerical results were compared with the analytical ones. In the multi layer case only the DRM-MD approach was used.
Keywords: axisymmetric laser-material interaction, dual reciprocity method.

1 Introduction

The dual reciprocity method (DRM) was applied for laser-material interaction analysis. Laser beams have a number of applications in different areas of science, technology, and medicine. In the present work, thermal models of interaction in case of cylindrical geometry and mono as well as multi layer structures were considered. The spatial and temporal distributions of temperature field were considered. The numerical model of laser-material interaction described here is restricted only to heating effects of the targeted material without destructive and

WIT Transactions on Modelling and Simulation, Vol 42, © 2006 WIT Press
www.witpress.com, ISSN 1743-355X (on-line)
doi:10.2495/BE06009

disintegration processes during interaction i.e. the incident intensity of laser radiation was considered to be equal to critical intensity.

In the present work the dual reciprocity method [1] is used to solve axisymmetric problems. The DRM has been used previously for axisymmetric problems, see for example [2,3,4]. The difference in this case is that the Laplace fundamental solution is used instead of the one for axisymmetric problems expressed in terms of Eliptic integrals. The present approach simplifies the DRM part and the construction of a suitable particular solution.

In order to estimate the accuracy of the numerical method, analytical results were compared to the results obtained using the boundary element DRM approach.

2 Mathematical model of the interaction

The heating process provoked by a laser beam during interaction was considered. It was assumed that absorption of the laser beam occurred in the thin surface layer of the bulk material.

The interaction with the material is modeled as an equivalent surface thermal source with appropriate spatial ant temporal distributions. The analysis is focused on cylindrical geometry and surface distributions of absorbed incoming laser beam fluxes, and accordingly the temperature field analysis was performed using the cylindrical coordinate system. Though this problem is a three-dimensional one, as there is axial symmetry, the temperature field is a function of the radial and axial coordinates only, i.e. the problem under consideration becomes a two dimensional one. In this work only mono and two layer structures, with ideal thermal contacts between adjacent layers, were considered, however the results can be applied to multi layer structures.

The geometry of the considered problem for a two-layer case is shown in Fig. 1. It was assumed that the spatial and temporal distributions of the laser beam intensity on the surface of the material specimens could be described by a product of two independent functions of the radial coordinate and time e.g. $q(r)$ and $\varphi(t)$, respectively.

It was also assumed that all the thermal parameters of the material of interest in the considered temperature range are constant and temperature independent. A linear temperature dependence of the material optical parameter, i.e. the absorption coefficient, was assumed [5]. The initial temperature inside the specimen is equal to the ambient temperature T_0. Heating of material, according to above assumptions, for a two layer cylindrical structure (Fig. 1.), with ideal thermal contact between layers, could be described by the following equations [5]:

$$\begin{aligned} \frac{1}{a_1}\frac{\partial T_1(z,r,t)}{\partial t} &= \Delta T_1(z,r,t), \quad 0 \le z \le h_1; \quad t \ge 0, \quad 0 \le r \le R \\ \frac{1}{a_2}\frac{\partial T_2(z,r,t)}{\partial t} &= \Delta T_2(z,r,t), \quad h_1 \le z \le h; \quad t \ge 0, \quad 0 \le r \le R \end{aligned} \tag{1}$$

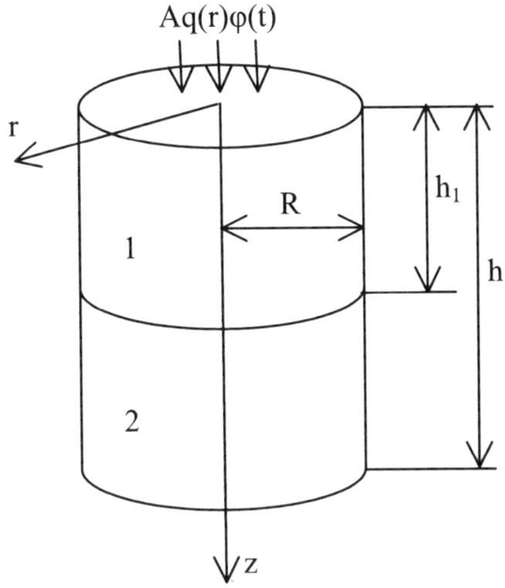

Figure 1: Geometry of the problem domain (R-radius of the structure; h_1-thickness of upper layer; h-height of whole structure; A-absorption coefficient).

Subscripts 1 and 2 correspond to the upper and to the lower layer, respectively. The corresponding boundary conditions are:

$$-\lambda_1 \frac{\partial T_1}{\partial z} = A(T)q(r)\varphi(t), \quad z=0, \quad 0 \le r \le R;$$

$$-\lambda_1 \frac{\partial T_1}{\partial r} = \alpha_1 T_1, \quad r=R, \quad 0 \le z \le h_1$$

$$-\lambda_2 \frac{\partial T_2}{\partial r} = \alpha_2 T_2, \quad r=R, \quad h_1 \le z \le h;$$

$$-\lambda_2 \frac{\partial T_2}{\partial z} = \alpha_2 T_2, \quad z=h, \quad 0 \le r \le R \tag{2}$$

$$T_1 = T_2, \quad \lambda_1 \frac{\partial T_1}{\partial z} = \lambda_2 \frac{\partial T_2}{\partial z}; \quad z=h_1, \quad 0 \le r \le R$$

$$\frac{\partial T_1}{\partial r} = 0; \quad r=0, \quad 0 \le z \le h_1; \quad \frac{\partial T_2}{\partial r} = 0; \quad r=0, \quad h_1 \le z \le h$$

where T_i is the temperature difference between the interior domain temperature and ambient one, λ is the coefficient of thermal conductivity, $a = \frac{\lambda}{\rho \cdot c}$ is the coefficient of thermal diffusivity, c is the specific heat, ρ is the material density, α is heat transfer coefficient which determines the rate of thermal losses on boundary surface, R and h are specimen's radius and length, respectively, $A(T)$ is absorption coefficient of the laser radiation by the material of the upper layer at temperature difference T.

The temperature dependence of the absorption coefficient is assumed to follow the following linear form

$$A(T) = A_0 + B \cdot T$$

where A_0 is of the absorption coefficient at ambient temperature T_0 and B is a constant whose value depends on the type of material [5]. For Al the above constants have the following numerical values [5]:

$$A_0 = 0.642; \quad B = -4.28 \cdot 10^{-4} \frac{1}{K}$$

The thermal losses, in axial and radial directions, were modeled by free thermal convection. Structures with three or more layers could also be described by the above model.

3 The boundary element formulation

For a mono-layer structures the governing equations (1) at n-th time step could be transformed for cylindrical coordinates into the following form:

$$\frac{\partial^2 T}{\partial r^2} + \frac{\partial^2 T}{\partial z^2} = \frac{1}{a}\frac{\partial T}{\partial t} - \frac{1}{r}\frac{\partial T}{\partial r} = b \quad 0 \le r \le R; \quad 0 \le z \le h$$

$$b = b_1 - b_2; \quad b_1 = \frac{1}{a}\frac{\partial T}{\partial t} \approx \frac{1}{a\Delta t}\Big(T\big(r, z, (n+1)\Delta t\big) - T\big(r, z, n\Delta t\big)\Big)$$

$$= \frac{1}{a\Delta t}\left(T_{n+1} - T_n\right); \quad b_2 = \frac{1}{r}\frac{\partial u}{\partial r} \tag{3}$$

where Δt is time step. Equation (3) is the main form of the equation which is solved in the present case. It is clear that the term with $1/r$ on the right hand side does not appear in the classical axisymmetric formulations which use fundamental solutions for axisymmetric problems. In the present case the Laplace fundamental solution is used and the term with $1/r$ is added to the non-homogeneous part of the Laplace equation. This term requires special care when $r \to 0$, as is explained further in the text.

By applying the Green's identity (3) can be transformed into the following integral form:

$$\chi(x)T(x) + \int_{\Gamma_y} q^*(x,y)T(y)d\Gamma_y - \int_{\Gamma_y} T^*(x,y)q(y)d\Gamma_y = -\int_{\Omega_y} T^*(x,y)b(y)d\Omega_y \tag{4}$$

where $x = (r_x, z_x)$, $y = (r_y, z_y)$, Ω is the problem domain Γ is the boundary of Ω, n is the direction of the normal to Γ, $q = \partial T/\partial n$ and $q^* = \partial T^*/\partial n$. The boundary conditions are given as:

$$q\big|_{z=0} = \frac{A(T)\cdot I(r,n\cdot\Delta t)}{\lambda},\quad 0\le r\le R;\quad \left.\frac{\partial T}{\partial r}\right|_{r=0} = 0,\ 0\le z\le h$$
$$q\big|_{r=R,z=H} = -\frac{\alpha\cdot T}{\lambda},\quad r=R, 0\le z\le h \vee z=h, 0\le r\le R \tag{5}$$

4 The dual reciprocity formulation

To avoid domain integration on the right hand side in expression (4) the DRM approximation is applied [1] yielding:

$$\chi(x)T(x) + \int_{\Gamma_y}\left(T(y)q^*(x,y) - T^*(x,y)q(y)\right)d\Gamma_y = \tag{6}$$
$$\sum_{j=1}^{N+L}\alpha_j\left(\chi(x)\hat{T}(x,y_j) + \int_{\Gamma_y}\left(\hat{T}(y,y_j)q^*(x,y) - \hat{q}(y,y_j)T^*(x,y)\right)d\Gamma_y\right)$$

In this work the DRM approximation function f was the 1+R radial basis function.

The thermal flux through the elementary surface S which encloses elementary volume dV during infinitesimally time period dt, see Figure 2, is represented using the following expressions:

$$dV\cdot\rho\cdot c\cdot dT = -\oint_S \vec{q}\cdot d\vec{s}\cdot dt;\quad dT = \frac{\partial T}{\partial t}dt + \vec{v}\nabla T, (\vec{v}=0);\quad dV = \Delta r^2\pi\cdot dz$$
$$\oint_S \vec{q}\cdot d\vec{s} = q_r\cdot 2\Delta r\pi\cdot dz + \left(q_{z_2} - q_{z_1}\right)\Delta r^2\pi; \Rightarrow \frac{\rho\cdot c}{2}\frac{\partial T}{\partial t} = -\left(\frac{q_r}{\Delta r} + \frac{1}{2}\frac{q_{z_2} - q_{z_1}}{dz}\right)$$
$$q_r = -\lambda\frac{\partial T}{\partial r};\quad q_z = -\lambda\frac{\partial T}{\partial z} \Rightarrow \lim_{\Delta r\to 0}\frac{1}{\Delta r}\frac{\partial T}{\partial r} = \frac{1}{2}\cdot\left(\frac{1}{a}\frac{\partial T}{\partial t} - \frac{\partial^2 T}{\partial z^2}\right) \tag{7}$$

where a, ρ and c have same meaning as in relations 1 and 2, $\vec{v}$ is velocity of element dV, $q_{1,2}$ is thermal flux in axial direction at point z and $z+dz$ respectively, and q_r is thermal flux in radial direction on boundary surface S. After discretization of the boundary Γ, the unknown temperature T is interpolated on elements on the boundary, the boundary integrals are evaluated and using collocation technique equation (6) is transformed into a system of linear equations.

The nodal values $T(x_i)$, $\hat{T}(x_i)$, $q(x_i)$, $\hat{q}(x_i)$, $b(x_i)$, $b_1(x_i)$, $b_2(x_i)$ and the coefficients α_i could be expressed in matrix form as:

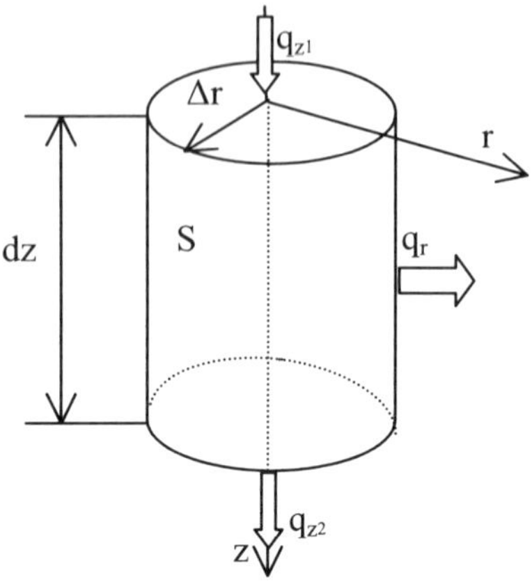

Figure 2: The thermal flux along z-axis.

$$\begin{aligned}
&\mathbf{b} = \mathbf{F}\boldsymbol{\alpha} = \left[b\left(y_1\right)\cdots b\left(y_K\right)\right]_{K\times 1}, \quad \mathbf{F} = \left[f\left(y_i, y_j\right)\right]_{K\times K}, \\
&\boldsymbol{\alpha} = \left[\alpha_1 \cdots \alpha_K\right]_{K\times 1} \Rightarrow \boldsymbol{\alpha} = \mathbf{F}^{-1}\mathbf{b} \\
&\mathbf{T} = \left[T\left(x_1\right)\cdots T\left(x_K\right)\right]_{K\times 1}, \quad \mathbf{q} = \left[q\left(x_1\right)\cdots q\left(x_K\right)\right]_{K\times 1}, \\
&\hat{\mathbf{T}} = \left[\hat{T}\left(x_1\right)\cdots \hat{T}\left(x_K\right)\right]_{K\times 1}, \quad \hat{\mathbf{q}} = \left[\hat{q}\left(x_1\right)\cdots \hat{q}\left(x_K\right)\right]
\end{aligned} \tag{8}$$

$$\mathbf{b} = \mathbf{b_1} - \mathbf{b_2}; \quad \mathbf{b_1} = \frac{1}{a\Delta t}\left(\mathbf{T} - \mathbf{T_0}\right) \tag{9}$$

$$\begin{aligned}
&b_{2i} = \begin{cases} \sum\limits_{j,n,m} \delta_{ij} \dfrac{1}{r_i} \dfrac{\partial f_{jn}}{\partial r} f_{nm}^{-1} T_m, & r_i \neq 0 \\ \dfrac{1}{2a\Delta t}\left(T_i - T_{0i}\right) - \sum\limits_{j,n} \dfrac{1}{2} \dfrac{\partial^2 f_{ij}}{\partial z^2} \cdot f_{in}^{-1} T_n, & r_i = 0 \end{cases}; \\
&\mathbf{b_2} = \left[b_{21} \cdots b_{2k}\right]_{K\times 1}; \quad K = N + L
\end{aligned} \tag{10}$$

$$\begin{aligned}
&\frac{\partial f_{jn}}{\partial r} = \left.\frac{\partial f\left(y, y_n\right)}{\partial r}\right|_{y=y_j}; \frac{\partial^2 f_{jn}}{\partial z^2} = \left.\frac{\partial^2 f\left(y, y_n\right)}{\partial z^2}\right|_{y=y_j}; \\
&f_{nm}^{-1} = \left[\mathbf{F}^{-1}\right]_{nm}; \quad \vec{y} = \left(r, z\right); \quad \vec{y}_n = \left(r_n, z_n\right)
\end{aligned} \tag{11}$$

where $\mathbf{T_0}$ is a vector obtained in the previous time step, N is the number of boundary nodes, L is the number of internal nodes, and δ_{ij} is the Kronecker delta symbol.

WIT Transactions on Modelling and Simulation, Vol 42, © 2006 WIT Press
www.witpress.com, ISSN 1743-355X (on-line)

Now the equation (6) can be expressed in the following matrix form:

$$\mathbf{Hu} - \mathbf{Gq} = (\mathbf{H\hat{u}} - \mathbf{G\hat{q}})\mathbf{F}^{-1}\mathbf{b} + \mathbf{I_0} \tag{12}$$

$$\mathbf{I_0} = \left[I_{0i}\right]; \quad I_{0i} = \frac{A_0}{\lambda}\int_0^{r_0} I\left(r_y, n \cdot \Delta t\right) \cdot u^*\left(\left|\vec{y} - \vec{y}_i\right|\right) \cdot dr_y; \qquad \vec{y} = \left(r_y, 0\right); \vec{y}_i = \left(r_i, z_i\right); \; i = 1, \ldots, K \tag{13}$$

where **H** and **G** are matrices whose matrix elements were evaluated from the contour integrals. The elements of the vector **q** over contour Γ could be expressed, according to the boundary conditions, by the elements of vector **u**. The elements of vector I_0 represent the equivalent thermal loads on upper surface of the specimens.

Sub-domain technique in the DRM, further referred to as DRM-MD [6] has been used in some examples in order to improve the accuracy.

5 Numerical results

The temperature field distributions in radial and axial direction, inside a mono-layer Al cylinder with radius 7 mm and length 5 mm, which were obtained using the DRM, the DRM-MD with four sub-domains and analytical solution [7] for t=1s are presented in Figures 3 and 4, respectively.

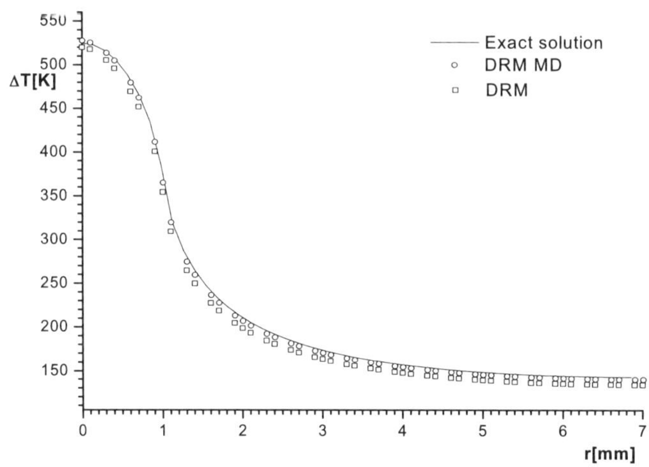

Figure 3: Temperature difference distribution along r-axis for the Al specimen on the upper surface (number of subdomains for DRM-MD=4).

The following properties of the incoming laser beam were considered: power-500W, radius of laser beam 1mm, top head profile and constant laser beam intensity with time duration of 1s.

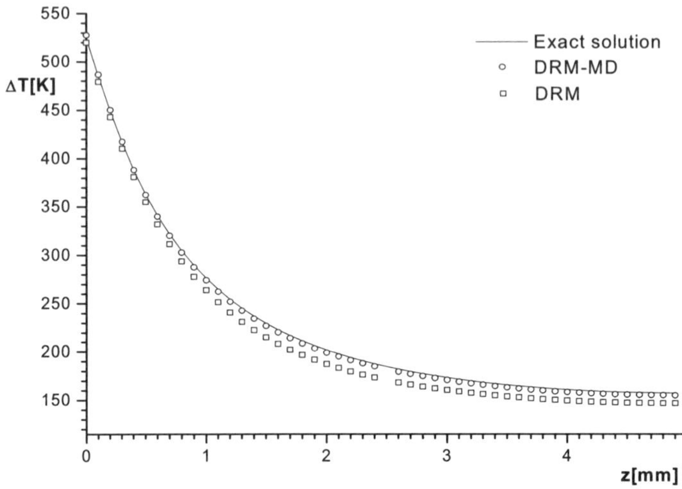

Figure 4: Temperature difference distribution along the z-axis inside the Al specimen obtained by using DRM, DRM-MD and analytical expression (number of subdomains for DRM-MD=4).

The DRM and the DRM-MD results compared to the analytical ones along radial direction for different number of boundary nodes are shown in Figure 5 for t=1s. It can be observed that the accuracy of the DRM-MD was higher than the one achieved using the DRM in all cases.

The distribution of temperature field at t=1 s, in axial directions, for the case of two layer cylindrical structures, is shown in Figure 6. The upper layer of the two-layer structure is made of Al and the lower layer is made of glass. The following dimensions of the structures were used: (i) Al-layer-0.5 mm, Glass layer- 4.5 mm; and (ii) Al-layer-0.7 mm, Glass layer- 4.3 mm thicknesses. In both cases the radius was 7 mm. The following properties of the laser beam were assumed: Power – 100W, radius of laser beam – 1mm, the laser beam has constant intensity with the top head profile and time duration of 1s. The presented results were obtained by using the DRM-MD procedure with nine sub domains. A linear temperature dependence of the absorption coefficient was assumed.

6 Conclusions

The boundary element dual reciprocity method (BE-DRM) was applied to the problem of interaction of laser-mono/multi layer structures with axial symmetry. The BEM formulation is based on the fundamental solution for the Laplace equation.

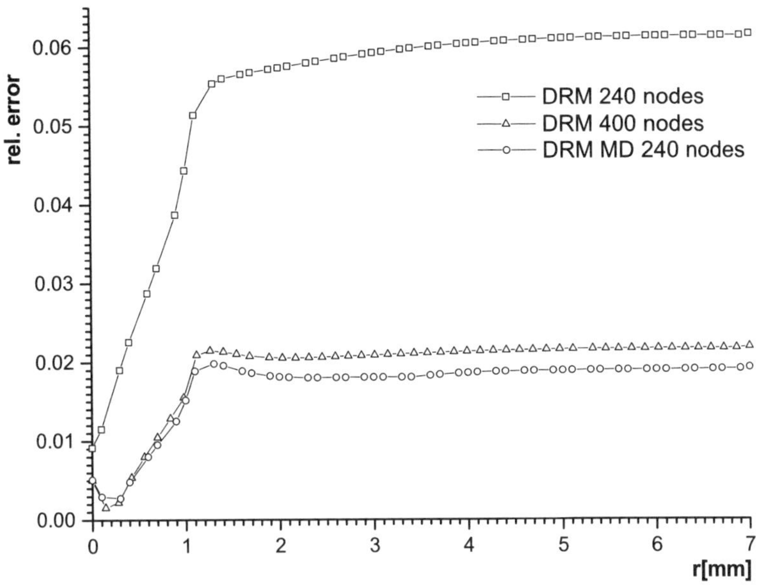

Figure 5: Relative error for the DRM and the DRM-MD along r direction on the upper surface for different number of boundary nodes (number of subdomains for DRM-MD=4).

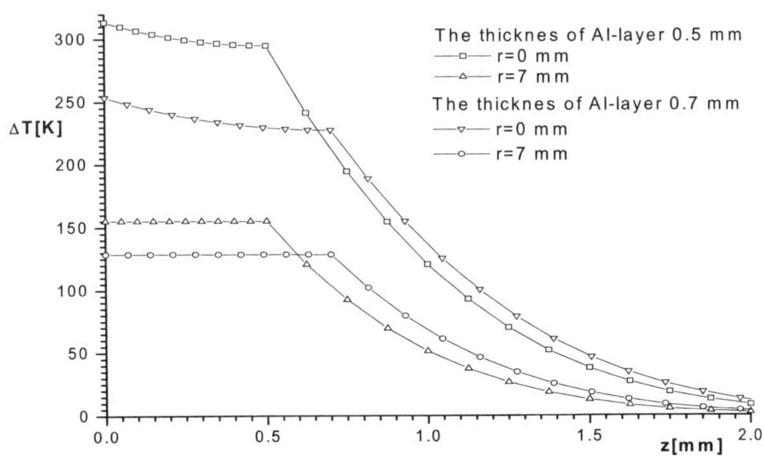

Figure 6: Temperature difference distribution along z-axis in case of two layer structures obtained by the DRM-MD with nine sub-domains at upper surface and at interfaces between layers.

The accuracy of the developed DRM formulation was first tested using analytical solution for a mono-layer case and then applied to a two-layer structure consisting of Al and glass. The results show that the formulation can provide accurate results for this type of problems. The results were compared for the DRM when the domain was kept as a single domain and when it was divided into sub-domains (DRM-MD). The sub-domain formulation showed increase in the accuracy. This behavior of the DRM formulation has already been reported in the past [8].

References

[1] Partridge PW, Brebbia CA, Wrobel LC. *The Dual Reciprocity Boundary Element Method.* Computational Mechanics Publications, Southampton & Elsevier Applied Science, New York, 1992.

[2] Wrobel LC, Telles JCF, Brebbia CA. A dual reciprocity boundary element formulation for axisymmetric diffusion problems. In: Brebbia CA, editor. Boundary elements VIII. Boston: Computational Mechanics Publications; 1986. p. 59–69.

[3] Fengwu Bai, Wen-Qiang Lu. The selection and assemblage of approximation functions and disposal of its singularity in axisymmetric DRBEM for heat transfer problems. Engineering Analysis with Boundary Elements 28 (2004) 955–965.

[4] Perrey-Debain E. Analysis of convergence and accuracy of the DRBEM for axisymmetric Helmhotz-type equation. Engng Anal Bound Elem 1999; 23:703–11.

[5] Rykalin N, Uglov A, Kokora A. Laser machining and welding. Mir Publishers, Moscow, 1979.

[6] Popov V, Power H. The DRM-MD integral equation method: An efficient approach for the numerical solution of domain dominant problems. Int. J. Num. Meth. Engrg, **44**, 327-353, 1999.

[7] Gospavic R, Sreckovic M, Popov V. Modelling of laser-material interaction using semi-analytical approach. Mathematics and Computers in Simulations 65, 211-219, 2004.

[8] Popov V, Power H. A domain decomposition in the dual reciprocity approach. Boundary Elements Communications, 7/1, 1-5, 1996.

Section 2
Advanced structural applications

Large deflection analysis of membranes containing rigid inclusions

M. S. Nerantzaki[1] & C. B. Kandilas[2]
[1]Department of Civil Engineering, National Technical University of Athens, Greece
[2]Department of Applied Mechanics and Marine Materials, Hellenic Naval Academy, Greece

Abstract

In this paper the deformation of membranes containing rigid inclusions is analyzed. These rigid inclusions can significantly change the entire stress distribution in the membrane and therefore create major difficulties for the design. The initially flat membrane, which may be prestretched by boundary in-plane tractions or displacements, is subjected to externally applied loads and to the weight of the rigid inclusions. The composite system is examined in cases where it's deformation reaches a state for which the undeformed and deformed shapes are substantially different. In such cases large deflections of membranes are considered, which result from nonlinear kinematic relations. The three coupled nonlinear equations in terms of the displacements governing the response of the membrane are solved using the Analog Equation Method (AEM), which reduces the problem to the solution of three uncoupled Poisson's equations with fictitious domain source densities. The problem is strongly nonlinear. In addition to the geometrical nonlinearity, the problem is itself nonlinear, because the membrane's reactions on the boundary of the rigid inclusions are not *a priori* known as they depend on the produced deflection surface. Iterative schemes are developed for the calculation of deformed membrane's configuration which converges to the final equilibrium state of the membrane with the given external applied loads. Several example problems are presented, which illustrate the method and demonstrate its accuracy and efficiency. The method has all the advantages of the pure BEM.
Keywords: rigid inclusion, elastic membranes, large deflections, nonlinear, boundary elements, analog equation method.

WIT Transactions on Modelling and Simulation, Vol 42, © 2006 WIT Press
www.witpress.com, ISSN 1743-355X (on-line)
doi:10.2495/BE06010

1 Introduction

In this paper elastic membranes of arbitrary shape containing rigid inclusions are studied. The initially flat membrane, which may be prestretched by boundary in-plane tractions or displacements, is subjected to externally applied loads and to the weight of the rigid inclusions. The equilibrium configuration of the membrane is reached when the reactions on the boundary with the given applied loads and the weight of the rigid inclusion are in equilibrium and the produced displacements field satisfy the equations governing the deformation response of the membrane and the boundary conditions of the problem. The problem is strongly nonlinear. Two types of geometric nonlinearity arise. One is due to nonlinear strain-displacement relations and is reflected in the three-coupled nonlinear partial differential equations governing the deformation response of the membrane. The other nonlinearity results from the rigid inclusion -structure interaction, namely from the fact that the reactions on the boundary of the rigid inclusion are not *a priori* known, as they depend on the deflection of the membrane and vice versa.

Analytical solutions for the problem of membranes with rigid inclusions as stated above are not available in the literature. The accuracy and efficiency of the method is examined as compared on the worked examples with other approximated numerical methods (FEM). It should be noted that, in contrast to the present method, the FEM method requires a very fine mesh to accurately capture the local stress concentration created around the rigid inclusion. The approach presented here requires the solution of two coupled problems, namely, the nonlinear analysis of membranes under a given load and the problem of finding the equilibrium configuration of the rigid inclusion through an iterative procedure. The membrane problem is solved using the AEM [3], while an iterative scheme is developed to determine the equilibrium state. Numerical examples are presented by analyzing membranes of various shapes with the rigid inclusion which illustrate the accuracy and efficiency of the method.

2 The nonlinear membrane problem

Consider a thin flexible initially flat elastic membrane consisting of homogeneous linearly elastic material occupying the two-dimensional domain Ω in x,y plane bounded by the K+1 curves $\Gamma_0,\Gamma_1,\ldots,\Gamma_K$ (see Fig. 1). The membrane is prestretched either by imposed displacement $\tilde{u}_n,\tilde{v}_t$ or by external forces $\tilde{T}_n,\tilde{T}_t$ acting along the boundary $\Gamma=\bigcup_{i=0}^{i=K}\Gamma_i$. Large deflections are considered resulting from nonlinear kinematic relations, where only the squares of the slopes of the deflection surface are retained in the strain components. Thus, the strain components are given as

$$\varepsilon_x = u,_x^o + u,_x + \frac{1}{2} w,_x^2 \tag{1a}$$

$$\varepsilon_y = v_{,y}^o + v_{,y} + \frac{1}{2} w_{,y}^2 \tag{1b}$$

$$\gamma_{xy} = u_{,y}^o + v_{,x}^o + u_{,y} + v_{,x} + w_{,x} w_{,y} \tag{1c}$$

where $u^o = u^o(x,y)$, $v^o = v^o(x,y)$ are the in-plane displacements components due to the prestress; $w = w(x,y)$ the transverse deflection produced when the membrane is subjected to the load $g = g(x,y)$ acting in the direction normal to its plane and to the weight of rigid inclusions; and $u = u(x,y)$, $v = v(x,y)$ are the additional membrane displacements due to the deflection.

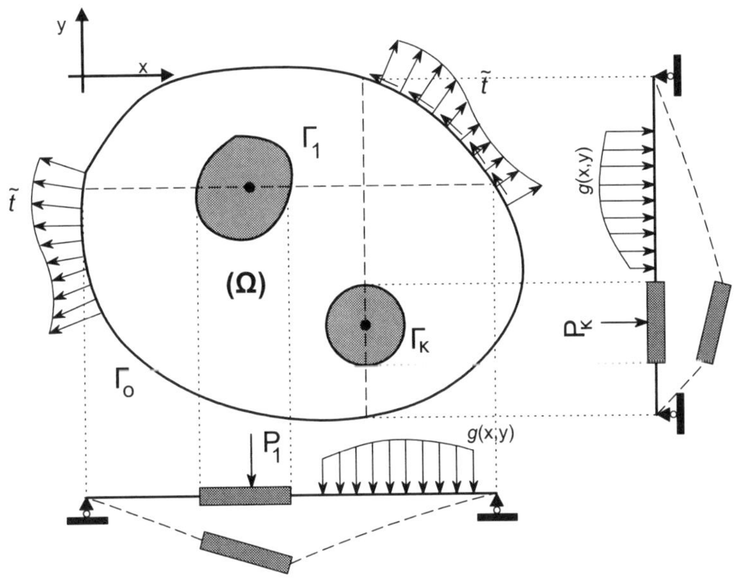

Figure 1: Domain Ω occupied by the membrane.

In the case of large deflection analysis of homogeneous elastic membranes the governing differential equations are [3]

$$N_{x,x} + N_{xy,y} = 0 \tag{2a}$$

$$N_{xy,x} + N_{y,y} = 0 \tag{2b}$$

$$(N_x^o + N_x) w_{,xx} + 2(N_{xy}^o + N_{xy}) w_{,xy} + (N_y^o + N_y) w_{,yy} + g = 0 \tag{2c}$$

in Ω; N_x^o, N_y^o, N_{xy}^o and N_x, N_y, N_{xy} are the membrane forces given as

$$N_x^o = C(u_{,x}^o + \nu v_{,y}^o) \tag{3a}$$

$$N_y^o = C(v_{,y}^o + \nu u_{,x}^o) \tag{3b}$$

$$N_{xy}^o = C\frac{1-\nu}{2}(u_{,y}^o + v_{,x}^o) \tag{3c}$$

$$N_x = C\left[\left(u,_x + \frac{1}{2}w,_x^2\right) + \nu\left(v,_y + \frac{1}{2}w,_y^2\right)\right] \tag{4a}$$

$$N_y = C\left[\left(v,_y + \frac{1}{2}w,_y^2\right) + \nu\left(u,_x + \frac{1}{2}w,_x^2\right)\right] \tag{4b}$$

$$N_{xy} = C\frac{1-\nu}{2}\left(u,_y + v,_x + w,_x\, w,_y\right) \tag{4c}$$

where $C = Et/(1-\nu^2)$ is the stiffness of the membrane, E the modulus of elasticity, ν the Poisson ratio and t its thickness. Attention should be paid to the resulting in plane tensile forces N_1, N_2 in the principal directions to avoid wrinkling of the membrane, namely

$$N_{1,2} = \frac{N_x + N_y}{2} \pm \sqrt{(\frac{N_x - N_y}{2})^2 + (N_{xy})^2}\ \rangle 0 \tag{5}$$

Substituting eqns (4) into eqns (2) yields the equilibrium equations in terms of the displacements

$$\frac{1-\nu}{2}\nabla^2 u + \frac{1+\nu}{2}(u,_x + v,_y),_x = -w,_x\,(w,_{xx} + \frac{1-\nu}{2}w,_{yy}) - \frac{1+\nu}{2}w,_y\, w,_{xy} \tag{6a}$$

$$\frac{1-\nu}{2}\nabla^2 v + \frac{1+\nu}{2}(u,_x + v,_y),_y = -w,_y\,(w,_{yy} + \frac{1-\nu}{2}w,_{xx}) - \frac{1+\nu}{2}w,_x\, w,_{xy} \tag{6b}$$

$$\begin{aligned}&\{N_x^o + (u,_x^2 + \frac{1}{2}w,_x^2) + \nu(v,_y + \frac{1}{2}w,_y^2)\}w,_{xx} + \\ &\{N_{xy}^o + (1-\nu)(u,_y + v,_x + w,_x\, w,_y)\}w,_{xy} \\ &+\{N_y^o + (v,_y + \frac{1}{2}w,_y^2) + \nu(u,_x + \frac{1}{2}w,_x^2)\}w,_{yy} = -g/C\end{aligned} \tag{6c}$$

For a fixed boundary the displacements should satisfy the following boundary conditions

$$\tilde{u} = \tilde{v} = \tilde{w} = 0 \tag{7a,b,c}$$

3 Implementation of the AEM for large deflections of membranes

The boundary value problem described by eqns (6) and (7) is solved using the Analog Equation Method (AEM). According to the concept of the analog equation, eqns (6) are replaced by three Poisson's equations

$$\nabla^2 u_i = b_i \qquad (i = 1,\ 2,\ 3) \tag{8}$$

where $b_i = b_i(x_1, x_2)$ are fictitious sources. Note that u_1, u_2, u_3 stand for the functions u, v, w, respectively.

The fictitious sources are established using the BEM. For this purpose b_i is approximated as

$$b_i = \sum_{j=1}^{M} a_j^{(i)} f_j \quad (i = 1, 2, 3) \tag{9}$$

where f_j are approximating radial basis functions and $a_j^{(i)}$ are $3M$ coefficients to be determined. We look for the solution as a sum of the homogeneous solution $\bar{u}_i$ and a particular one u_i^p, $u_i = \bar{u}_i + u_i^p$. The particular solution is obtained as

$$u_i^p = \sum_{j=1}^{M} a_j^{(i)} \hat{u}_j \tag{10}$$

where $\hat{u}_j$ is a particular solution of

$$\nabla^2 \hat{u}_j = f_j \tag{11}$$

The homogenous solution is obtained from the boundary value problem

$$\nabla^2 \bar{u}_i = 0 \quad \text{in } \Omega \tag{12a}$$

$$\bar{u}_i = \tilde{u}_i - \sum_{j=1}^{M} a_j^{(i)} \hat{u}_j \quad \text{on } \Gamma \tag{12b}$$

Writing the solution of eqn (12a) in integral form, we have

$$c\bar{u}_i = -\int_{\Gamma} (u^* \bar{u}_{i,n} - \bar{u}_i u^*_{,n}) ds \;, \qquad i = 1, 2, 3 \tag{13}$$

with $u^* = \ell n r / 2\pi$, $r = | P - Q |$, $Q \in \Gamma$ being the fundamental solution of the Laplace equation and $c = 1, 1/2, 0$ depending on whether $P \in \Omega$, $P \in \Gamma$, $P \notin \Omega \cup \Gamma$, respectively. On the base of eqns (10) and (13), the solution of eqn (8) for points inside Ω $(c = 1)$ is written as

$$u_i = -\int_{\Gamma} (u^* \bar{u}_{i,n} - \bar{u}_i u^*_{,n}) ds + \sum_{j=1}^{M} a_j^{(i)} \hat{u}_j \;, \qquad i = 1, 2, 3 \tag{14}$$

The first and second derivatives for points inside Ω are obtained by direct differentiation of eqn (14). Thus, we have

$$u_{i,k} = -\int_{\Gamma} (u^*_{,k} \bar{u}_{i,n} - \bar{u}_i u^*_{,nk}) ds + \sum_{j=1}^{M} a_j^{(i)} \hat{u}_{j,k} \qquad (k = 1, 2) \tag{15a}$$

$$u_{i,kl} = -\int_{\Gamma} (u^*_{,kl} \bar{u}_{i,n} - \bar{u}_i u^*_{,nkl}) ds + \sum_{j=1}^{M} a_j^{(i)} u_{j,k\ell} \qquad (k, l = 1, 2) \tag{15b}$$

Using the BEM with N constant boundary elements, discretizing eqn (13) and applying it to the N boundary nodal points yields

$$\mathbf{C}\bar{\mathbf{u}}_i = \mathbf{H}\bar{\mathbf{u}}_i - \mathbf{G}\bar{\mathbf{u}}_{i,n} \qquad (i = 1,\ 2,\ 3) \tag{16}$$

with $\mathbf{C}$ being a $N \times N$ diagonal matrix including the values of the coefficient c at the N boundary nodal points and $\mathbf{H}, \mathbf{G}$ are $N \times N$ matrices originating from the integration of the kernels on the boundary elements. Eqns (14) and (15) are subsequently applied to M points inside the domain $(c = 1)$ (see Fig.2). This yields after eliminating $\bar{\mathbf{u}}$ and $\bar{\mathbf{u}}_n$ by virtue of the boundary conditions (12b) and eqn (16)

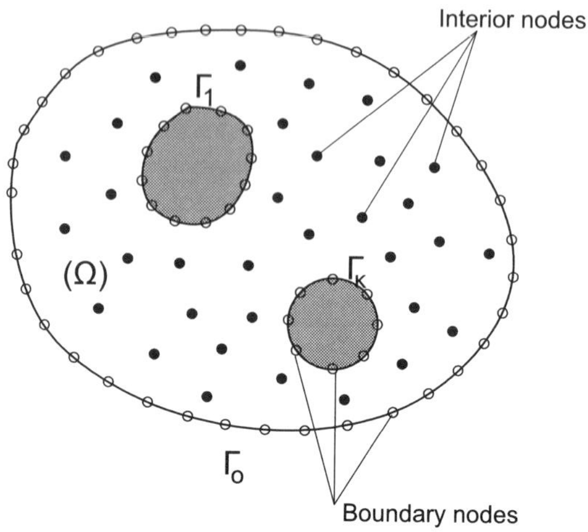

Figure 2: Boundary discretization and domain nodal points.

$$\mathbf{u}_i = \mathbf{D}\mathbf{a}^{(i)} + \mathbf{E}\tilde{\mathbf{u}}_i \tag{17}$$

$$\mathbf{u}_{i,k} = \mathbf{D}_k\mathbf{a}^{(i)} + \mathbf{E}_k\tilde{\mathbf{u}}_i \tag{18}$$

$$\mathbf{u}_{i,kl} = \mathbf{D}_{kl}\mathbf{a}^{(i)} + \mathbf{E}_{kl}\tilde{\mathbf{u}}_i \tag{19}$$

where $\mathbf{D}$, $\mathbf{E},\ldots,\mathbf{E}_{kl}$ are known matrices and $\mathbf{a}^{(i)}$ is the vector of the unknown coefficients.

The final step of AEM is to apply eqns (6) to the M points inside Ω and replace the derivatives of u_i using eqns (18) and (19). This yields

$$\mathbf{F}_1(\mathbf{a}^{(1)},\mathbf{a}^{(2)},\mathbf{a}^{(3)},\mathbf{u}^{in}) = 0 \tag{20a}$$

$$\mathbf{F}_2(\mathbf{a}^{(1)},\mathbf{a}^{(2)},\mathbf{a}^{(3)},\mathbf{u}^{in}) = 0 \tag{20b}$$

$$\mathbf{F}_3(\mathbf{a}^{(1)},\mathbf{a}^{(2)},\mathbf{a}^{(3)},\mathbf{u}^{in}) = 0 \tag{20c}$$

where $\mathbf{u}^{in}$ is the array of the unknown displacements on the boundary of each rigid inclusion. The solution of the problem posed by eqns (20) cannot be approached directly and a form of iteration procedure is required. Since the membrane's deformed configuration is originally unknown, the iterative technique used proceeds by examining the equilibrium of the rigid inclusion.

We assume $\mathbf{u}_o$ and θ_o an initial guess of displacement and rotation of the center of the rigid inclusion. Then the compatibility equations $\mathbf{u}^{\mathbf{in}} = \mathbf{u}_o + \theta_o \times \mathbf{r}$ must be satisfied for all selected nodes on the rigid inclusion boundary where $\mathbf{r}$ the distance a node of the interior boundary from the center of the rigid inclusion. For the displacements of the interior boundary, the membrane's externally applied load and the boundary conditions, the system of non linear algebraic equations is solved and a deflection surface $w^{(1)}$ is determined. Once the vectors $\mathbf{a}^{(i)}$ ($i = 1,2,3$) are evaluated the displacements and their derivatives are computed from eqns (17)-(19). Finally, the stress resultants are computed from eqns (4) and the boundary reactions from [3]

$$T_x = N_x \cos a + N_{xy} \sin a \tag{21a}$$

$$T_y = N_{xy} \cos a + N_y \sin a \tag{21b}$$

$$V = T_x w,_x + T_y w,_y \tag{21c}$$

where $a = angle(x, \mathbf{n})$.

The procedure continues checking the six equilibrium equations on each rigid inclusion in the deformed configuration. If these are not satisfied, the procedure iterates with new values of $\mathbf{u}_o$ and θ_o.

4 Numerical examples

On the basis of the procedure described in previous section a FORTRAN program was written and a rectangular membrane with a circular rigid inclusion was analyzed. Radial basis functions of multiquadric type, i.e. $f = (r^2 + c^2)^{1/2}$, have been employed; c is an appropriately chosen arbitrary constant. Computationally, the method was checked with other numerical procedures (FEM) and found very accurate and efficient. The numerical results were obtained using the MS Fortran PowerStation 4.0 on a Pentium III PC. For all calculations the solution of non linear algebraic equations converged in less than 10 iterations with a tolerance $\varepsilon_w = 0.0001$. The solution of eqns (20) was obtained using the subroutine DNEQNF of IMSL.

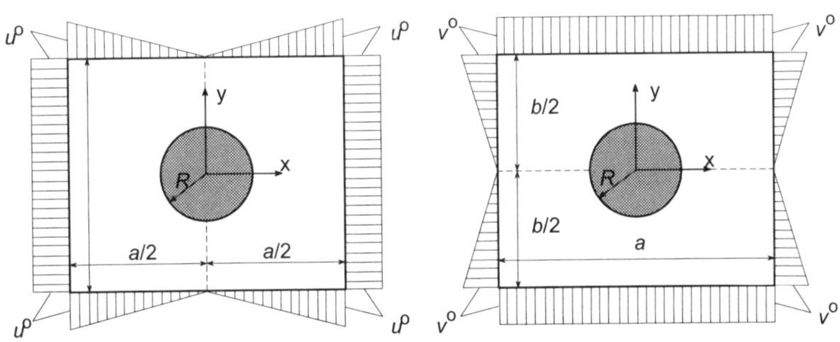

Figure 3: Imposed boundary displacements.

4.1 Rectangular membrane with a circular inclusion

A rectangular membrane with a circular rigid inclusion, subjected to a concentrate load P or a moment M_x applied to the center of it has been analyzed. The membrane was prestretched by imposed boundary displacements as shown in Fig. 3, which produce approximately uniform prestress. The employed data are: $a = 6.0\,m$, $b = 4.0\,m$, $R = 1.0\,m$, $E = 110\,MPa$:

$\nu = 0.3$, $\tilde{u} = \tilde{v} = 2\,cm$. The membrane's thickness is $t = 5\,mm$, or $t = 10\,mm$, in the case where a moment M_x is also applied. The membrane was analysed using $N = 110$ constant boundary elements and $M = 72$ interior collocation points. The FEM code ANSYS is employed using 184 triangular finite elements capable to large deflection analysis to compare the obtained results. In Fig. 4 and 5 the in-plane displacement u and the computed profile of the deflection w along the x axis are presented for various values of applied load P. Also in Fig. 7 the stress resultant N_x in polar coordinates along the boundary of rigid inclusion is presented for various values of P. The deflection at the center of the rigid inclusion is in very good agreement with those calculated from the FEM solution, as shown in Fig. 6. The computed profile of the deflection w along the x axis shown in Fig. 8 for a load $P = 5kN$ and a moment $M_x = 5kNm$.

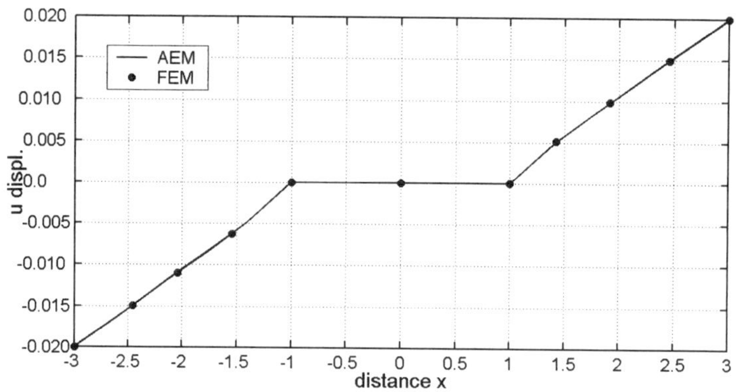

Figure 4: In-plane displacement u at $y = 0$ for various values of applied load P.

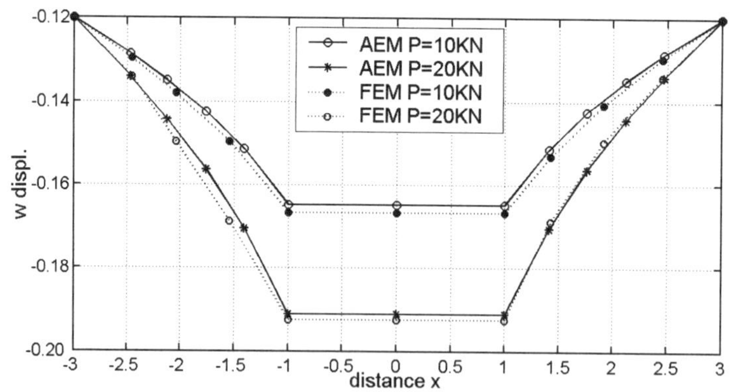

Figure 5: Membrane's profile at $y = 0$ for different values of applied load P.

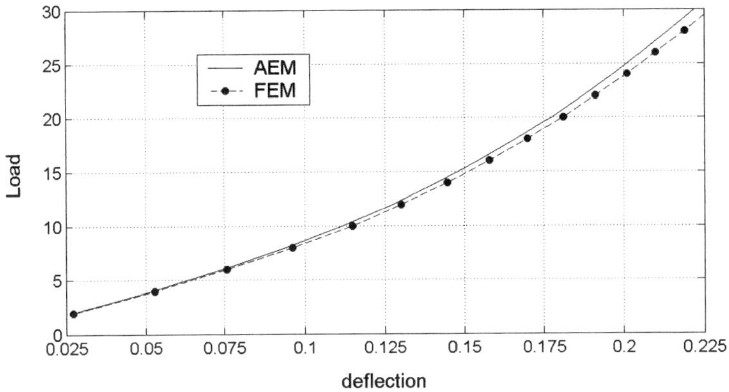

Figure 6: Deflection w at the center of rigid inclusion for various values of applied load P.

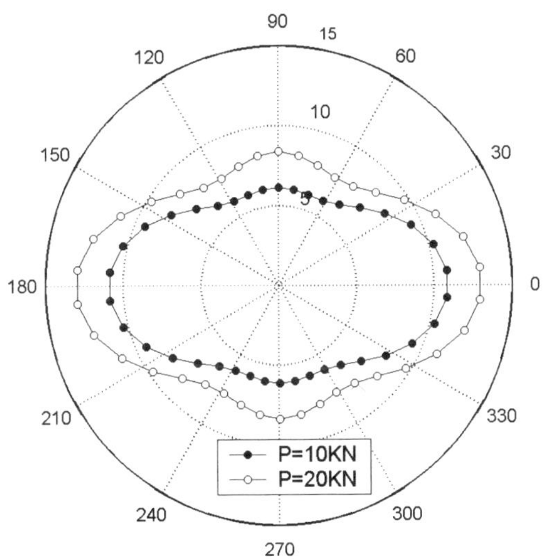

Figure 7: Membrane forces N_x in polar coordinates along the boundary of rigid inclusion for various values of P.

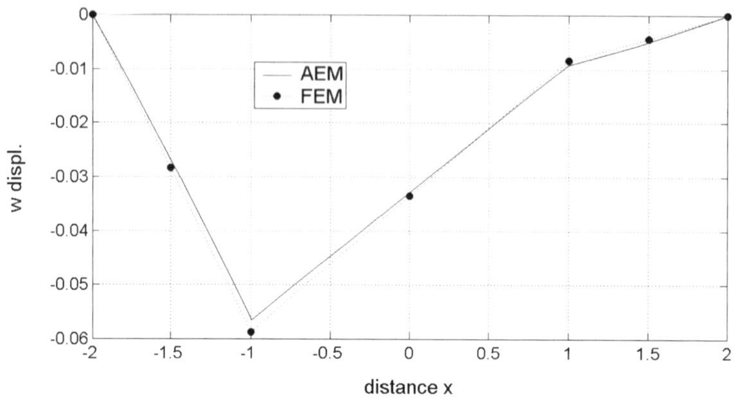

Figure 8: Membrane's profile at $y = 0$ for a load P and a moment M_x

References

[1] Haughton D. M. 1991. An Exact Solution for the Stretching of Elastic Membranes Containing a Hole or Inclusion, *Mechanics Research Communications*, 18(1), 29-39.

[2] Chen, D. and Cheng, S., 1996. Nonlinear Analysis of Prestretched Circular Membrane and a Modified Iteration Technique, *International Journal of Solids and Structures*, 33, 545-553.

[3] Katsikadelis, J.T., Nerantzaki, M.S. & Tsiatas, G.C., 2001. The analog equation method for large deflection analysis of membranes. A boundary-only solution. *Computational Mechanics,* 27(6), pp. 513-523.

Shear deformation effect in nonlinear analysis of spatial beams subjected to variable axial loading by BEM

E. J. Sapountzakis & V. G. Mokos
School of Civil Engineering, National Technical University, Zografou Campus, Athens, Greece

Abstract

In this paper a boundary element method is developed for the nonlinear analysis of beams of arbitrary doubly symmetric simply or multiply connected constant cross section, taking into account shear deformation effect. The beam is subjected in an arbitrarily concentrated or distributed variable axial loading, while the shear loading is applied at the shear center of the cross section, avoiding in this way the induction of a twisting moment. To account for shear deformations, the concept of shear deformation coefficients is used. Five boundary value problems are formulated with respect to the transverse displacements, the axial displacement and to two stress functions and solved using the Analog Equation Method, a BEM based method. Application of the boundary element technique yields a system of nonlinear equations from which the transverse and axial displacements are computed by an iterative process. The evaluation of the shear deformation coefficients is accomplished from the aforementioned stress functions using only boundary integration. Numerical examples with great practical interest are worked out to illustrate the efficiency, the accuracy and the range of applications of the developed method. The influence of both the shear deformation effect and the variableness of the axial loading are remarkable.
Keywords: transverse shear stresses, shear center, shear deformation coefficients, beam, second order analysis, nonlinear analysis, boundary element method.

doi:10.2495/BE06011

1 Introduction

An important consideration in the analysis of the components of plane and space frames or grid systems is the influence of the action of axial, lateral forces and end moments on the deformed shape of a beam. Lateral loads and end moments generate deflection that is further amplified by axial compression loading. The aforementioned analysis becomes much more accurate and complex taking into account that the axial force is nonlinearly coupled with the transverse deflections, avoiding in this way the inaccuracies arising from a linearized second – order analysis.

Over the past twenty years, many researchers have developed and validated various methods of performing a linearized second-order analyses on structures. Early efforts led to methods based on accounting for the aforementioned effect by using magnification factors applied to the results obtained from first-order analyses. Consequently, due to the demand of more rigorous and accurate second-order analysis of structural components several research papers have been published including a non-linear incremental stiffness method, closed-form stiffness methods, the analysis of non-linear effects by treating every element as a "beam-column" one, a design method for space frames using stability functions to capture second-order effects associated with P-δ and P-Δ effects and the finite element method using linear and cubic shape functions. Recently, Katsikadelis and Tsiatas [1] presented a BEM-based method for the nonlinear analysis of beams with variable stiffness. In all these studies shear deformation effect is ignored.

In this paper a boundary element method is developed for the nonlinear analysis of beams of arbitrary doubly symmetric simply or multiply connected constant cross section, taking into account shear deformation effect. The essential features and novel aspects of the present formulation compared with previous ones are summarized as follows.

i. The beam is subjected in an arbitrarily concentrated or distributed variable axial loading.
ii. The beam is supported by the most general boundary conditions including elastic support or restrain.
iii. The analysis is not restricted to a linearized second – order one but is a nonlinear one arising from the fact that the axial force is nonlinearly coupled with the transverse deflections (additional terms are taken into account).
iv. Shear deformation effect is taken into account.
v. The shear deformation coefficients are evaluated using an energy approach, instead of Timoshenko and Goocher's [2] and Cowper's [3] definitions, for which several authors have pointed out that one obtains unsatisfactory results or definitions given by other researchers, for which these factors take negative values.
vi. The effect of the material's Poisson ratio v is taken into account.
vii. The proposed method employs a pure BEM approach (requiring only boundary discretization) resulting in line or parabolic elements instead of

area elements of the FEM solutions (requiring the whole cross section to be discretized into triangular or quadrilateral area elements), while a small number of line elements are required to achieve high accuracy.

Numerical examples with great practical interest are worked out to illustrate the efficiency, the accuracy and the range of applications of the developed method. The influence of both the shear deformation effect and the variableness of the axial loading are remarkable.

2 Statement of the problem

Consider a prismatic beam of length L with a doubly symmetric cross section of arbitrary shape, occupying the two dimensional multiply connected region Ω of the y,z plane bounded by the $K+1$ curves $\Gamma_1, \Gamma_2, \ldots, \Gamma_K, \Gamma_{K+1}$, as shown in Fig.1. These boundary curves are piecewise smooth, i.e. they may have a finite number of corners. The material of the beam, with shear modulus G and Poisson's ratio v is assumed homogeneous, isotropic and linearly elastic. Without loss of generality, it may be assumed that the x-axis of the beam principal coordinate system is the line joining the centroids of the cross sections. The beam is subjected to an arbitrarily distributed axial loading p_x and to torsionless bending arising from arbitrarily distributed transverse loading p_y, p_z and bending moments m_y, m_z along y and z axes, respectively (Fig.1a).

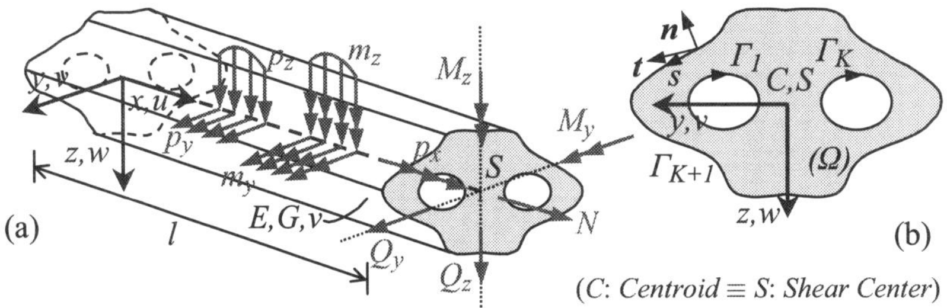

Figure 1: Prismatic beam in torsionless bending (a) with an arbitrary doubly symmetric cross-section occupying the two dimensional region Ω (b).

According to the linear theory of beams (small deflections), the angles of rotation of the cross-section in the x-z and x-y planes of the beam subjected to the aforementioned loading and taking into account shear deformation effect satisfy the following relations

$$\cos\omega_y \simeq 1 \qquad \cos\omega_z \simeq 1 \tag{1a,b}$$

$$\sin\omega_y \simeq \omega_y = -\frac{dw}{dx} = \theta_y - \gamma_z \qquad \sin\omega_z \simeq \omega_z = -\frac{dv}{dx} = -\theta_z - \gamma_y \tag{1c,d}$$

where $w = w(x)$, $v = v(x)$ are the beam transverse displacements with respect to *z*, *y* axes, respectively, while the corresponding curvatures are given as

$$k_y = \frac{d\theta_y}{dx} = -\frac{d^2 w}{dx^2} + \frac{d\gamma_z}{dx} = -\frac{d^2 w}{dx^2} - \frac{p_z}{GA_z} \tag{2a}$$

$$k_z = \frac{d\theta_z}{dx} = \frac{d^2 v}{dx^2} - \frac{d\gamma_y}{dx} = \frac{d^2 v}{dx^2} + \frac{p_y}{GA_y} \tag{2b}$$

where γ_y, γ_z are the additional angles of rotation of the cross-section due to shear deformation and GA_y, GA_z are its shear rigidities of the Timoshenko's beam theory, where

$$A_z = \kappa_z A = \frac{1}{a_z} A \qquad A_y = \kappa_y A = \frac{1}{a_y} A \tag{3a,b}$$

are the shear areas with respect to y, z axes, respectively with κ_y, κ_z the shear correction factors, a_y, a_z the shear deformation coefficients and A the cross section area.

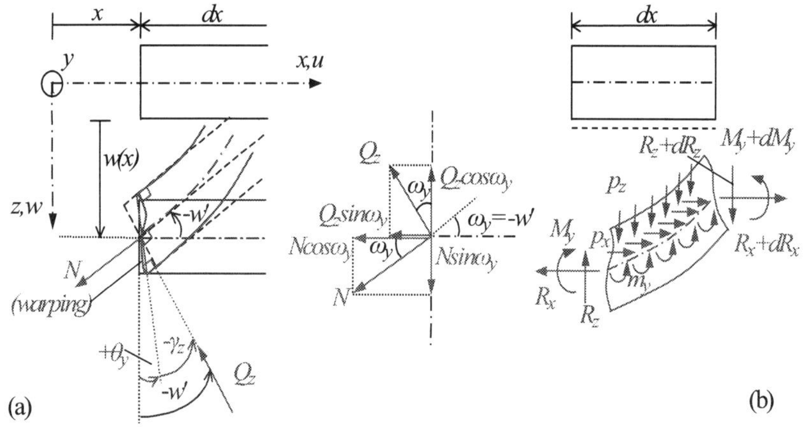

Figure 2: Displacements (a) and forces (b) acting on the deformed element in the xz plane.

Referring to Fig. 2(b), the stress resultants R_x, R_z acting in the x, z directions, respectively, are related to the axial N and shear Q_z forces as

$$R_x = N\cos\omega_y + Q_z \sin\omega_y \qquad R_z = Q_z\cos\omega_y - N\sin\omega_y \tag{4a,b}$$

which by virtue of eqns. (1) become

$$R_x = N - Q_z \frac{dw}{dx} \qquad R_z = Q_z + N\frac{dw}{dx} \tag{5a,b}$$

The second term in the right hand side of eqn. (5a), expresses the influence of the shear force Q_z on the horizontal stress resultant R_x. However, this term can be neglected since $Q_z w'$ is much smaller than N and thus eqn. (5a) is written as

$$R_x \simeq N \tag{6}$$

Similarly, the stress resultant R_y acting in the y direction is related to the axial N and shear Q_y forces as

$$R_y = Q_y + N\frac{dv}{dx} \tag{7}$$

The governing equation of the beam transverse displacement $w = w(x)$ will be derived by considering the equilibrium of the deformed element in the *x-z* plane. Thus, referring to Fig. 2 we obtain

$$\frac{dR_x}{dx} + p_x = 0 \qquad \frac{dR_z}{dx} + p_z = 0 \qquad \frac{dM_y}{dx} - Q_z + m_y = 0 \tag{8a,b,c}$$

Substituting eqns. (6), (5b) into eqns. (8a,b), using eqn. (8c) to eliminate Q_z, employing the well-known relation

$$M_y = EI_y k_y \tag{9}$$

and utilizing eqn. (2a) we obtain the expressions of the angle of rotation due to bending θ_y and the stress resultants M_y, R_z as

$$\theta_y = -\frac{dw}{dx} + \frac{1}{GA_z}\left(-EI_y\frac{d^3w}{dx^3} - \frac{EI_y}{GA_z}\left(\frac{dp_z}{dx} + N\frac{d^3w}{dx^3} - 2p_x\frac{d^2w}{dx^2} - \frac{dp_x}{dx}\frac{dw}{dx}\right) + m_y\right) \tag{10}$$

$$M_y = -EI_y\frac{d^2w}{dx^2} - \frac{EI_y}{GA_z}\left(p_z + \frac{dN}{dx}\frac{dw}{dx} + N\frac{d^2w}{dx^2}\right) \tag{11a}$$

$$R_z = -EI_y \frac{d^3 w}{dx^3} - \frac{EI_y}{GA_z}\left(\frac{dp_z}{dx} + N\frac{d^3 w}{dx^3} - 2p_x \frac{d^2 w}{dx^2} - \frac{dp_x}{dx}\frac{dw}{dx} \right) + m_y + N\frac{dw}{dx} \quad (11b)$$

and the governing differential equation as

$$EI_y\left(1 + \frac{N}{GA_z} \right)\frac{d^4 w}{dx^4} = p_z - p_x \frac{dw}{dx} + N\frac{d^2 w}{dx^2} + \frac{dm_y}{dx} - \\ - \frac{EI_y}{GA_z}\left(\frac{d^2 p_z}{dx^2} - 3p_x \frac{d^3 w}{dx^3} - 3\frac{dp_x}{dx}\frac{d^2 w}{dx^2} - \frac{d^2 p_x}{dx^2}\frac{dw}{dx} \right) \quad (12)$$

Moreover, the pertinent boundary conditions of the problem at the beam ends $x = 0,l$ are given as

$$\alpha_1^z w(x) + \alpha_2^z R_z(x) = \alpha_3^z \qquad \beta_1^z \theta_y(x) + \beta_2^z M_y(x) = \beta_3^z \quad (13a,b)$$

where $\alpha_i^z, \beta_i^z \ (i = 1,2,3)$ are given constants, while the angle of rotation θ_y and the stress resultants M_y, R_z at the beam ends $x = 0,l$ are given as

$$\theta_y = -\frac{EI_y}{GA_z}\left(1 + \frac{N}{GA_z} \right)\frac{d^3 w}{dx^3} - \frac{dw}{dx} \quad (14a)$$

$$M_y = -EI_y\left(1 + \frac{N}{GA_z} \right)\frac{d^2 w}{dx^2} \qquad R_z = -EI_y\left(1 + \frac{N}{GA_z} \right)\frac{d^3 w}{dx^3} + N\frac{dw}{dx} \quad (14b,c)$$

Eqns. (13) describe the most general boundary conditions associated with the problem at hand and can include elastic support or restrain. It is apparent that all types of the conventional boundary conditions (clamped, simply supported, free or guided edge) can be derived form these equations by specifying appropriately the functions α_i^z and β_i^z (e.g. for a clamped edge it is $\alpha_1^z = \beta_1^z = 1$, $\alpha_2^z = \alpha_3^z = \beta_2^z = \beta_3^z = 0$).

Similarly, considering the beam in the *x-y* plane we obtain the boundary value problem of the beam transverse displacement $v = v(x)$ as

$$EI_z\left(1 + \frac{N}{GA_y} \right)\frac{d^4 v}{dx^4} = p_y - p_x \frac{dv}{dx} + N\frac{d^2 v}{dx^2} - \frac{dm_z}{dx} - \\ - \frac{EI_z}{GA_y}\left(\frac{d^2 p_y}{dx^2} - 3p_x \frac{d^3 v}{dx^3} - 3\frac{dp_x}{dx}\frac{d^2 v}{dx^2} - \frac{d^2 p_x}{dx^2}\frac{dv}{dx} \right) \quad \text{inside the beam} \quad (15)$$

$$\alpha_1^y v(x) + \alpha_2^y R_y(x) = \alpha_3^y \qquad \beta_1^y \theta_z(x) + \beta_2^y M_z(x) = \beta_3^y \qquad \text{at } x = 0, l \tag{16a,b}$$

where α_i^y, β_i^y $(i = 1,2,3)$ are given constants and the expressions of the angle of rotation θ_z and the stress resultants M_z, R_y inside the beam are given as

$$\theta_z = \frac{dv}{dx} - \frac{1}{GA_y}\left(-EI_z \frac{d^3 v}{dx^3} - \frac{EI_z}{GA_y}\left(\frac{dp_y}{dx} + N\frac{d^3 v}{dx^3} - 2p_x \frac{d^2 v}{dx^2} - \frac{dp_x}{dx}\frac{dv}{dx} \right) - m_y \right) \tag{17a}$$

$$M_z = EI_z \frac{d^2 v}{dx^2} + \frac{EI_z}{GA_y}\left(p_y + \frac{dN}{dx}\frac{dv}{dx} + N\frac{d^2 v}{dx^2} \right) \tag{17b}$$

$$R_y = -EI_z \frac{d^3 v}{dx^3} - \frac{EI_z}{GA_y}\left(\frac{dp_y}{dx} + N\frac{d^3 v}{dx^3} - 2p_x \frac{d^2 v}{dx^2} - \frac{dp_x}{dx}\frac{dv}{dx} \right) - m_y + N\frac{dv}{dx} \tag{17c}$$

In both of the aforementioned boundary value problems the axial force N inside the beam or at its boundary is given from the following relation

$$N = EA\left[\frac{du}{dx} + \frac{1}{2}\left(\frac{dw}{dx} \right)^2 + \frac{1}{2}\left(\frac{dv}{dx} \right)^2 \right] \tag{18}$$

where $u = u(x)$ is the bar axial displacement, which can be evaluated from the solution of the following boundary value problem

$$EA\left[\frac{d^2 u}{dx^2} + \frac{d^2 w}{dx^2}\frac{dw}{dx} + \frac{d^2 v}{dx^2}\frac{dv}{dx} \right] = -p_x \qquad \text{inside the beam} \tag{19}$$

$$c_1 u(x) + c_2 N(x) = c_3 \qquad \text{at the beam ends } x = 0, l \tag{20}$$

where c_i $(i = 1,2,3)$ are given constants.

The solution of the boundary value problems prescribed from eqns (12), (13a,b) and (15), (16a,b) presumes the evaluation of the shear deformation coefficients a_z, a_y corresponding to the principal centroidal system of axes Cyz. These coefficients are established equating the approximate formula of the shear strain energy per unit length

$$U_{appr.} = \frac{a_y Q_y^2}{2AG} + \frac{a_z Q_z^2}{2AG} \tag{21}$$

with the exact one given from

$$U_{exact} = \int_{\Omega} \frac{(\tau_{xz})^2 + (\tau_{xy})^2}{2G} d\Omega \tag{22}$$

and are obtained as [4]

$$a_y = \frac{1}{\kappa_y} = \frac{A}{\Delta^2} \int_{\Omega} \left[(\nabla \Theta) - \boldsymbol{e} \right] \cdot \left[(\nabla \Theta) - \boldsymbol{e} \right] d\Omega \tag{23a}$$

$$a_z = \frac{1}{\kappa_z} = \frac{A}{\Delta^2} \int_{\Omega} \left[(\nabla \Phi) - \boldsymbol{d} \right] \cdot \left[(\nabla \Phi) - \boldsymbol{d} \right] d\Omega \tag{23b}$$

where $(\tau_{xz}), (\tau_{xy})$ are the transverse (direct) shear stress components, $(\nabla) \equiv \boldsymbol{i_y} (\partial/\partial y) + \boldsymbol{i_z} (\partial/\partial z)$ is a symbolic vector with $\boldsymbol{i_y}, \boldsymbol{i_z}$ the unit vectors along y and z axes, respectively, Δ is given from

$$\Delta = 2(1+\nu) I_y I_z \tag{24}$$

$\boldsymbol{e}$ and $\boldsymbol{d}$ are vectors defined as

$$\boldsymbol{e} = \left(\nu I_y \frac{y^2 - z^2}{2} \right) \boldsymbol{i_y} + (\nu I_y yz) \boldsymbol{i_z} \qquad \boldsymbol{d} = (\nu I_z yz) \boldsymbol{i_y} + \left(\nu I_z \frac{z^2 - y^2}{2} \right) \boldsymbol{i_z} \tag{25a,b}$$

and $\Theta(y,z)$, $\Phi(y,z)$ are stress functions, which are evaluated from the solution of the following Neumann type boundary value problems [4]

$$\nabla^2 \Theta = -2 I_y y \quad \text{in } \Omega \,, \qquad \frac{\partial \Theta}{\partial n} = \boldsymbol{n} \cdot \boldsymbol{e} \quad \text{on } \Gamma = \bigcup_{j=1}^{K+1} \Gamma_j \tag{26a,b}$$

$$\nabla^2 \Phi = -2 I_z z \quad \text{in } \Omega \,, \qquad \frac{\partial \Phi}{\partial n} = \boldsymbol{n} \cdot \boldsymbol{d} \quad \text{on } \Gamma = \bigcup_{j=1}^{K+1} \Gamma_j \tag{27a,b}$$

where $\boldsymbol{n}$ is the outward normal vector to the boundary Γ. In the case of negligible shear deformations $a_z = a_y = 0$. It is also worth here noting that the boundary conditions (26b), (27b) have been derived from the physical consideration that the traction vector in the direction of the normal vector $\boldsymbol{n}$ vanishes on the free surface of the beam.

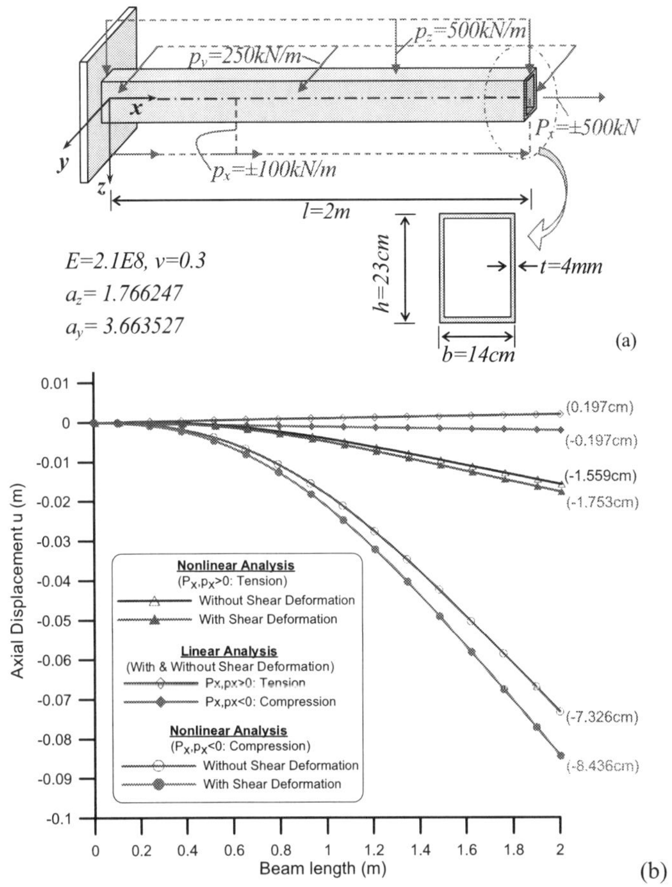

Figure 3: Thin-walled beam and axial displacement.

3 Integral representations: numerical solution

According to the precedent statement of the problem, the non-linear analysis of a beam including shear deformation reduces in establishing the displacements w, v, u and the stress functions Φ, Θ. The numerical solution of the problems described by eqns (12), (13a,b) and (15), (16a,b) for the displacements w, v is similar and is accomplished using the AEM [5], as this is developed for partial differential equations [6, 7]. Moreover, the evaluation of the stress functions is carried out using BEM as this is presented in [4], while the numerical solution of the problem described by eqns (19), (20) for the axial displacement u is also achieved using BEM.

The numerical solution of the aforementioned problems leads to a nonlinear coupled system of equations the solution of which is accomplished employing iterative methods.

4 Numerical examples

The cantilever thin-walled beam of Fig.3a subjected in distributed axial p_x, transverse p_y, p_z and concentrated P_x loading has been studied. In Fig.3b the axial displacement of points along the beam are presented performing both a linear and a nonlinear analysis. From this figure the influence of both the axial loading and the shear deformation effect in the nonlinear analysis is remarkable.

Acknowledgements

Financial support provided by the "*HRAKLEITOS Research Fellowships with Priority to Basic Research*", an EU funded project in the special managing authority of the Operational Program in Education and Initial Vocational Training. The Project "*HRAKLEITOS*" is co-funded by the European Social Fund (75%) and National Resources (25%).

References

[1] Katsikadelis, J.T. & Tsiatas, G.C., Large deflection analysis of beams with variable stiffness, *Acta Mechanica*, **164**, pp. 1-13, 2003.

[2] Timoshenko, S.P. & Goodier, J.N., *Theory of Elasticity,* 3rd edn, McGraw-Hill, New York, 1984.

[3] Cowper, G.R., The shear coefficient in Timoshenko's beam theory, *Journal of Applied Mechanics*, ASME, **33(2)**, pp. 335-340, 1966.

[4] Sapountzakis, E.J. & Mokos, V.G., A BEM solution to transverse shear loading of beams, *Computational* Mechanics, **36**, pp. 384-397, 2005.

[5] Katsikadelis, J.T., The analog equation method. A boundary-only integral equation method for nonlinear static and dynamic problems in general bodies, *Theoretical and Applied Mechanics*, **27**, pp. 13-38, 2002.

[6] Katsikadelis, J.T. & Tsiatas, G.C., Nonlinear dynamic analysis of beams with variable stiffness, *Journal of Sound and Vibration* 270, pp. 847-863, 2004.

[7] Sapountzakis, E.J. & Katsikadelis, J.T., Dynamic analysis of elastic plates reinforced with beams of doubly-symmetrical cross section, *Computational Mechanics*, **23**, pp. 430-439, 1999.

High rate continuum modeling mesh reduction methodologies and advanced applications

E. L. Baker, D. Pfau, J. M. Pincay, T. Vuong & K. W. Ng
U.S. Army Armament Research, Development and Engineering Center
Picatinny, USA

Abstract

A variety of mesh reduction methodologies (MRM) have been developed for use in high rate continuum modeling. An adaptive mesh refinement (AMR) technique has been implemented for use in the CTH high rate Eulerian finite difference model. This new implementation allows increased rectilinear mesh refinement in localized areas of interest. We have applied this AMR to successfully resolve dominating physical phenomena involved in concrete wall impact modeling, as well as physical phenomena observed at the material interface of explosively welded metals. In addition, a variety of MRM relaxation algorithms have been developed for high rate continuum Arbitrary Lagrangian-Eulerian (ALE) modeling. These relaxation algorithms are now routinely used to provide the high resolution simulation of explosively produced metal jetting using the CALE computer program. Finally, a multi-mesh MRM technique has been implemented in the ALE-3D computer model. This MRM technique has been used to provide the modeling of fragment impact for the development of safer munitions. These new MRM techniques are now allowing the high rate continuum modeling of physical phenomena that was not previously simulated.
Keywords: mesh reduction methodologies, high rate continuum modeling, impact physics, high explosives.

1 Introduction

High rate continuum modeling is used for the modeling of high rate events including high explosive detonation and high velocity impact. These models typically provide explicit second order integration in time and space of the

WIT Transactions on Modelling and Simulation, Vol 42, © 2006 WIT Press
www.witpress.com, ISSN 1743-355X (on-line)
doi:10.2495/BE06012

conservation equations. Due to the extreme deformations often observed, Eulerian and ALE modeling are often chosen over traditional Lagrangian methodology. As the requirements for increased problem size and increased resolution constantly push existing computational platforms to their limits, methods to reduce computational time and memory requirements can be extremely beneficial. In particular, a variety of mesh reduction methodologies (MRM) have been developed and implemented for use in high rate continuum modeling. These new MRM techniques are now being applied for the modeling of high rate physical phenomena not previously successfully modeled.

2 Eulerian adaptive mesh refinement

Adaptive mesh refinement (AMR) has been investigated as a method for improving computational resolution, reducing memory requirements and increasing computational efficiency for high rate continuum modeling by a number of researchers [1, 2, 3]. The implementation of such an adaptive mesh refinement capability has been recently completed by Sandia National Laboratories in the multi-materials high rate continuum computer model, CTH [4]. In order to achieve a practical implementation with good parallel performance, a block-based approach has been implemented with refinement and un-refinement occurring in an isotropic 2:1 manner. Crawford et al. [5] showed that practical speed-up from 3 to 10 times and at least 3 times memory requirement reductions can be achieved on multiprocessor and massively parallel platforms. His studies also indicated further improvements for larger problems. As the block refinement and un-refinement are very computationally expensive serial procedures, parallel performance of the adaptivity required that CTH use a super-cycling capability for refinement and un-refinement. In the current implementation, refinement occurs every two or three cycles and un-refinement will occur every six cycles. This is accomplished by using a two cell thick trigger region at the boundary of each block. When a refinement indicator reaches its threshold, then the adjacent block is signaled for refinement at least two cycles before the refinement is needed. Although somewhat ad-hoc in nature, the performance of this procedure has proven improved parallel efficiency, particularly when coupled with message passing consolidation and an efficient load balancing strategy. To date, the entire mesh is cycled with a single time step, determined by smallest time increment that meets stability based criteria, normally associated with the highest resolution mesh. Currently implemented refinement indicators include velocity, pressure, temperature, materials and material boundaries as indicated by mixed cells. Refinement indicator operators include absolute value, gradient magnitude and maximum difference.

We have used this new capability to produce increased rectilinear mesh refinement in localized areas of interest. In particular, AMR has been used to successfully resolve physical phenomena observed at the material interface of explosively welded metals, as well as dominating physical phenomena involved in masonry wall impact modeling. An ongoing computational investigation of explosive cladding of Ta based materials is being conducted in support of the U.S. Army Future Combat System Durable Gun Barrels Manufacturing Technology Objective (MTO) and the Chromium Elimination SERDP program. The objective of the effort is to develop and demonstrate physics based modeling for explosive barrel cladding process design and optimization. Actual explosive welding of tantalum based liners to the inside of gun barrels is being undertaken by TPL Inc. Figures 1 and 2 present CTH calculations of the tantalum gun barrel cladding process.

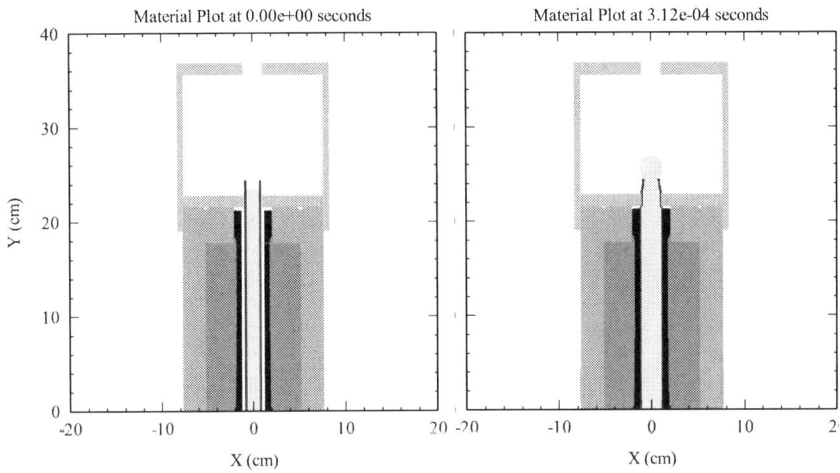

Figure 1: Material plots of gun barrel cladding.

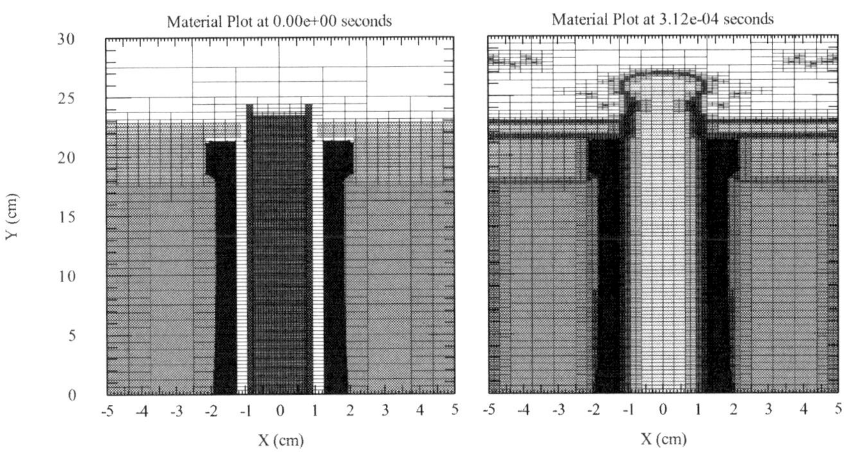

Figure 2: Material plots of gun barrel cladding in the muzzle region displaying mesh blocks.

We have also used the new AMR routines in CTH to successfully resolve dominating physical phenomena involved in concrete wall impact modeling. In previous modeling efforts of modeling projectile penetration of concrete walls, the steel reinforcement bars were not successfully resolved, due to the large memory requirements to resolve these relatively fine structures. The steel reinforcement bars are an important part of the physical process, as they provide both significant strengthening for the wall and a higher density material within the concrete wall that can significantly damage the projectile during the concrete wall penetration process. Figure 3 presents initial and late time plots of the projectile concrete wall penetration process. Figure 4 presents the initial state material region plot if the mesh refinement is not used. In this example, refinement indicators are used in order to refine everywhere individual volume fractions are greater than 0 (anywhere those materials exist), except for concrete. Three levels of refinement are used, with a minimum cell size of 1.6mm. In this way, the projectile and concrete reinforcement bars are highly refined, whereas the surrounding air (modeled as void) and concrete are less refined.

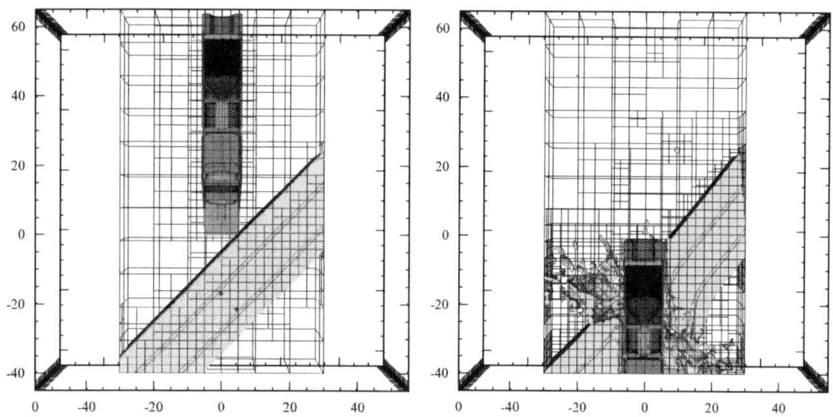

Figure 3: Initial and late time plots of the projectile concrete wall penetration process showing mesh blocks (dimensions in cm).

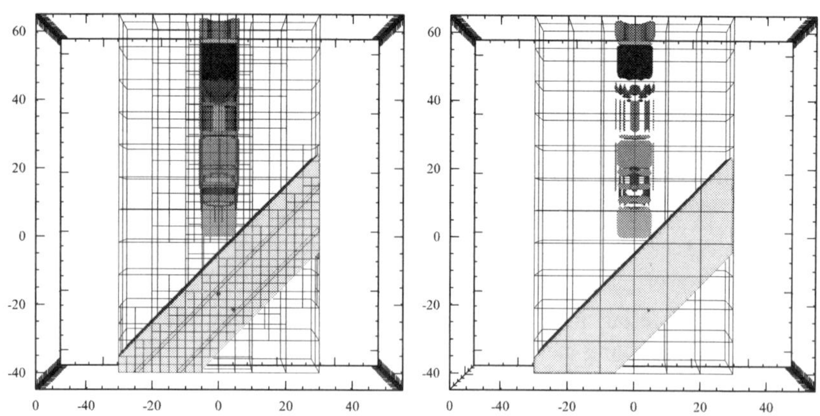

Figure 4: Initial state material region plot if the mesh refinement is not used.

3 Arbitrary Lagrange-Eulerian mesh relaxation

In addition, a variety of MRM relaxation algorithms have been developed for high rate continuum Arbitrary Lagrangian-Eulerian (ALE) modeling and implemented by R. Tipton in the CALE computer program [6]. These relaxation algorithms have been successfully used to provide high resolution simulation of explosively produced jetting. This jetting is produced in shaped charge warheads where an axisymmetric hollow metal liner, typically copper, is imploded onto the axis. The imploding metal liner undergoes an extremely high rate jetting deformation process and projects a high velocity metal jet forward. Figures 5 and 6 present material region and mesh plots of the jetting process of a shaped charge. A weighted potential algorithm [7] based on material fractions is used in the CALE program for this modeling. In this way, the higher weighted materials result in increased resolution mesh areas. This can clearly been seen in the figures.

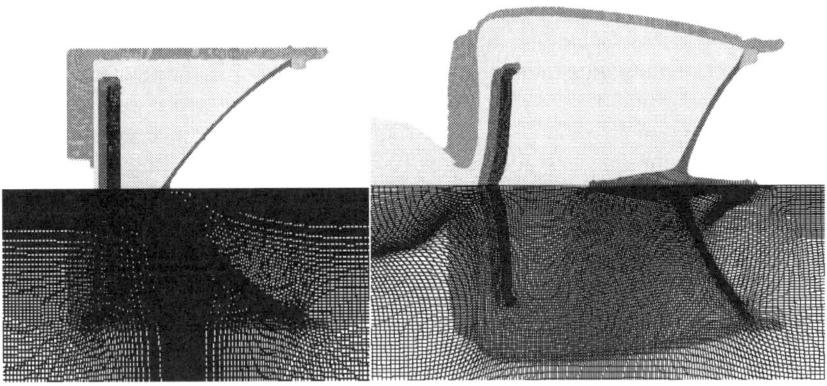

Figure 5: Material and mesh plots of the initial shaped charge configuration.

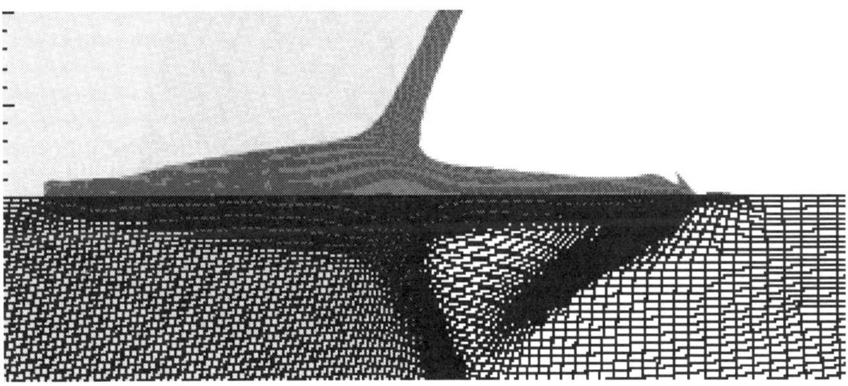

Figure 6: Material and mesh plots of the jetting shaped charge liner.

4 Multi-mesh technique

Finally, a multi-mesh MRM technique has been developed for ALE modeling and has been implemented in the ALE-3D computer program [8, 9]. In this implementation, separate meshes can be attached to form a larger "multi-mesh configuration. This mesh reduction methodology can provide significant mesh reduction by only meshing regions of interest. We have used the MRM technique has been used to provide modeling of bullet impact for development of safer munitions. Reduced mesh sizes are achieved with required mesh resolutions, by meshing the munition that is being attacked and than using a small attached sub-mesh that included the bullet that is impacting the munition being attacked. Figure 7 presents an example bullet impact configuration in which the to mesh setup configuration is used. Figure 8 presents material plots of the resulting bullet impact response of the attacked munition.

5 Conclusions

A variety of MRM techniques have been investigated and successfully applied for high rate continuum modeling of ballistic events. In particular, Eulerian adaptive mesh refinement has been used for 2D and 3D modeling using the CTH computer program to reduce memory requirements and reduce computational times. This technique has allowed extremely resolved modeling of explosive welding of tantalum materials to steel that was not previously possible. Additionally, this MRM technique now allows the resolution of masonry wall penetration details that was not previously possible due to memory restrictions. Steel reinforcement materials have now been resolved, that are physically important to munition wall penetration process.

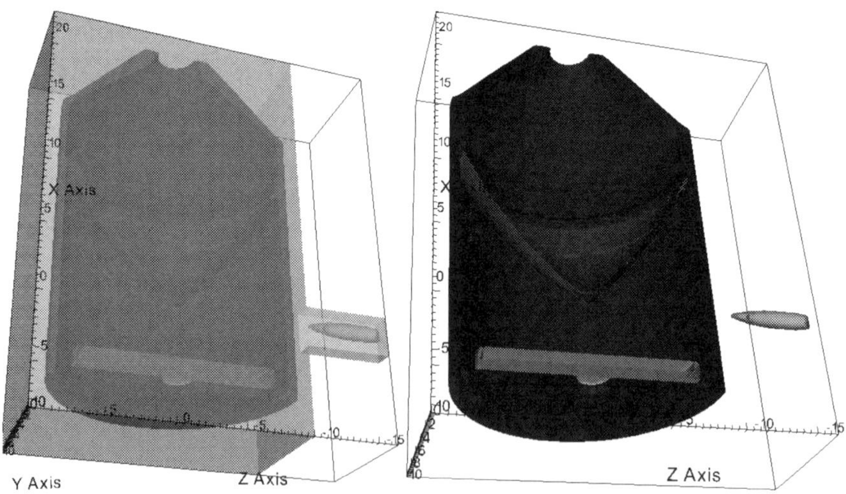

Figure 7: Initial bullet impact configuration showing the two mesh setup.

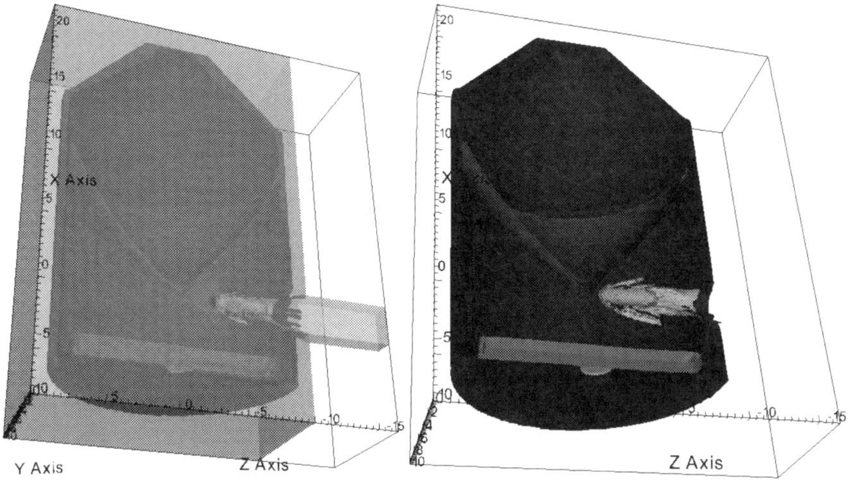

Figure 8: Material plots of the resulting bullet impact response.

Weighted potential algorithms have also been successfully used in shaped charge jet formation modeling to produce high resolution physics of the jet formation process. Finally, a multi-mesh technique has been used in the ALE-3D computer program to reduce memory requirements and reduce computational time required to perform bullet impact modeling in support of safer munitions development. These new MRM techniques are currently available in several advanced high rate continuum modeling programs and are now allowing the high rate continuum modeling of physical phenomena that was not previously simulated.

References

[1] Berger, M.J. & Colella, P., Local adaptive mesh refinement for shock hydrodynamics, *J. Comp. Phys.*, 82, 64-84, 1989.

[2] Jones, B., SHAMROCK – an adaptive, multi-material hydrocode. *Proc. of the International Workshop on new Models and Numerical Codes for Shock Wave Processes in Condensed Media*, St. Catherines College, Oxford, UK, 15-19 September 1997.

[3] Tang, P.K & Scannapieco, A.J., Modeling cylinder test, *Proc. of the Conference of the American Physical Society Topical Group on Shock Compression of Condensed Matter*, Seattle, WA, USA, August 13-18, 1995, 449-452.

[4] Crawford, D.A., Adaptive mesh refinement in CTH, *Proc. of the 15th U.S. Army Symposium on Solid Mechanics*, Myrtle Beach, South Carolina, USA, April 12-14, 1999.

[5] Crawford, D.A., Taylor, P.A., Bell, R.L. & Hertel, E.S., Adaptive mesh refinement in the CTH shock physics hydrocode, *Proc. of New Models*

and Hydrocodes for Shock Wave Processes in Condensed Matter, Edinburgh, U.K., May 19-24, 2002.

[6] Tipton, R.E., CALE users manual, Lawrence Livermore National Laboratory, Oct. 1995.

[7] Winslow, A. 1963, Equipotential zoning of two dimensional meshes, LLNL Report UCRL-7312, 1963.

[8] Couch, R., Sharp, R., Otero, I., Neely, R., Futral, S., Dube, E., McCallen, R., Maltby, J., & and Nichols, A., ALE hydrocode development, Joint DOD/DOE Munitions Technology Development Progress Report, UCRL-ID-103482-95, January 1996.

[9] Dube, E. & Rodrigue, G., A geometric weighted elliptic grid regeneration method for 3D unstructured ALE hydrodynamics, *Proc. of the 5th International Symposium on Computational Fluid Dynamics*, Sendai, Japan, August 31-September 3, 1993.

Boundary element analysis of strain fields in InAs/GaAs quantum wire structures

F. Han[1], E. Pan[1] & J. D. Albrecht[2]
[1]*Department of Civil Engineering, University of Akron, USA*
[2]*Air Force Research Laboratory, Wright-Patterson Air Force Base, USA*

Abstract

This paper studies the elastic fields in InAs/GaAs quantum wire (QWR) structures arising from the mismatch lattice between InAs and GaAs. The present treatment is different from recent analyses based on the Eshelby inclusion approach where the QWR material, for simplicity, is assumed to be the same as the matrix/substrate. Here, a more complete treatment is developed taking into account the structural inhomogeneity using the boundary integral equation method. We implement our model using discrete boundary elements at the interface between the QWR and its surrounding matrix. The coefficients of the algebraic equations are derived exactly for constant elements using our recent Green's function solutions in the Stroh formalism. For both (001) and (111) growth directions, our results show that while the elastic fields far from the QWR are approximated well by the homogeneous inclusion approach, for points within or close to the QWR, the differences between the fields computed with the simplified inclusion and complete inhomogeneity models can be as large as 10% for the test system. These differences in the strain fields will have strong implications for the modeling of the quantized energy states of the quantum wire nanostructures. Since the strain fields inside and close to the wire are more important than the exterior strain fields from the standpoint of the confined electronic states, we suggest that in the vicinity of the QWR, the inhomogeneity model be used with proper elastic constants, whilst the simple exact inclusion model be used in the bulk of surrounding medium.
Keywords: inclusion, inhomogeneity, exact closed-form solution, full/half plane Green's function, boundary integral equation.

WIT Transactions on Modelling and Simulation, Vol 42, © 2006 WIT Press
www.witpress.com, ISSN 1743-355X (on-line)
doi:10.2495/BE06013

1 Introduction

Quantum wire (QWR) semiconductor nanostructures and their quantum mechanical properties have been the subjects of many investigations over the past decade [1–4]. Similar to quantum-dot (QD) structures, the electronic energy features are closely related to and in some cases can be controlled by the strain field induced by the lattice and/or thermal mismatches among component materials. As a first approximation to fabricated structures, the intrinsic material differences (mostly the differences in elastic compliances) between the QWR/QD and its substrate were ignored [5–7]. The crucial simplification of these approaches is to impose a strain field arising from lattice mismatches without allowing for variations in the material constants from material to material. This approximation was recently shown to be reliable for computing the strain fields sufficiently away from the QWR where the system is sensitive to the QWR geometry and interfacial mismatch but relatively unaffected by the internal properties of the QWR [8, 9]. Whether this approximation is valid for points inside and close to the QWR (the crucial locations from an electronic device standpoint) is suspect, and is the motivation for the present investigation.

In previous treatments, many researchers use the established Eshelby inclusion method [10, 11] to solve the QWR induced field. By convention, the so-called Eshelby inclusion problem is that of an embedded subdomain, such as a QWR, consisting of material which is modeled as a strained region but with identical mechanical properties as the matrix (i.e. substrate, buffer, or overgrowth layers). We contrast this with the structural inhomogeneity problem where the materials are different in all respects except for their lattice type [11]. Investigations on homogeneous polygonal inclusions have been carried out for both isotropic and anisotropic elastic cases [9, 12, 13]. More recently, general solutions to anisotropic Eshebly problems that account for electro-mechanical coupling have been derived. These solutions are based either on the analytical continuation and conformal mapping method [8, 14, 15] or on the Green's function method using the equivalent body-force concept [16, 17]. Numerically, finite element (FEM) and finite difference (FDM) methods have been successfully applied to the strained QD or QWR problems [1, 18], so is the boundary integral equation method (i.e., BEM) with its advantages in dealing with singularities and comparatively smaller computational demands [19, 20].

In this paper, we address the issue of the homogeneous inclusion vs. structural inhomogeneity in the context of QWR semiconductor structures. The model is based on the formalism recently proposed by one of the authors [16] and treated computationally using an accurate BEM. First, we convert the contribution of the eigenstrain to an integral along the interface of the QWR and its matrix/substrate. After this conversion, the boundary integral equation is applied to both the QWR and substrate. Using the constant-element discretization the integration of the Green's function is carried out exactly and the resulting system of algebraic equations is solved for the interface quantities. The elastic fields inside and outside the QWR are finally evaluated using the solved interfacial values.

As numerical examples, InAs/GaAs QWR structures are analyzed for both (001) and (111) growth directions. QWRs with trapezoidal cross-sections are considered. The results of this work show several features which should be useful for numerical modeling when accounting for strain fields in device designs. *(i)* In the substrate and far away from the QWR, both inclusion and inhomogeneity models give similar results. In other words, we have validated that if one is only interested in the elastic far fields, the simplified homogeneous inclusion model can be safely applied. *(ii)* For points within or near the QWR, the difference between the inclusion and inhomogeneity models can be as high as 10% for these materials. Using a linear bandgap deformation potential relation, this would correspond to a 10% modification to the local electronic energy bandgap.

2 Problem description and basic equations

Let us suppose that there is a misfit elastic strain γ_{ij}^{*}(i, j=1,2,3) inside an arbitrarily shaped polygon QWR domain V, which is embedded in the z<0 half-plane substrate (fig. 1 for trapezoid shape). Assume also that the misfit strain within the QWR is uniform and is zero outside. The interface between the QWR and matrix is denoted by S. We also denote C_{ijkl}^{w} and C_{ijkl}^{m} as the elastic moduli of the QWR and the matrix materials, respectively. For the homogeneous inclusion problem, C_{ijkl}^{w} is equal to C_{ijkl}^{m}.

We define γ_{ij} as the total elastic strain, which is related to the total elastic displacement u_i as

$$\gamma_{ij} = (u_{i,j} + u_{j,i})/2 \tag{1}$$

The total strain can be written as

$$\gamma_{ij} = \gamma_{ij}^{e} + \gamma_{ij}^{*} \tag{2}$$

where γ_{ij}^{e} is the elastic strain that appears in the constitutive relation

$$\sigma_{ij} = C_{ijkl}\gamma_{kl}^{e} \tag{3}$$

which can be written as

$$\sigma_{ij} = C_{ijkl}(\gamma_{kl} - \chi\gamma_{kl}^{*}) \tag{4}$$

In eqn. (4), χ is equal to 1 if the field point is within the QWR domain V and to 0 in the substrate. σ_{ij} is the stress, and C_{ijkl} is the material elastic modulus, replaced with either C_{ijkl}^{w} or C_{ijkl}^{m} depending on the problem domain. We further define the traction

$$t_i = \sigma_{ji}n_j \quad (i, j = 1,2,3) \tag{5}$$

where n_j are the direction cosines of the outward normal $\boldsymbol{n}$ along the interface S.

Substituting the stress in eqn. (4) into the equilibrium equation for the stress

$$\sigma_{ij,i} + f_j = 0 \tag{6}$$

results in the expression

$$C^{\rm w}_{\rm ijkl}u_{\rm k,li}-C^{\rm w}_{\rm ijkl}\gamma^{*}_{\rm kl,i}=0 \tag{7}$$

for the QWR domain. It is clear that the second term in eqn. (7) is equivalent to a body force defined as

$$f_j^{(w)}=-C^{\rm w}_{\rm ijkl}\gamma^{*}_{\rm kl,i} \tag{8}$$

which is also called the equivalent body force of the eigenstrain [11, 16]. This equivalent body force will be employed in the next section to convert the contribution of the eigenstrain to a boundary integral along the interface of the QWR and its substrate.

3 Boundary integration equations and constant element discretization

To solve the problem in fig. 1, we apply the BEM to both the QWR and its matrix/substrate. The boundary integral formulation can be derived as [19]

$$b_{\rm ij}(\boldsymbol{X})u_j^{\rm (m)}(\boldsymbol{X})=\int_S\left[U_{\rm ij}^{\rm (m)}(\boldsymbol{X},\boldsymbol{x})t_{\rm j}^{\rm (m)}(\boldsymbol{x})-T_{\rm ij}^{\rm (m)}(\boldsymbol{X},\boldsymbol{x})u_{\rm j}^{\rm (m)}(\boldsymbol{x})\right]dS(\boldsymbol{x}) \tag{9}$$

for the matrix, and

$$b_{\rm ij}(\boldsymbol{X})u_j^{\rm (w)}(\boldsymbol{X})=\int_S\left[U_{\rm ij}^{\rm (w)}(\boldsymbol{X},\boldsymbol{x})\left(t_{\rm j}^{\rm (w)}(\boldsymbol{x})+f_{\rm j}^{\rm (w)}(\boldsymbol{x})\right)-T_{\rm ij}^{\rm (w)}(\boldsymbol{X},\boldsymbol{x})u_{\rm j}^{\rm (w)}(\boldsymbol{x})\right]dS(\boldsymbol{x}) \tag{10}$$

for the QWR. The superscripts (m) and (w) denote quantities associated with the matrix and wire, respectively. In eqns. (9) and (10) $t_{\rm j}$ and $u_{\rm j}$ are the traction and displacement components, $\boldsymbol{x}$ and $\boldsymbol{X}$ are the coordinates of the field and source points, respectively. The coefficients $b_{\rm ij}$ is equal to $\delta_{\rm ij}$ if $\boldsymbol{X}$ is an interior point and $1/2\delta_{\rm ij}$ at a smooth boundary point. For point at complicated geometry locations, these coefficients can be determined by the rigid-body motion method [19]. Furthermore, in eqn. (10), $f_{\rm j}^{\rm (w)}$ is the traction induced by the misfit elastic eigenstrain inside the QWR, which is given by eqn. (8).

The Green's functions $U_{\rm ij}$ and $T_{\rm ij}$ in eqns. (9) and (10) are taken to be the special two-dimensional Green's functions for full/half plane, which are described in details in Pan [16]. The indexes i and j indicate the j-th Green's elastic displacement/traction (at $\boldsymbol{x}$) in response to a line-force in the i-th direction (applied at $\boldsymbol{X}$). Note also that the Green's functions [16] are in exact closed form, and thus their integration over constant elements can be carried out exactly. This is computationally desirable as it is very efficient and accurate for the simulation.

Employing the constant element, we divide the boundary (interface) into N elements with the n-th element being labeled as $\Gamma_{\rm n}$. The constant values $u_{\rm jn}$ and $t_{\rm jn}$ on the n-th element equal to those at the center of the element. Under this assumption, the boundary integral equations (9) and (10) for the matrix and QWR domains are reduced to the following algebraic equations

$$b_{\rm ij}u_j+\sum_{n=1}^{N}\left(\int_{\Gamma_{\rm n}}T_{\rm ij}^{\rm (m)}d\Gamma\right)u_{j{\rm n}}=\sum_{n=1}^{\rm N}\left(\int_{\Gamma_{\rm n}}U_{\rm ij}^{\rm (m)}d\Gamma\right)t_{j{\rm n}} \tag{11}$$

and

$$b_{\mathrm{ij}}u_j + \sum_{\mathrm{n}=1}^{\mathrm{N}}\left(\int_{\Gamma_\mathrm{n}} T_{\mathrm{ij}}^{(\mathrm{w})} d\Gamma\right)u_{j\mathrm{n}} = \sum_{\mathrm{n}=1}^{\mathrm{N}}\left(\int_{\Gamma_\mathrm{n}} U_{\mathrm{ij}}^{(\mathrm{w})} d\Gamma\right)(t_{j\mathrm{n}} + C_{\mathrm{ijkl}}^{(\mathrm{w})}\gamma_{\mathrm{kl}}^{*}n_j) \tag{12}$$

Obviously, the difference between eqns. (11) and (12) is that there is a traction induced by the misfit elastic eigenstrain inside the QWR in eqn. (12).

Given the eqns. (11) and (12), the problem now is to find the suitable Green's functions U_{ij} and T_{ij}, as well as their integrals on each element Γ_n, which are the kernel functions in these equations.

In this paper, we utilized our recently derived exact closed-form Green's function in Pan [16]. By assuming constant element, the line integral can be carried out analytically. After the derivation of the Green's displacement and its integral, all the other physical quantities, such as the strain, stress, traction, etc. can be derived uniquely from the displacement. Finally, the full- and half-plane traction and displacement Green's function can be obtained in a concise format, which is also easy to program and calculate.

Eventually, using the constant element discretization the two boundary integral equations (11) and (12) for the QWR and matrix/substrate can be cast into a system of algebraic equations for the interface points. In matrix form, they can be expressed as

$$\hat{\boldsymbol{U}}^{(w)}\boldsymbol{t}^{(w)} - \hat{\boldsymbol{T}}^{(w)}\boldsymbol{u}^{(w)} = \boldsymbol{f}^{(w)} \tag{13}$$

$$\hat{\boldsymbol{U}}^{(m)}\boldsymbol{t}^{(m)} - \hat{\boldsymbol{T}}^{(m)}\boldsymbol{u}^{(m)} = 0 \tag{14}$$

where the coefficients $\hat{\boldsymbol{U}}$ and $\hat{\boldsymbol{T}}$ are the exact integrals of Green's functions on each constant element given in Pan's paper [16], and $\boldsymbol{u}$ and $\boldsymbol{t}$ are the displacement and traction vectors in the middle of each constant element. The term $\boldsymbol{f}^{(\mathrm{w})}$ on the right-hand side of eqn. (13) is the equivalent force corresponding to the misfit eigenstrain within the QWR.

We assume that the matrix and QWR are perfectly bonded along the interface S, that is, the continuity conditions $\boldsymbol{u}^{(\mathrm{m})} = \boldsymbol{u}^{(\mathrm{w})}$ and $\boldsymbol{t}^{(\mathrm{m})} = -\boldsymbol{t}^{(\mathrm{w})}$ hold. Then the number of unknowns is identical to the number of equations. Therefore, all the nodal displacements and tractions can be determined. Furthermore, making use of the Somigliana's identity, the displacement at any location within either the QWR or the matrix can be easily obtained as

$$b_{\mathrm{ij}}u_j = \sum_{\mathrm{n}=1}^{N}\left(\int_{\Gamma_\mathrm{n}} U_{\mathrm{ij}}^{(\mathrm{m,w})} d\Gamma\right)t_{j\mathrm{n}} - \sum_{\mathrm{n}=1}^{N}\left(\int_{\Gamma_\mathrm{n}} T_{\mathrm{ij}}^{(\mathrm{m,w})} d\Gamma\right)u_{j\mathrm{n}} + \sum_{\mathrm{n}=1}^{N}\left(\int_{\Gamma_\mathrm{n}} U_{\mathrm{ij}}^{(\mathrm{w})} d\Gamma\right)f_{jn}^{(w)} \tag{15}$$

where the last force term exists for the QWR domain only. Furthermore, utilizing the basic equations (1)-(5), all the internal elastic response in the matrix and QWR can also be calculated.

In summary, we have derived the exact boundary integral equations for the QWR and matrix domains by assuming constant elements along their interface. These equations can be used to find the elastic response along the interface and at any location within the QWR and its matrix. Before applying our exact closed-form solutions to a buried InAs QWR in GaAs, we have first checked these solutions against available results [21, 22] for different QWR inclusions in full-plane/half-plane systems and our solutions are found to be the same as previously published results.

4 Numerical analysis

In this section, we study a realistic trapezoidal InAs QWR structure model within a half plane substrate, and the effect of the traction-free surface in this situation is considered. As shown in fig. 1, the trapezoidal QWR was chosen to have crystallographically allowed sidewall angles for the (111) orientation so that the length along the top is 5.1716nm and along the bottom 8nm. The centroid *c* is located at (0, -4nm) with a height of 2nm. Uniform misfit hydrostatic strain is assumed within the QWRs, i.e., $\gamma_{xx}^* = \gamma_{yy}^* = \gamma_{zz}^* = 0.07$.

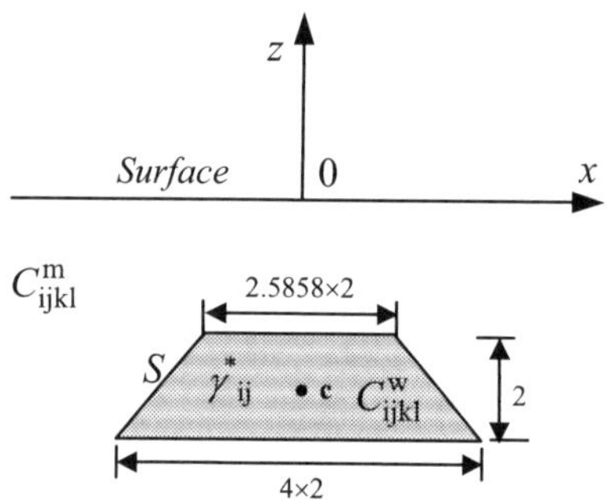

Figure 1: A trapezoid QWR inclusion/inhomogeneity in GaAs half plane (z<0): An γ_{ij}^* extended eigenstrain within the QWR.

In the homogeneous inclusion model, the QWR is assumed to have the same elastic properties as its matrix (i.e., GaAs), whilst in the complete inhomogeneity model the elastic properties are assumed to be those of the bulk InAs. Two orientations are considered: One is InAs/GaAs (001) in which the global coordinates *x*, *y*, and *z* are coincident with the crystalline axes [100], [010], and [001], and the other is InAs/GaAs (111) where the *x*-axis is along [11-2], *y*-axis along [-110], and *z*-axis along [111] directions of the crystal. The bulk elastic constants of InAs and GaAs for the two different orientations (i.e., (001) and (111)) are given in [16] and [18]. We further emphasize that the material orientations for both the QWR and its matrix are assumed to be the same in all the numerical examples, and that the boundary condition on the surface of the substrate is assumed to be traction-free.

Figs. 2 and 3 show the contour of hydrostatic strain inside the half plane with (a) and (b) corresponding to the simplified inclusion and complete inhomogeneity models, respectively. As expected, the strain field is symmetric about the *z*-axis for the (001) structure (fig. 2) and asymmetric for the (111) orientation (fig. 3).

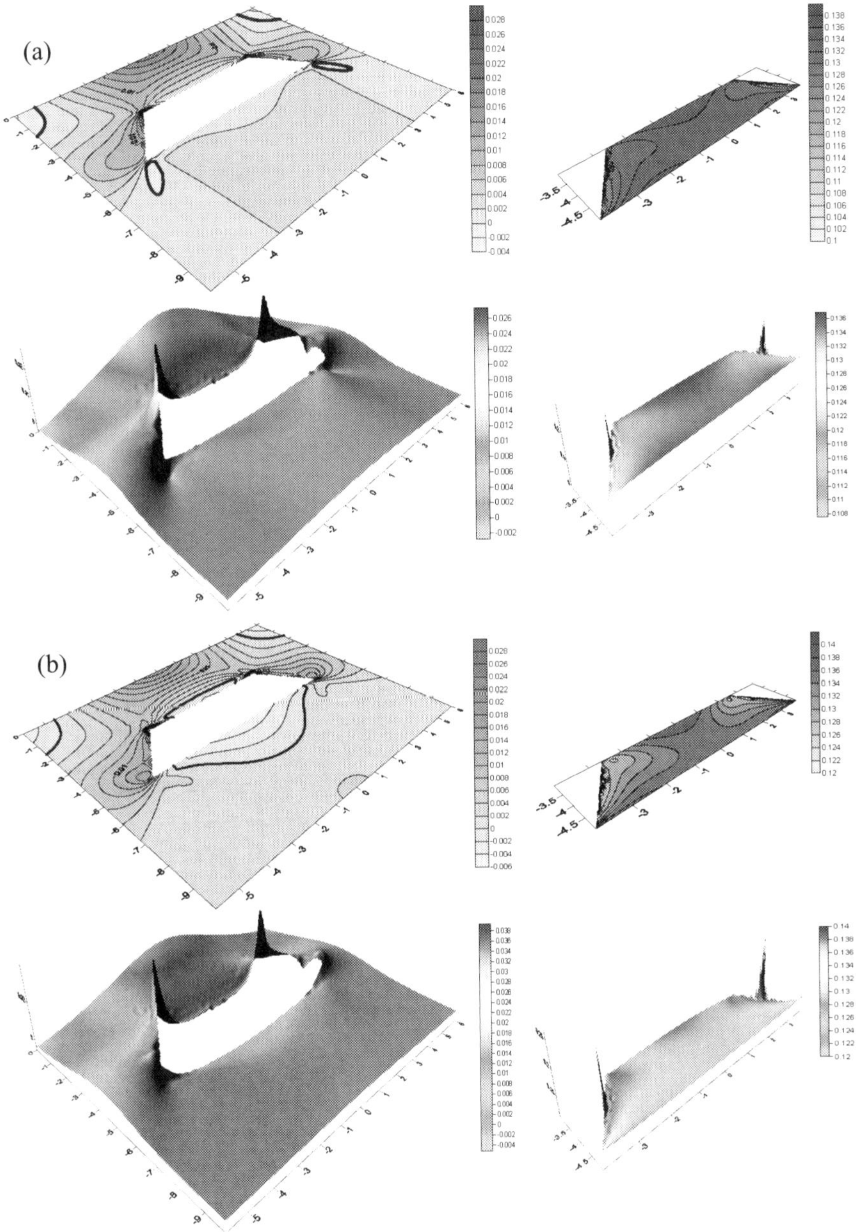

Figure 2: Contour of hydrostatic strain ($\gamma_{xx}+\gamma_{zz}$) of trapezoid QWR inside GaAs (001) half plane. GaAs (001) inclusion in (a), and InAs (001) inhomogeneity in (b).

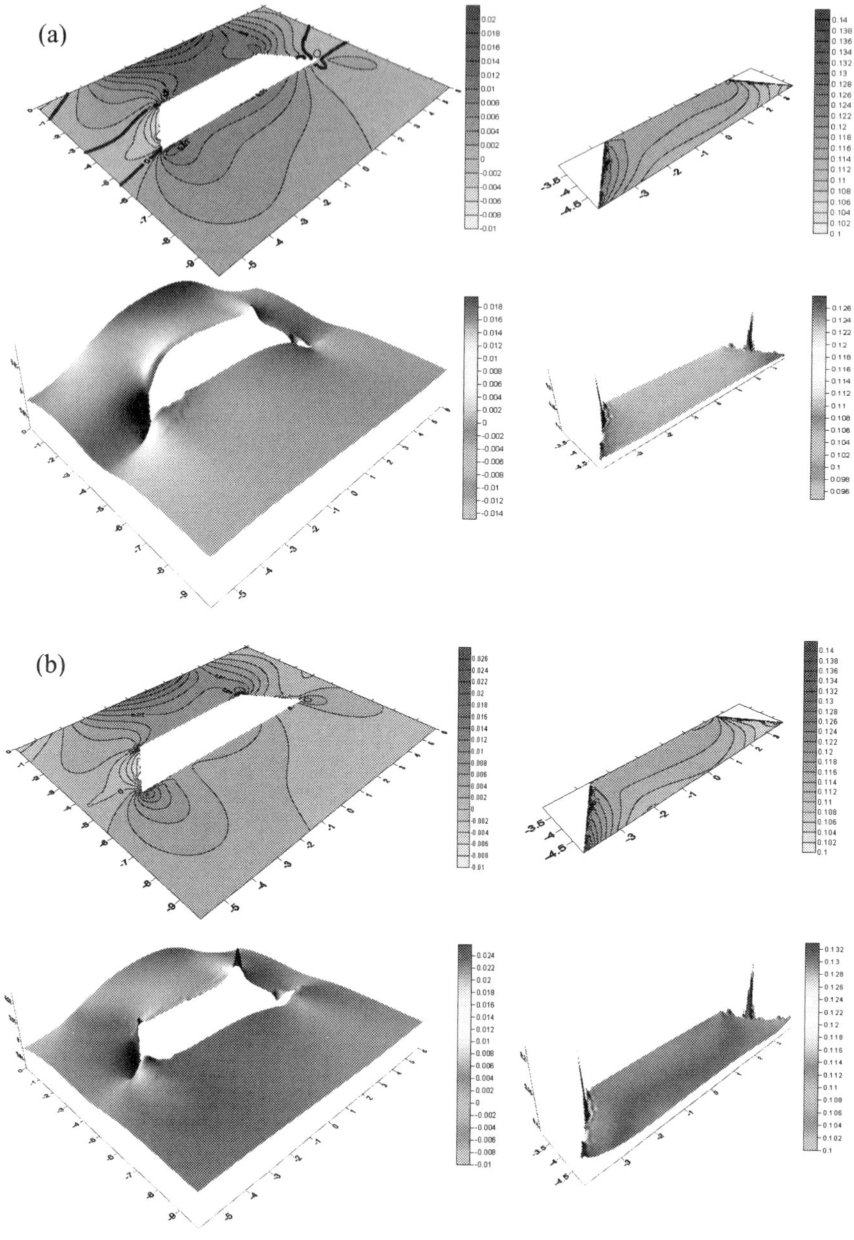

Figure 3: Contour of hydrostatic strain ($\gamma_{xx}+\gamma_{zz}$) of trapezoid QWR inside GaAs (111) half plane. GaAs (111) inclusion in (a), and InAs (111) inhomogeneity in (b).

We further notice that the hydrostatic strain value inside the QWR is much larger than that outside, whilst the concentration near the center of the free surface can be also observed from figs. 2 and 3. We marked the zero value with a thick line, which is the transition from tensile (positive values plotted as solid lines) to compressive (negative values plotted as dashed lines) domains. In general, different models (inclusion and inhomogeneity) affect mostly the region inside or nearby but exterior to the QWR.

5 Concluding remarks

In the paper, we proposed an accurate BEM for the quantum nanostructure model of a QWR embedded in a substrate. By utilizing half-plane Green's functions, their analytic integrals, as well as interface conditions, the elastic response at any location can be predicted based on the inclusion and inhomogeneity models. From our study, some important features are observed: 1). In the substrate and far away from the QWR, both the inclusion and inhomogeneity models predict very close results. In other words, if one is only interested in the elastic fields far away from the QWR interface, the simplified homogeneous inclusion model can be safely applied. 2). For points inside or close to the QWR, the difference between the two models can be as high as 10% for the test structures, which can result in strong variations of the confined electronic states.

Acknowledgements

This work was carried out while the second author (EP) was attached to the School of Mechanical and Production Engineering at Nanyang Technological University under the Tan Chin Tuan Exchange Fellowship in the summer of 2004. He would like to thank Prof. K. M. Liew and the School for their hospitality during his stay there.

References

[1] Grundmann, M., Stier, O. & Bimberg, D., InAs/GaAs pyramidal quantum dots: Strain distribution, optical phonons, and electronic structure. *Phys. Rev. B*, **52**, pp. 11969–11981, 1995.

[2] Caro, L.O. & Tapfer, L., Strain and piezoelectric fields in arbitrarily oriented semiconductor heterostructures. II. Quantum wires. *Phys. Rev. B*, **51**, pp. 4381–4387, 1995.

[3] Goldoni, G., Rossi, F. & Molinari, E., Strong exciton binding in hybrid GaAs-based nanostructures. *Physica B*, **272**, pp. 518–521, 1999.

[4] Bimberg, D., Grundmann, M. & Ledentsov, N.N., *Quantum Dot Heterostructures*, Wiley: New York, 1998.

[5] Andreev, A.D. & O'Reilly, E.P., Theory of the electronic structure of GaN/AlN hexagonal quantum dots. *Phys. Rev. B*, **62**, pp. 15851–15870, 2000.

[6] Pan, E. & Yang, B., Elastostatic fields in an anisotropic substrate due to a buried quantum dot. *J. Appl. Phys.*, **90**, pp. 6190–6196, 2001.

[7] Pan, E., Elastic and piezoelectric fields in substrates GaAs (001) and GaAs (111) due to a buried quantum dot. *J. Appl. Phys.*, **91**, pp. 6379–6387, 2002.

[8] Ru, C.Q., Eshelby's problem for two-dimensional piezoelectric inclusions of arbitrary shape. *Proc. R. Soc. Lond.*, **A456**, pp. 1051–1068, 2000.

[9] Glas, F., Elastic relaxation of truncated circular cylinder with uniform dilatational eigenstain in a half-space. *Phys. Stat. Sol. B*, **237**, pp. 599–610, 2003.

[10] Ting, T.C.T., *Anisotropic Elasticity*, Oxford University Press: Oxford, 1996.

[11] Mura, T., *Micromechanics of Defects in Solids*, Kluwer Academic Publishers: Dordrecht, 1987.

[12] Nozaki, N. & Taya, M., Elastic fields in a polyhedral inclusion with uniform eigenstrains and related problems. *J. Appl. Mech.*, **68**, pp. 441–452, 2001.

[13] Faux, D.A., Downes, J.R. & O'Reilly, E.P., Analytic solutions for strain distribution in quantum-wire structures. *J. Appl. Phys.*, **82**, pp. 3754–3762, 1997.

[14] Ru, C.Q., Analytical solution for Eshelby's problem of an inclusion of arbitrary shape in a plane or half-plane. *J. Appl. Mech.*, **66,** pp. 315–322, 1999.

[15] Wang, X., & Shen, Y.P., Inclusions of arbitrary shape in magnetoelectroelastic composite materials. *Int. J. Eng. Sci.*, **41**, pp. 85–102, 2003.

[16] Pan, E., Eshelby Problem of polygonal inclusion in anisotropic piezoelectric full and half-planes. *J. Mech. Phys. Solids*, **52**, pp. 567–589, 2004.

[17] Jiang, X. & Pan, E., Exact solution for 2D polygonal inclusion problem in anisotropic magnetoelectroelastic full-, half-, and biomaterial-planes. *Inter. J. Solids Struct.*, **41**, 4361–4382, 2004.

[18] Jogai, B., Three dimension strain field calculations in coupled InAs/GaAs quantum dots. *J. Appl. Phys.*, **88**, pp. 5050–5055, 2000.

[19] Brebbia, C.A. & Dominguez, J., *Boundary Elements An Introductory Course*, Computational Mechanics Publications: Southampton, 1992.

[20] Yang, B. & Pan, E., Elastic analysis of an inhomogeneous quantum dot in multilayered semiconductors using a boundary element method. *J. Appl. Phys.* **92**, pp. 3084–3088, 2002.

[21] Glas, F., Analytical calculation of the strain field of single and periodic misfitting polygonal wires in a half-space. *Phil. Mag.* **A82,** pp. 2591–2608, 2002.

[22] Pan, E. & Jiang, X., Effect of QWR shape on the induced elastic and piezoelectric fields. *Computer Model. Eng. Sci.*, **6,** pp. 77–89, 2004.

Section 3
Heat and mass transfer

A meshless solution procedure for coupled turbulent flow and solidification in steel billet casting

B. Šarler[1], R. Vertnik[1] & G. Manojlović[2]
[1]*Laboratory for Multiphase Processes, Nova Gorica Polytechnic, Nova Gorica, Slovenia*
[2]*Technical Development, Štore-Steel, Štore, Slovenia*

Abstract

This paper introduces a physical model and a mesh-free computational procedure for simulation of coupled heat, mass and momentum transport in the continuous casting of steel. The governing equations are based on the mixture continuum formulation of the solid and the liquid phase. The model takes into account the pure liquid, nucleation and movement of the globulitic solid phase, formation of the rigid porous solid matrix, and complete solid. The zero-equation Prandtl mixing-length theory turbulence model is incorporated into the momentum equation. The domain and boundary of interest are divided into overlapping influence areas. On each of them, the fields are represented by the multiquadrics radial basis function collocation on a related sub-set of nodes. Time-stepping is performed in an explicit way. The governing equations are solved in its strong form, i.e. no integrations are performed. The polygonisation is not present and the method is practically independent on the problem dimension. The strategy for the interconnected solution of velocity, grain movement, pressure, and temperature fields is described in detail. The geometry discretisation of the Štore-Steel billet caster is presented.
Keywords: billet casting, steel, solidification, turbulence, meshless method, local radial basis function collocation method, moving boundary problem, multiquadrics.

1 Introduction

Continuous casting is currently the most common [1] casting practice in production of steel. The process involves molten metal being feed through a

WIT Transactions on Modelling and Simulation, Vol 42, © 2006 WIT Press
www.witpress.com, ISSN 1743-355X (on-line)
doi:10.2495/BE06014

submerged entry nozzle into a water-cooled mould where it is sufficiently solidified around the outer surface that it takes the shape of the mould and acquires sufficient mechanical strength to contain the molten core at the centre. As the strand, pulled by a system of rolls, emerges from the mould, water impinges directly onto the surface through a system of sprays, flows over the cast surface and completes the solidification. Related transport, solid mechanics, and phase change kinetics phenomena are extensively studied [2]. Continuous casting technology has advanced due to innovations resulting from knowledge gained from several different tools. These include expensive plant trials, laboratory experiments, and mathematical models [3]. The computational modelling of the process is becoming increasingly important in improving the cast quality, as well as the caster safety and yield.

A common complication of the classical numerical methods is the need to create a polygonisation, either in the domain and/or on its boundary. This type of (re)meshing is often the most time consuming part of the solution process and is far from being fully automated, particularly in moving and/or deforming boundary problems. In recent years, a new class of methods is developed which do not require polygonisation but use only a set of nodes to approximate the solution. The rapid development of these types of meshless (polygon-free) methods and their classification is elaborated in the very recent monographs [4,5]. A broad class of meshfree methods in development today are based on Radial Basis Functions (RBFs) [6]. The global RBF collocation method (RBFCM) or Kansa method [7] is the simplest of them. The method has been extended to cope with large-scale diffusion problems in [8], and to convective-diffusive solid liquid phase change problems in [9] by developing its local version (LRBFCM). The Navier-Stokes equations have been solved by the RBFCM in [10] and by the LRBFCM in [11]. This paper upgrades the recently developed LRBFCM to solve the coupled mass, momentum and heat transfer in continuous cast steel billets.

2 Governing equations

The governing equations are based on the turbulence-extended version of the physical model for solid-liquid systems that include nucleation phenomena and solid phase movement, developed in [12]. The model copes with columnar and equiaxed solidification within one equation and properly describes the Newtonian fluid with variable viscosity and density in the pure laminar liquid limit, which was not the case in previous models. The model for the first time incorporates the nucleation phenomena into mixture continuum model of turbulent solidification. Consider a connected domain Ω with boundary Γ occupied by a liquid-solid phase change material described with the temperature dependent density $\rho_\wp$ of the phase $\wp$, temperature dependent specific heat at constant pressure $c_\wp$, thermal conductivity $k_\wp$, and the specific latent heat of the solid-liquid phase change h_m. The liquid phase properties are additionally described by the viscosity $\mu_{L\ell}$ and turbulent eddy viscosity $\mu_{L\tau}$. We seek the solution of the temperature, velocity and pressure fields at time $t_0 + \Delta t$, where t_0 represents initial time and Δt the positive time increment, based on the coupled set of governing equations that follows. The macroscopic mass conservation of the assumed system is governed by

$$\frac{\partial}{\partial t}\rho + \nabla \cdot (\rho \vec{v}) = 0; \; \rho = f_S^V \rho_S + f_L^V \rho_L, \; \rho\vec{v} = f_S^V \rho_S \vec{v}_S + f_L^V \rho_L \vec{v}_L \,. \qquad (1,2,3)$$

where $f_\wp^V$ represents the volume fraction of the phase $\wp$ which add to unity $f_S^V + f_L^V = 1$. The liquid volume fraction is assumed to vary from 0 to 1 between solidus T_S and liquidus temperature T_L. ρ and $\vec{v}$ represent the mixture density and velocity. The macroscopic energy conservation for the assumed system is governed by

$$\frac{\partial}{\partial t}(\rho h) + \nabla \cdot (\rho \vec{v} h) = \nabla \cdot (k_{\text{eff}} \nabla T) + \nabla \cdot \left(\rho \vec{v} h - f_S^V \rho_S \vec{v}_S h_S - f_L^V \rho_L \vec{v}_L h_L\right), \qquad (4)$$

with the mixture enthalpy and effective thermal conductivity defined as

$$h = f_S^V h_S + f_L^V h_L, \quad k_{\text{eff}} = f_S^V k_S + f_L^V (k_{L\ell} + k_{L\tau}), \qquad (5,6)$$

where $k_{L\tau}$ represents the turbulent correction of the liquid thermal conductivity $k_{L\ell}$. The constitutive temperature-enthalpy relationships are

$$h_S = \int_{T_{ref}}^{T} c_S dT, \qquad h_L = h_S(T) + \int_{T_S}^{T} (c_L - c_S) dT + h_m, \qquad (7,8)$$

with T_{ref} standing for the reference temperature. The macroscopic momentum conservation of the system is governed by

$$\begin{aligned} &\frac{\partial}{\partial t}(\rho\vec{v}) + \nabla \cdot (\rho\vec{v}\vec{v}) = -\nabla P + {}^V f_L \nabla \cdot \boldsymbol{\tau}_L + \nabla \cdot \left(\rho\vec{v}\vec{v} - {}^V f_S \rho_S \vec{v}_S \vec{v}_S - {}^V f_L \rho_L \vec{v}_L \vec{v}_L\right) + \\ &+ \left({}^V f_S \nabla \cdot \boldsymbol{\tau}_S + {}^V f_S \rho_S \vec{\mathrm{f}}_S + {}^V f_L \rho_L \vec{\mathrm{f}}_L\right) c^+ \left\{{}^V f_S^C - {}^V f_S\right\} + \\ &+ \left(\rho_L \vec{\mathrm{f}}_L + \vec{\mathrm{g}}_L\right) c^+ \left\{{}^V f_S - {}^V f_S^C\right\}, \\ &\nabla \cdot \boldsymbol{\tau}(\mu, \vec{v}) = \mu \nabla^2 \vec{v} + \nabla \cdot (2\mu \mathbf{v}) - \frac{2}{3} \nabla \cdot \left[\mu (\nabla \cdot \vec{v}) \cdot \mathbf{I}\right], \end{aligned} \qquad (9,10)$$

Or in the compact form

$$\frac{\partial}{\partial t}(\rho\vec{v}) + \nabla \cdot (\rho\vec{v}\vec{v}) = -\nabla P + \vec{F} \,. \qquad (11)$$

The interphase force is modelled by the Darcy law with the Kozeny-Karman permeability with δ_K denoting a small number

$$\vec{g}_L = -\frac{\mu_L}{K}\frac{\rho}{\rho_L}(\vec{v} - \vec{v}_{SYS}), \quad K = K_0 \frac{\left(1 - {}^V f_S\right)^3}{{}^V f_S^2 + \delta_K}. \tag{12}$$

The body force is modelled by the Bussinesq approximation with $\vec{a}$ denoting acceleration, $\rho_{0\wp}$ and T_0 the reference density and temperature and $\beta_{\wp T}$ the respective thermal expansion coefficients

$$\rho_\wp \vec{f}_\wp = \rho_{0\wp}\vec{a} \cdot \left[1 - \beta_{\wp T}(T - T_0)\right], \tag{13}$$

where P represents pressure, $\boldsymbol{\tau}_\wp$ the extra stress tensor, $\vec{f}_\wp$ the body force, $\vec{g}_\wp$ the inter-phase force on phase $\wp$, and c^+ the linearised Heaviside function over interval $2\delta x$

$$c_+\{x\} = \begin{cases} 1; & x \geq +\delta x \\ (x + \delta x)/(2\delta x) & ;\delta x > x > -\delta x \\ 0; & x \leq -\delta x \end{cases}. \tag{14}$$

Note that the second row in Equation (9) sets-in in the so called slurry region with ${}^V f_S < {}^V f_S^C$ where solid grains are free to move, and the third row in Equation (9) sets-in in the mushy zone, where solid phase with ${}^V f_S > {}^V f_S^C$ moves with the system velocity $\vec{v}_S = \vec{v}_{SYS}$. ${}^V f_S^C$ represents the rigid solid fraction limit. The stress tensors are calculated from the following logic

$$\boldsymbol{\tau}_S = \boldsymbol{\tau}_S(\mu_S, \vec{v}_S), \boldsymbol{\tau}_L = \boldsymbol{\tau}_L(\mu_L, \vec{v}_L(\vec{v}, \vec{v}_S)), \mu_S = \frac{\mu_L}{{}^V f_S}\left(1 - \frac{{}^V f_S}{{}^V f_S^C}\right)^{-2.5\,{}^V f_S^C} - \frac{{}^V f_L}{{}^V f_S}\mu_L. \tag{15,16}$$

The liquid viscosity is modeled as

$$\mu_L = \mu_{L\ell} + \mu_{L\tau}, \tag{17}$$

with the $\mu_{L\ell}$ and $\mu_{L\tau}$ representing the laminar and the turbulent viscosity. Based on the determination of $\mu_{L\tau}$, the available turbulent models are classified as zero-equation, one-equation, and two-equation models. One- and two-equation models are more sophisticated than the zero-equation models, but they lack the simplicity and ease with which the zero equation model can be used. Therefore, a zero equation model based on the Prandtl mixing-length hypothesis is used in the current basic model. By introducing a two dimensional Cartesian coordinate system with base vectors $\vec{i}_\varsigma; \varsigma = x, y$ and coordinates $p_\varsigma; \varsigma = x, y$, i.e. $\vec{p} = \vec{i}_\varsigma p_\varsigma;\ \varsigma = x, y$ can the eddy turbulent viscosity be expressed with the Prandtl mixing length ℓ_m with

$$\mu_L^T = \rho_L \ell_m^2 \left[\left(\frac{\partial v_x}{\partial p_x} \right)^2 + \left(\frac{\partial v_x}{\partial p_y} \right)^2 + \left(\frac{\partial v_y}{\partial p_x} \right)^2 + \left(\frac{\partial v_y}{\partial p_y} \right)^2 \right]^{1/2} . \tag{18}$$

The relative movement of the solid phase with respect to the liquid phase is described by assuming a linear drag force

$$\vec{v}_S = \begin{cases} \vec{v}_L + \dfrac{1 - {}^V f_S}{18\mu_{eff}} (\rho_S - \rho_L) d_S^2 \vec{a};\ {}^V f_S < {}^V f_S^C \\ \vec{v}_{SYS};\ {}^V f_S \geq {}^V f_S^C \end{cases} . \tag{19}$$

The solid grains move with the grain transport equation and grow as

$$\frac{\partial}{\partial t} n + \nabla \cdot (\vec{v}_S n) = \frac{n_{\max}}{\sqrt{2\pi} \Delta T_\sigma} \exp\left(-\frac{1}{2} \left(\frac{\Delta T - \Delta T_n}{\Delta T_\sigma} \right)^2 \right), \quad d_S = \left(\frac{3\, {}^V f_S}{4\pi n} \right)^{\frac{1}{3}}, \tag{20,21}$$

where ΔT_N represents the mean nucleation under-cooling corresponding to the maximum of the distribution, ΔT_σ is the standard deviation of the distribution, $n_{\max}$ is the maximum density of nuclei given by the integral of the total distribution from zero under-cooling to infinite under-cooling, and $\Delta T = T_L - T$, with T_L standing for the liquidus temperature. The three parameters strongly depend on the melt composition, inclusions and solidification conditions.

The pressure field is solved by taking the divergence of the momentum equation and considering the mass conservation

$$\nabla \cdot \frac{\partial}{\partial t} (\rho \vec{v}) = \frac{\partial}{\partial t} \nabla \cdot (\rho \vec{v}) = -\frac{\partial^2 \rho}{\partial t^2} = -\nabla \cdot [\nabla \cdot (\rho \vec{v} \vec{v})] - \nabla^2 P + \nabla \cdot \vec{F} . \tag{22}$$

The pressure is calculated from a false transient of the following equation towards the steady-state

$$\frac{\partial}{\partial t} P = -\nabla^2 P - \nabla \cdot [\nabla \cdot (\rho \vec{v} \vec{v})] + \nabla \cdot \vec{F} + \frac{\partial^2 \rho}{\partial t^2} . \tag{23}$$

with Neumann boundary conditions obtained by multiplication of the momentum equation with the normal derivative

$$\frac{\partial P}{\partial n_\Gamma} = \left[-\frac{\partial}{\partial t} (\rho \vec{v}) - \nabla \cdot (\rho \vec{v} \vec{v}) + \nabla \cdot \vec{F} \right] \cdot \vec{n}_\Gamma . \tag{24}$$

Let us assume the initial velocity and pressure fields are known. The initial pressure is calculated from a three-level time-step procedure

$$P_0 = P_0^{'} + \left[-\nabla^2 P_0^{'} - \nabla \cdot \left[\nabla \cdot \left(\rho_0 \vec{v}_0 \vec{v}_0 \right) \right] + \nabla \cdot \vec{F}_0 + \frac{\rho_{-2} - 2\rho_{-1} + \rho_0}{\Delta t^2} \right] \Delta t \,. \quad (25)$$

where $P_0^{'}$ represents the value from previous iteration, and ρ_{-1}, ρ_{-2} represent the value of density at time $t_0 - \Delta t, t_0 - 2\Delta t$. The pressure equation boundary conditions are discretised as

$$\frac{\partial P}{\partial n_\Gamma} = \left[\frac{-\rho_0 \vec{v}_0 + \rho_{-1} \vec{v}_{-1}}{\Delta t} - \nabla \cdot \left(\rho_0 \vec{v}_0 \vec{v}_0 \right) + \nabla \cdot \vec{F}_0 \right] \cdot \vec{n}_\Gamma \,. \quad (26)$$

After calculation of the pressure field at time t_0, the new velocity field at time $t_0 + \Delta t$ is calculated from

$$\vec{v} = \left[\frac{\rho_0 v_0}{\Delta t} - \nabla \cdot \left(\rho \vec{v}_0 \vec{v}_0 \right) - \nabla P_0 + \vec{F}_0 \right] \frac{\Delta t}{\rho} \,. \quad (27)$$

Afterwards, the solid phase movement (19) is calculated, followed by the solution of the grain transport (20,21), and the solution of the energy equation. After a sufficiently small difference between results of the two successive internal timestep iterations is achieved, the next timestep is attempted. The meshless computational details regarding solution of the scalar transport equation are given as follows.

3 Solution procedure

The solution of the problem is elaborated on a general transport equation, defined on a fixed domain Ω with boundary Γ, standing as a general type of equation representing all the introduced governing equations in previous chapter as a special case

$$\frac{\partial}{\partial t}\left[\rho \mathrm{C}(\Phi) \right] + \nabla \cdot \left[\rho \vec{v}\, \mathrm{C}(\Phi) \right] = -\nabla \cdot \left(-\mathbf{D} \nabla \Phi \right) + S \quad (28)$$

with ρ, Φ, t, $\mathbf{D}$, and S standing for density, transport variable, time, velocity, diffusion tensor and source, respectively. Scalar function C stands for possible more involved constitutive relations between the conserved and the diffused quantities. The solution of the governing equation for the transport variable at the final time $t_0 + \Delta t$ is sought. The solution is constructed by the initial and boundary conditions that follow. The initial value of the transport variable $\Phi(\vec{p},t)$ at a point with position vector $\vec{p}$ and time t_0 is defined through the known function Φ_0

$$\Phi(\vec{p},t) = \Phi_0(\vec{p}); \; p \in \Omega + \Gamma \,. \quad (29)$$

The boundary Γ is divided into not necessarily connected parts $\Gamma = \Gamma_D \cup \Gamma_N \cup \Gamma_R$ with Dirichlet, Neumann and Robin type boundary conditions, respectively. These boundary conditions are at the boundary point $\vec{p}$ with normal $\vec{n}_\Gamma$ and time $t_0 \le t \le t_0 + \Delta t$ defined through known functions Φ_Γ^D, Φ_Γ^R, $\Phi_{\Gamma ref}^R$

$$\Phi = \Phi_\Gamma^D;\ \vec{p} \in \Gamma_D, \qquad \frac{\partial}{\partial n_\Gamma}\Phi = \Phi_\Gamma^N;\ \vec{p} \in \Gamma_N, \qquad \frac{\partial}{\partial n_\Gamma}\Phi = \Phi_\Gamma^R\left(\Phi - \Phi_{\Gamma ref}^R\right);\ \vec{p} \in \Gamma_R. \tag{30,31,32}$$

The involved parameters of the governing equation and boundary conditions are assumed to depend on the transport variable, space and time. The solution procedure is based on combined explicit-implicit scheme. The discretisation in time can be written as

$$\frac{\partial}{\partial t}\left(\rho \mathrm{C}(\Phi)\right) \approx \frac{\rho \mathrm{C} - \rho_0 \mathrm{C}_0}{\Delta t} \approx \frac{\bar{\rho}\bar{\mathrm{C}} + \bar{\rho}\dfrac{d\bar{C}}{d\Phi}\left(\Phi - \bar{\Phi}\right) - \rho_0 \mathrm{C}_0}{\Delta t}, \tag{33}$$

by using the two-level time discretisation and the Taylor expansion of the function $\mathrm{C}(\Phi)$. The known quantities are denoted with overbar. The source term can be expanded as

$$S(\Phi) \approx \bar{S} + \frac{d\bar{S}}{d\Phi}\left(\Phi - \bar{\Phi}\right). \tag{34}$$

The unknown Φ can be directly calculated from the equation

$$\Phi = \frac{\dfrac{\rho_0}{\Delta t}\mathrm{C}_0 - \dfrac{\bar{\rho}}{\Delta t}\bar{\mathrm{C}} + \dfrac{\bar{\rho}}{\Delta t}\dfrac{d\bar{C}}{d\Phi}\bar{\Phi} + \nabla\cdot\left(\mathbf{D}_0 \nabla \Phi_0\right) - \nabla\cdot\left(\rho_0 \vec{v}_0 \mathrm{C}_0\right) + \bar{S} - \dfrac{d\bar{S}}{d\Phi}\bar{\Phi}}{\dfrac{\bar{\rho}}{\Delta t}\dfrac{d\bar{C}}{d\Phi} - \dfrac{d\bar{S}}{d\Phi}}. \tag{35}$$

The value of the transport variable Φ_n is solved in a set of nodes $\vec{p}_n; n = 1,2,...,N$ of which N_Ω belong to the domain and N_Γ to the boundary. The iterations over one time-step are completed when the equation (36) is satisfied, and the steady-state is achieved when the equation (37) is achieved

$$\max\left|\Phi_n - \bar{\Phi}_n\right| \le \Phi_{itr}, \qquad \max\left|\Phi_n - \Phi_0\right| \le \Phi_{ste}. \tag{36,37}$$

The representation of function Φ over a set of l (in general) non-equally spaced ${}_lN$ nodes ${}_l\vec{p}_n;\ n = 1,2,...,{}_lN$ is made in the following way

$$\Phi(\vec{p}) \approx \sum_{k=1}^{{}_lK} {}_l\psi_k(\vec{p})\,{}_l\alpha_k. \tag{38}$$

${}_l\psi_k$ stands for the shape functions, ${}_l\alpha_k$ for the coefficients of the shape functions, and ${}_lK$ represents the number of the shape functions. The left lower index on entries of expression (38) represents the sub-domain ${}_l\omega$ on which the coefficients ${}_l\alpha_k$ are determined. The sub-domains ${}_l\omega$ can in general be contiguous (overlapping) or non-contiguous (non-overlapping). Each of the sub-domains ${}_l\omega$ includes ${}_lN$ grid-points of which ${}_lN_\Omega$ are in the domain and ${}_lN_\Gamma$ are on the boundary. The coefficients can be calculated from the sub-domain nodes in two distinct ways. The first way is collocation (interpolation) and the second way is approximation by the least squares method. Only the simpler collocation version for calculation of the coefficients is considered in this text. Let us assume the known function values ${}_l\Phi_n$ in the nodes ${}_l\vec{p}_n$ of the subdomain ${}_l\omega$. The collocation implies

$$\Phi\left({}_l\vec{p}_n\right)=\sum_{k=1}^{{}_lN}{}_l\psi_k\left({}_l\vec{p}_n\right){}_l\alpha_k \,. \tag{39}$$

For the coefficients to be computable, the number of the shape functions has to match the number of the collocation points ${}_lK={}_lN$, and the collocation matrix has to be non-singular. The system of equations (39) can be written in a matrix-vector notation

$${}_l\underline{\Psi}\,{}_l\vec{\alpha}={}_l\vec{\Phi};\; {}_l\underline{\psi}_{kn}={}_l\psi_k\left({}_l\vec{p}_n\right),\; {}_l\Phi_n=\Phi\left({}_l\vec{p}_n\right). \tag{40}$$

The coefficients ${}_l\vec{\alpha}$ can be computed by inverting the system (40)

$${}_l\vec{\alpha}={}_l\underline{\Psi}^{-1}\,{}_l\vec{\Phi}\,. \tag{41}$$

By taking into account the expressions for the calculation of the coefficients ${}_l\vec{\alpha}$, the collocation representation of function $\Phi(\vec{p})$ on subdomain ${}_l\omega$ can be expressed as

$$\Phi(\vec{p})\approx\sum_{k=1}^{{}_lN}{}_l\psi_k(\vec{p})\sum_{n=1}^{{}_lN}{}_l\underline{\psi}_{kn}^{-1}\,{}_l\Phi_n \,. \tag{42}$$

The first partial spatial derivatives of $\Phi(\vec{p})$ on subdomain ${}_l\omega$ can be expressed as

$$\frac{\partial}{\partial p_\varsigma}\Phi(\vec{p})\approx\sum_{k=1}^{{}_lN}\frac{\partial}{\partial p_\varsigma}{}_l\psi_k(\vec{p})\sum_{n=1}^{{}_lN}{}_l\underline{\psi}_{kn}^{-1}\,{}_l\Phi_n;\; \varsigma=x,y \,. \tag{43}$$

The second partial spatial derivatives of $\Phi(\vec{p})$ on subdomain ${}_l\omega$ can be expressed as

$$\frac{\partial^2}{\partial p_\varsigma p_\xi}\Phi(\vec{p}) \approx \sum_{k=1}^{{}_lN} \frac{\partial^2}{\partial p_\varsigma p_\xi} {}_l\psi_k(\vec{p}) \sum_{n=1}^{{}_lN} {}_l\underline{\psi}_{kn}^{-1} {}_l\Phi_n;\ \varsigma,\xi = x,y. \tag{44}$$

The radial basis functions, such as multiquadrics, can be used for the shape function

$$ {}_l\psi_k(\vec{p}) = \left[1 + c^2\, {}_lr_k^2 / {}_lr_0^2\right]^{1/2};\ {}_lr_k^2 = (\vec{p} - {}_l\vec{p}_k)\cdot(\vec{p} - {}_l\vec{p}_k). \tag{45}$$

where c represents the shape parameter and ${}_lr_0$ the scaling constant, set to the maximum distance between two nodes in a subdomain. The complete solution procedure follows the below defined steps 1-6. **Step 1:** First, the initial conditions are set in the domain and boundary nodes and the required derivatives are calculated from the known nodal values. **Step 2:** The equation (35) is used to calculate the new values of the variable ${}_l\Phi_n$ at time $t_0 + \Delta t$ in the domain nodes. **Step 3:** What follows in the steps 3 and 4 defines variable ${}_l\Phi_n$ at time $t_0 + \Delta t$ in the Dirichlet, Neumann, and Robin boundary nodes of the new boundary at time $t_0 + \Delta t$. For this purpose, in the step 3, the coefficients ${}_l\vec{\alpha}$ have to be determined from the new values in the domain $\Omega(t_0 + \Delta t)$ and from the information on the boundary conditions at $\Gamma(t_0 + \Delta t)$. Let us introduce domain, Dirichlet, Neumann, and Robin boundary indicators for this purpose. These indicators are defined as

$$\Upsilon_{\Omega n} = \begin{cases} 1; \vec{p}_n \in \Omega \\ 0; \vec{p}_n \notin \Omega \end{cases},\ \Upsilon_{\Gamma n}^D = \begin{cases} 1; \vec{p}_n \in \Gamma^D \\ 0; \vec{p}_n \notin \Gamma^D \end{cases},\ \Upsilon_{\Gamma n}^N = \begin{cases} 1; \vec{p}_n \in \Gamma^N \\ 0; \vec{p}_n \notin \Gamma^N \end{cases},\ \Upsilon_{\Gamma n}^R = \begin{cases} 1; \vec{p}_n \in \Gamma^R \\ 0; \vec{p}_n \notin \Gamma^R \end{cases}. \tag{46}$$

The coefficients ${}_l\vec{\alpha}$ are calculated from the system of linear equations

$$\begin{aligned} &\sum_{k=1}^{{}_lN} {}_l\Upsilon_{\Omega n}\, {}_l\psi_k({}_l\vec{p}_n)\, {}_l\alpha_k + \sum_{k=1}^{{}_lN} {}_l\Upsilon_{\Gamma n}^D\, {}_l\psi_k({}_l\vec{p}_n)\, {}_l\alpha_k \\ &+ \sum_{k=1}^{{}_lN} {}_l\Upsilon_{\Gamma n}^N \frac{\partial}{\partial n_\Gamma} {}_l\psi_k({}_l\vec{p}_n)\, {}_l\alpha_k + \sum_{k=1}^{{}_lN} {}_l\Upsilon_{\Gamma n}^R \frac{\partial}{\partial n_\Gamma} {}_l\psi_k({}_l\vec{p}_n)\, {}_l\alpha_k \\ &= {}_l\Upsilon_{\Omega n}\, {}_lT_n + {}_l\Upsilon_{\Gamma n}^D\, {}_lT_n^D + {}_l\Upsilon_{\Gamma n}^N\, {}_lT_n^N + {}_l\Upsilon_{\Gamma n}^R\, {}_lT_{\Gamma n}^R \left(\sum_{k=1}^{{}_lN} {}_l\psi_k({}_l\vec{p}_n)\, {}_l\alpha_k - {}_lT_{\Gamma ref\, n}^R \right). \end{aligned} \tag{47}$$

The system (47) can be written in a compact form

$$ {}_l\underline{\Psi}\, {}_l\vec{\alpha} = {}_l\vec{b}, \tag{48}$$

with the following system matrix entries

$$\begin{aligned} {}_l\underline{\Psi}_{nk} &= {}_l\Upsilon_{\Omega n}\, {}_l\psi_k({}_l\vec{p}_n) + {}_l\Upsilon_{\Gamma n}^D\, {}_l\psi_k({}_l\vec{p}_n) \\ &+ {}_l\Upsilon_{\Gamma n}^N \frac{\partial}{\partial n_\Gamma} {}_l\psi_k({}_l\vec{p}_n) + {}_l\Upsilon_{\Gamma n}^R \left[\frac{\partial}{\partial n_\Gamma} {}_l\psi_k({}_l\vec{p}_n) - {}_lT_{\Gamma n}^R \sum_{k=1}^{{}_lN} {}_l\psi_k({}_l\vec{p}_n) \right]. \end{aligned} \tag{49}$$

and with the following explicit form of the augmented right hand side vector

$$ {}_{l}\vec{b}_{n} = {}_{l}\Upsilon_{\Omega n} T_{n} + {}_{l}\Upsilon_{\Gamma n}^{D} T_{\Gamma n}^{D} + {}_{l}\Upsilon_{\Gamma n}^{N} T_{\Gamma n}^{N} - {}_{l}\Upsilon_{\Gamma n}^{R}\, {}_{l}T_{\Gamma n}^{R}\, {}_{l}T_{\Gamma ref\, n}^{R} \tag{50} $$

Step 4: The unknown boundary values of $\Gamma(t_0 + \Delta t)$ are set from equation (39).

Step 5: The under-relaxation $T_n = \underline{T}_n + \varepsilon(T_n - \underline{T}_n)$ might be required in the general case for all the computational nodes $\vec{p}_n; n = 1, 2, ..., N$ with ε standing for the under-relaxation factor. The iterations over one timestep are completed when the iteraton criterion (36) is satisfied in all computational nodes. The steady-state is achieved when the criterion (37) is met. The parameter Φ_{ste} is defined as the steady-state convergence margin. In case the steady-state criterion is achieved or the time of calculation exceeds the foreseen time of interest, the calculation is stopped.

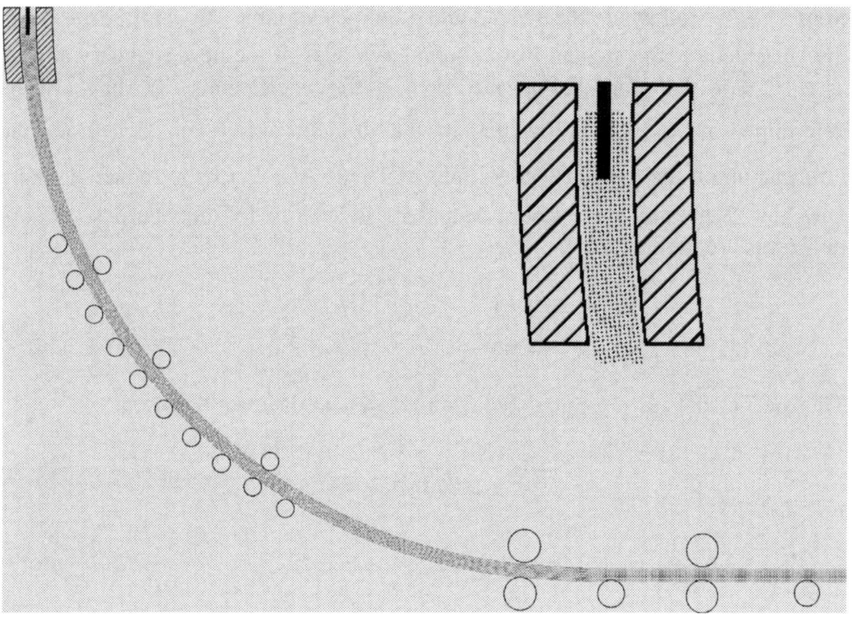

Figure 1: Schematics of meshless 2D discretization (pointisation) of the Štore-Steel billet caster. Right upper corner: detail of the mould with the submerged entry nozzle.

4 Conclusions

The present paper introduces the LRBFCM for numerical evaluation of highly complex solidification model, associated with the continuous casting of steel billets. Probably for the first time, it copes with the solution of the momentum and pressure equations in the turbulent solidification context as occurs in the continuous casting of steel billets. Attempts of solving such models were previously made only by more established numerical methods. The developments are almost independent on the problem dimension. The complicated geometry is easy to cope with. No polygonisation is needed. No integrations are needed. The method appears efficient, because it does not require a solution of the large systems of equations like the original Kansa method. Instead, small systems of linear equations have to be solved in each timestep for each node and associated sub-domain, representing the most natural and automatic domain decomposition. The energy part of the presented model has been already solved by three different mesh reduction or meshless techniques. The dual reciprocity boundary element method has been used in [13], the polynomial variant (diffuse approximate method) of the present method in [14], and in [15] the LRBFCM. The numerical examples incorporating the solution of the complete model will be shown in one of our future publications.

Acknowledgements

The author would like to acknowledge the support of the Štore-Steel company, Štore, Slovenia and Slovenian Ministry for High Education, Science and Technology in the framework of the project L2-5387-1540-03 Modelling and Optimisation for Competitive Continuous Casting and L2-7204-1540-05 Modelling and Optimisation of Continuous Casting.

References

[1] Irving, W.R. *Continuous Casting of Steel*, The University Press, Cambridge, 1993.

[2] Beckermann, C. Modelling of macrosegregation, *International Materials Reviews*, 2002, **47**, 243-261.

[3] Šarler, B., Vertnik, R., Šaletić, S., Manojlović, G., Cesar, J. Application of continuous casting simulation at Štore-Steel, *Berg- und Hüttenmännische Monatshefte*, 2005, **150**, 300-306.

[4] Liu, G.R. *Mesh Free Methods,* CRC Press, Boca Ratorn, 2003.

[5] Šarler, B. *Chapter 9: Meshless Methods,* in: Nowak, A.J. (Ed.) *Advanced Numerical Methods in Heat Transfer*, Silesian Technical University Press, Poland, 2004.

[6] Buhmann, M.D. *Radial Basis Function: Theory and Implementations,* Cambridge University Press, Cambridge, 2003.

[7] Kansa, E.J. Multiquadrics – a scattered data approximation scheme with applications to computational fluid dynamics-II. Solutions to parabolic, hyperbolic and elliptic partial differential equations, *Computers and Mathematics with Application*, 1990, **19**, 147-161.

[8] Šarler, B. & Vertnik, R. Meshfree explicit local radial basis function collocation method for diffusion problems, *Computers and Mathematics with Application*, (in print).

[9] Vertnik, R. & Šarler, B. Meshless local radial basis function collocation method for convective diffusive solid-liquid phase change problems,

International Journal of Numerical Methods in Heat and Fluid Flow, (in print).

[10] Šarler, B. A radial basis function collocation approach in computational fluid dynamics, *Computer Modelling in Engineering and Sciences*, 2005, **7**, 185-193.

[11] Divo, E. & Kassab, A.J., An efficient localized RBF meshless method applied to fluid flow and conjugate heat transfer, *ASME Paper Number IMECE2005-82150* (expanded version submitted *ASME J. of Heat Transfer*, under review).

[12] Šarler, B. & Pepper, D.W. Momentum transport modelling including solid phase movement and nucleation for direct chill casting processes, in: Bergles, A.E., Globič, I., Amon, C. and Bejan, A. (Eds.). *Thermal Sciences 2004 (ASME-ZSIS International Thermal Sciences Seminar)*, University of Ljubljana, Ljubljana, 2004.

[13] Šarler, B. & Perko, J. DRBEM solution of temperatures and velocities in DC cast aluminium slabs, in: Brebbia, C.A., Kassab, A.J., Chopra, M.B., & Divo, E. (Eds.). *Boundary element technology XIV*, WIT Press, Southampton, 2001, 357-369.

[14] Šarler, B., Vertnik, R. & Perko, J. Application of diffuse approximate method in convective-diffusive solidification problems, *Computers, Materials, Continua*, 2005, **2**, 77-83.

[15] R. Vertnik, B. Šarler, Solution of the heat transfer model of the start-up phase of the direct chill casting process by a meshless numerical method, in: Nowak, A.J., Weçel, G., Bialecki, R.A. *Proceedings of the Eurotherm Seminar 82: numerical heat transfer 2005,* Institute of Thermal Technology, Silesian University of Technology, 2005, 367-376.

Conduction heat transfer with nonzero initial conditions using the Boundary Element Method in the frequency domain

N. Simões[1], A. Tadeu[1] & W. Mansur[2]
[1]Department of Civil Engineering, University of Coimbra, Portugal
[2]Department of Civil Engineering,
COPPE/Federal University of Rio de Janeiro, Brazil

Abstract

The diffusion equation with non-zero initial condition is solved using the Boundary Element Method in the frequency domain. Complex frequencies are used in order to avoid aliasing phenomena and to allow the computation of the steady state response. One numerical example is given to illustrate the effectiveness of the proposed approach to solve diffusion equations in two dimensions.
Keywords: *transient heat transfer; conduction; Boundary Element Method; 2.5D Green's functions; frequency domain; Fourier transform; nonzero initial conditions.*

1 Introduction

Analytical solutions are only known for solving simple problems, while the Boundary Element Method (BEM) can be implemented to deal with general problems. The BEM is an alternative to the Finite Element Method (FEM) and the Finite Difference Method (FDM) for solving physical problems, such as that of heat diffusion (Brebbia et al. [1], Pina and Fernandez [2], Godinho et al. [3])

Most of the BEM implementations to solve transient heat transfer problems are based on either "time marching schemes" (Chang et al. [4], Shaw [5], Wrobel and Brebbia [6], Carini et al. [7], Lesnic et al. [8]) or Laplace transforms (Sutradhar et al. [9], Sutradhar and Paulino [10], Chen et al. [11]). The Laplace

WIT Transactions on Modelling and Simulation, Vol 42, © 2006 WIT Press
www.witpress.com, ISSN 1743-355X (on-line)
doi:10.2495/BE06015

transform technique removes the time dependent derivative using a transform variable. However, this process requires then an inverse transform to achieve the solution in the time domain that may lead to loss of accuracy and amplification of small truncation errors.

To improve the efficiency of the BEM and its applicability to more general problems different numerical schemes have been proposed such as the Dual Reciprocity Boundary Method (DRBEM) (see Nardini and Brebbia [12] and Satravaha and Zhu ([13], [14])) and the coupling of the FEM and the BEM (Guven and Madenci [15]).

The present work solves the problem of heat conduction in the frequency domain after the application of Fourier transform in the time domain. Initial conditions are assumed and dealt with source density distribution (Mansur et al. ([16], [17]). To avoid the aliasing phenomena and the calculation of the static response, complex frequencies are used. Their effect is removed in the time domain by rescaling the response.

This paper describes first the BEM formulation used and how the time solutions are obtained. The proposed approach is then used to solve the case of heat conduction in a finite rectangle with a unit initial temperature and maintaining null temperatures along the boundary, for which analytical solutions are known.

2 BEM formulation

The transient heat transfer by conduction in a homogeneous domain Ω, bounded by a surface C can be modelled by

$$\left(\frac{\partial^2}{\partial x^2}+\frac{\partial^2}{\partial y^2}\right)T=\frac{1}{K}\frac{\partial T}{\partial t}, \tag{1}$$

in which t is time, $T(t,x,y)$ is temperature, $K=k/\rho\, c$ is the thermal diffusivity, k is the thermal conductivity, ρ is the density and c is the specific heat.

To solve this equation, we move from the time domain to the frequency domain by applying a Fourier transformation in the time domain to eqn. (1). Performing the integration by parts, one obtains

$$\left(\frac{\partial^2}{\partial x^2}+\frac{\partial^2}{\partial y^2}+\lambda^2\right)\hat{T}(\omega,x,y)=-\frac{T_0}{K}, \tag{2}$$

where $\hat{T}(\omega,x,y)=\int_0^\infty T(t,x,y)\,dt$, $\lambda=\sqrt{\frac{-\mathrm{i}\omega}{K}}$, $\mathrm{i}=\sqrt{-1}$, T_0 is the initial tempearure at $t=0$ and ω is the frequency.

The boundary integral equation can be constructed by applying the reciprocity theorem, leading to

$$p\hat{T}(x_0,y_0,\omega)=\int_C q(x,y,\nu_n,\omega)G(x,y,x_0,y_0,\omega)\,dc - \int_C H(x,y,\nu_n,x_0,y_0,\omega)\hat{T}(x,y,\omega)\,dc+\int_\Omega \frac{T_0}{K}G(x,y,x_0,y_0,\omega)\,d\Omega . \quad (3)$$

In this equation, ν_n is the unit outward normal along the boundary, G and H are respectively the fundamental solutions (Green's functions) for the temperature (T) and heat flux (q), at (x,y) due to a virtual point heat source at (x_0,y_0). The factor p is a constant defined by the shape of the boundary, taking the value $1/2$ if $(x_0,y_0)\in C$ and the boundary is smooth.

The required Green's functions for temperature and heat flux in Cartesian co-ordinates are given by

$$G(x,y,x_0,y_0,\omega)=\frac{-\mathrm{i}}{4k}H_0(\lambda r), \quad (4)$$

$$H(x,y,\nu_n,x_0,y_0,\omega)=\frac{\mathrm{i}}{4k}\lambda H_1(\lambda r)\frac{\partial r}{\partial \nu_n}, \quad (5)$$

in which $r=\sqrt{(x-x_0)^2+(y-y_0)^2}$ and $H_n(\;)$ are Hankel functions of the second kind and order n.

Equation (3) is solved by discretizing the boundary C into N straight boundary elements, with one nodal point in the middle of each element. The integrations along the elements are evaluated using a Gaussian quadrature scheme, when they are not performed along the loaded element. For the loaded element, the existing singular integrands of the Green's functions are calculated in closed form.

The integral $\int_0^{L/2} H_0(\lambda r)dr$ is evaluated as

$$\int_0^{L/2} H_0(\lambda r)dr=\frac{L}{2}H_0\left(\lambda\frac{L}{2}\right)+ \pi\frac{L}{4}\left[H_1\left(\lambda\frac{L}{2}\right)S_0\left(\lambda\frac{L}{2}\right)-H_0\left(\lambda\frac{L}{2}\right)S_1\left(\lambda\frac{L}{2}\right)\right], \quad (6)$$

where $S_n(\ldots)$ are Struve functions of order n and L is the size of the boundary elements.

The integral $\int_\Omega \frac{T_0}{K}G(x,y,x_0,y_0,\omega)\,d\Omega$ is evaluated by discretizing the domain Ω into small sub-domains. In the vicinity of the element being loaded a semicircle, Ω_1, is defined, whose centre coincides with the position of the virtual load and with a radius whose length is equal to half of the boundary size, as shown in Figure 1. In this domain, the integration is performed analytically following the expression

$$\int_{\Omega_1} H_0(\lambda r) d\Omega_1 = \int_0^{L/2}\int_0^{\pi} H_0(\lambda r) r \, dr \, d\theta = \frac{\pi L}{2\lambda}\left[H_1\left(\lambda \frac{L}{2}\right)\right] - \frac{2\mathrm{i}}{\lambda^2}, \tag{7}$$

while for the other sub-domains the integrations are evaluated using a Gaussian quadrature scheme.

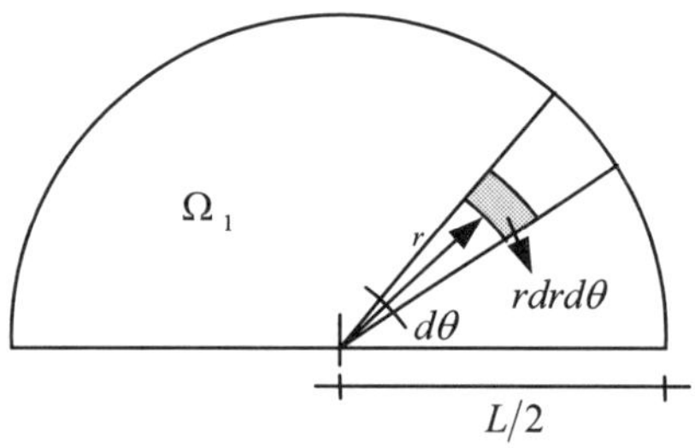

Figure 1: Integration scheme.

3 Responses in the time domain

After the frequency results responses have been obtained, the time responses in the temporal domain are computed by means of an inverse Fourier transform. Aliasing phenomena are prevented by using complex frequencies with a small imaginary part of the form $\omega_c = \omega - \mathrm{i}\eta$ (with $\eta = 0.0005\Delta\omega$, and $\Delta\omega$ being the frequency step). The constant η cannot be arbitrarily large, since this leads either to severe loss of numerical precision, or to underflows and overflows in the evaluation of the exponential windows (see Kausel and Roësset [18]). The response needs to be computed from 0.0Hz to very high frequencies. An intrinsic characteristic of this problem is that the heat responses decay very rapidly as the frequency increases, which allows us to limit the upper frequency where the solution is required. The static response can be computed when the frequency is zero, since the use of complex frequencies leads to arguments of the Hankel function other than zero ($\omega_c = -\mathrm{i}\eta$ for 0.0Hz).

4 Verification of the solution

To verify the formulation described above the heat propagation was computed in a finite rectangle $-a \le x \le a$, $-b \le y \le b$ ($a = b = 0.2$ m) with a unit initial temperature across the full domain, maintaining null temperatures at the boundaries. The analytical solution for the temperature distribution has been given by Carslaw and Jaeger [19]:

$$T(t,x,y)=\frac{16}{\pi^2}\sum_{n=0}^{\infty}\sum_{m=0}^{\infty}L_{n,m}\cos\frac{(2n+1)\pi x}{2a}\cos\frac{(2m+1)\pi y}{2b}\exp\left(-D_{n,m}\,t\right)\,, \tag{8}$$

where $L_{n,m}=\dfrac{(-1)^{n+m}}{(2n+1)(2m+1)}$ and $D_{n,m}=\dfrac{K\pi^2}{4}\left[\dfrac{(2n+1)^2}{a^2}+\dfrac{(2m+1)^2}{b^2}\right]$. In the present example, the thermal properties of the homogeneous unbounded medium are $k=0.06\,\mathrm{W.m^{-1}.{}^{\circ}C^{-1}}$, $\rho=80.0\,\mathrm{Kg.m^{-3}}$ and $c=1300.0\,\mathrm{J.Kg^{-1}.{}^{\circ}C^{-1}}$, which defines a thermal diffusivity of $K=5.8\times10^{-7}\,\mathrm{m^2.s^{-1}}$.

The calculations were obtained in the frequency range $[0,1024\times10^{-5}]\mathrm{Hz}$ with an increment of $\Delta\omega=10^{-5}\,\mathrm{Hz}$, which defines a time window of $t=27.78\mathrm{h}$. The heat responses were computed over along a grid of 39×39 receivers equally spaced along the x and y directions. The application of an inverse Fourier transform in the frequency domain allows the time domain response to be obtained.

Figure 2 presents the time responses at different times ($t=207.52\mathrm{s}$, $t=512.7\mathrm{s}$, t=9008.8 s, $t=25012.2\mathrm{s}$), computed using the formulation proposed in the present work. This Figure also illustrates the amplitude differences between the numerical and the analytical solutions (eqn. 8). The major errors can be found at the beginning of the heat propagation phenomenon.

In order to illustrate the time evolution of the heat propagation, Figure 3 shows the temperature curves at the receivers placed at $(0.0,0.0)\mathrm{m}$, $(-0.1,-0.1)\mathrm{m}$ and $(-0.19,-0.19)\mathrm{m}$. The results were computed using equation (8) and the proposed formulation. The solid lines identify the analytical responses, while the marks indicate the temperatures calculated by the proposed formulation. There is good agreement between both numerical and analytical results.

Notice that, at $t=0.0\mathrm{s}$, the numerical solutions do not coincide with the exact time values, because $T(t,x,y)$ obtained from a Fourier transform is discontinuous at this instant.

Figure 4 shows the temperature values obtained at different times ($t=512.7\mathrm{s}$, $t=9008.8\mathrm{s}$, $t=25012.2\mathrm{s}$, $t=50012.2\mathrm{s}$) for a line of receivers placed at $y=0.0\mathrm{m}$ and $y=0.1\mathrm{m}$ crossing the rectangle. The solid lines represent the solution given by equation (8) while the marks identify the responses provided by the proposed formulation. These results are very similar. As time elapses the temperatures fall very quickly, to establish equilibrium with the boundary conditions (null temperatures).

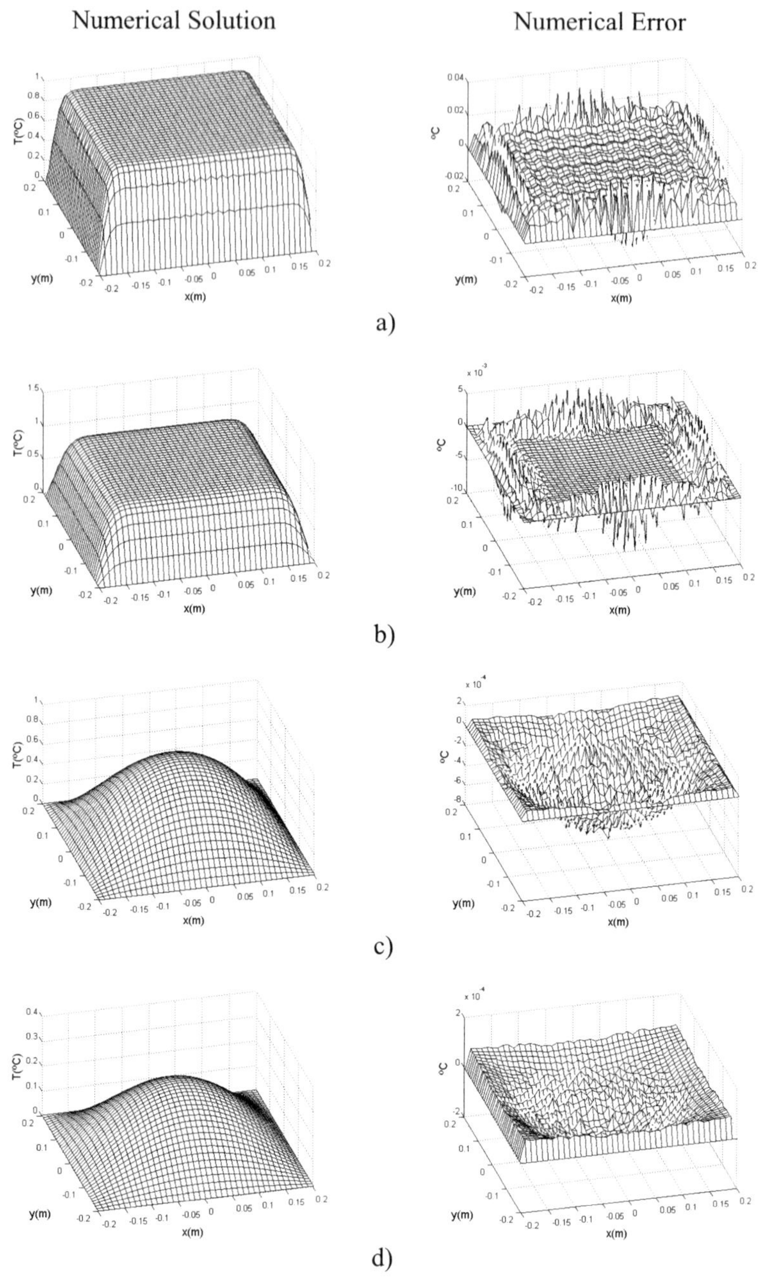

Figure 2: Temperature responses and numerical errors at different times: a) 207.52s b) 512.7s ; c) 9008.8 s ; d) 25012.2 s.

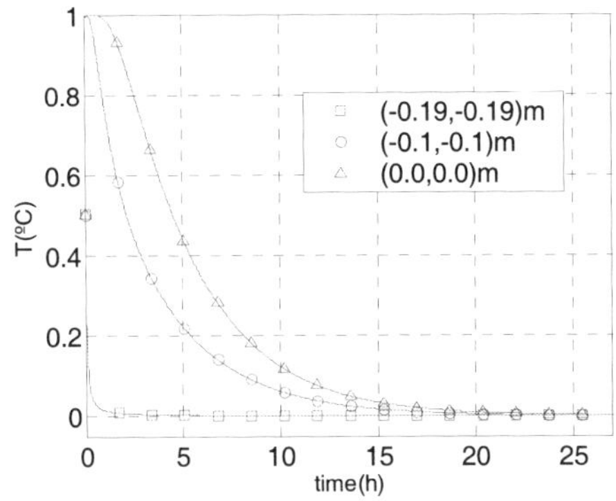

Figure 3: Time domain responses and error for different receivers placed at $(0.0, 0.0)\text{m}$, $(-0.1, -0.1)\text{m}$ and $(-0.19, -0.19)\text{m}$.

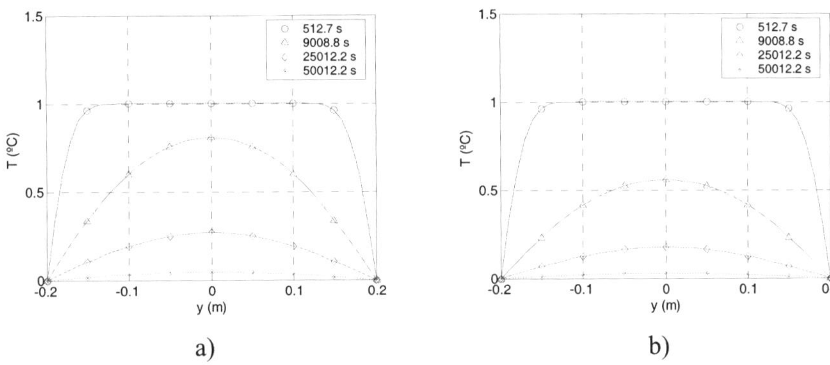

Figure 4: Analytical versus numerical heat responses at different instants, for two vertical line of receivers: a) $y = 0.0\text{m}$; b) $y = 0.1\text{m}$.

5 Conclusions

This paper has illustrated how the heat diffusion time history due to initial temperature conditions can be obtained in the frequency domain by means of a direct BEM formulation. Analytical solutions have been used for verification purposes, to confirm the formulation's accuracy in the frequency domain, transforming the initial conditions into equivalent source density distribution. Good results were obtained for a finite rectangle with unit initial conditions subjected to null temperatures at the boundaries.

References

[1] Brebbia, C.A., Telles, J.C.F., Wrobel L.C., *Boundary Element Techniques: Theory and Applications in Engineering*, Springer, Berlin, 1984.

[2] Pina, M.L.G., Fernandez, J.L.M., Applications in heat conduction by BEM, *In Topics in Boundary Element Research*, C.A Brebbia (ed.), Springer: Berlin, 1984.

[3] Godinho, L., Tadeu, A., Simões, N., Study of transient heat conduction in 2.5D domains using the Boundary Element Method, *Engineering Analysis with Boundary Elements*, **28**, pp. 593-606, 2004.

[4] Chang, Y.P., Kang, C.S., Chen, D.J., The use of fundamental Green's functions for solution of problems of heat conduction in anisotropic media, *International Journal of Heat and Mass Transfer*, **16**, pp. 1905-1918, 1973.

[5] Shaw, R.P., An integral equation approach to diffusion, *International Journal of Heat and Mass Transfer*, **17**, pp. 693-699, 1974.

[6] Wrobel, L.C., Brebbia C.A., A formulation of the Boundary Element Method for axisymmetric transient heat conduction, *International Journal of Heat and Mass Transfer*, **24**, pp. 843-850, 1981.

[7] Carini, A., Diligenti, M., Salvadori, A., Implementation of a symmetric Boundary Element method in transient heat conduction with semi-analytical integrations, *Int. J. for Numer. Meth. in Engng.*, **46**, pp. 1819-1843, 1999.

[8] Lesnic, D., Elliot, L., Ingham, D.B., Treatment of singularities in time-dependent problems using the boundary element method, *Engineering Analysis with Boundary Elements-EABE*, **16**, pp. 65-70, 1995.

[9] Sutradhar, A., Paulino, G.H., Gray, L.J., Transient heat conduction in homogeneous and non-homogeneous materials by the Laplace transform Galerkin Boundary Element Method, *Engineering Analysis with Boundary Elements-EABE*, **26 (2)**, pp. 119-132, 2002.

[10] Sutradhar, A., Paulino, G.H., The simple boundary element method for transient heat conduction in functionally graded materials. *Computer Methods in Applied Mechanics and Engineering*, **193**, pp. 4511-4539, 2004.

[11] Chen, C.S., Golberg, M.A., Rashed, Y.F., A mesh free method for linear diffusion equations, *Numerical Heat Transfer*, Part B, **33**, pp. 469-486, 1998.

[12] Nardini, D., Brebbia, C.A., A new approach to free vibration analysis using boundary elements, *In Boundary Element Methods in Engineering*, Computational Mechanics Publications, Southampton/Springer-Verlag, Berlin, 1982.

[13] Zhu, S.P., Satravaha, P., An efficient computational method for nonlinear transient heat conduction problems, *Applied Mathematical Modeling,* **20,** pp. 513-522, 1996.

[14] Satravaha, P., Zhu, S., An application of the LTDRM to transient diffusion problems with nonlinear material properties and nonlinear boundary conditions, *Applied Mathematics and Computation*, **87**, pp. 127-160, 1997.

[15] Guven, I., Madenci, E., Transient heat conduction analysis in a piecewise homogeneous domain by a coupled boundary and finite element method, *Int. J. for Numer. Meth. in Engng.*, **56**, pp. 351-380, 2003.

[16] Mansur, W.J., Soares, D, Ferro, M.A.C., Initial conditions in frequency-domain analysis: the FEM applied to the scalar wave equation, J*ournal of Sound and Vibration*, **270 (4-5)**, pp. 767-780, 2004.

[17] Mansur, W. J., Abreu, A. I., Carrer, J.A.M. Initial conditions contribution in frequency-domain BEM analysis, *Computer Modeling in Engineering & Sciences – CMES,* **6(1)**, pp. 31-42, 2004.

[18] Kausel, E., Roësset, J.M., Frequency domain analysis of undamped systems, *Journal of Engineering Mechanics, ASCE*, **118(4)**, pp. 721-734, 1992.

[19] Carslaw H.S., Jaeger J.C., *Conduction of heat in solids*, second edition, Oxford University Press, London, 1959.

The heat release rate of the fire predicted by sequential inverse method

W. S. Lee[1] & S. K. Lee[2]
[1]*Department of Mechanical Engineering, National Cheng Kung University, Taiwan*
[2]*Department of Safety, Health and Environmental Engineering, National Kaohsiung First University of Science and Technology, Taiwan*

Abstract

The sequential inverse method of locating and sizing accidental fire in a compartment is developed in this paper. The prediction procedure in this study includes two parts: the first is to simulate fire plume and ceiling jet flow by using the FDS model, then the result of temperature field at specified measurement sensors are utilized to predict the heat release rate and location of fire by the proposed method. A few examples are used to demonstrate the efficiency of the proposed method. The predictions show that the proposed method is capable of estimating the transient heat release rate of fire. The accuracy of the heat release rate of the fire is sensitive to random error and systematic error. Furthermore, the proposed method is noniterative and cost effective.
Keywords: sequential inverse method, Large Eddy Simulation, fire detection system.

1 Introduction

In the semiconductor manufacturing cleanroom, the processing facilities such as wet bench, diffusion furnace may exist significant fire risks. When accidental fire occurs in these areas the time before the fire is detected plays a crucial role in extinguishing fire. As a result, it is necessary to automatically monitor the working places and quickly determine the location of the fire in order to decrease the threat to life and property. The inverse heat conduction problem (IHCP) was used to estimate the surface heat flux, internal heat source and thermal properties from one or more measured temperature sensors inside a heat-conducting body

WIT Transactions on Modelling and Simulation, Vol 42, © 2006 WIT Press
www.witpress.com, ISSN 1743-355X (on-line)
doi:10.2495/BE06016

during the past two decades. Recently, the inverse method has been widely applied to solve many practical engineering problems [1]. Most studies utilized the numerical methods to determine the unknown conditions in the inverse problem [2, 3].

The solution of inverse problem can be divided into two steps. Firstly, the prediction of the transient temperature fields given the specified heat source by using numerical model is called the forward problem. And then, the minimization of the residuals between measured and predicted temperatures to determinate the location and heat release rate of fire. In order to investigate the buoyant plume and ceiling jet flow in detail, we adapted the FDS (Fire Dynamics Simulator) code [4] developed by Building and Fire Research Laboratory in National Institute of Standards and Technology (NIST) as research tool to find the solution of the forward problem. The purpose of this study is to develop the sequential inverse method in order to estimate the heat release rate and location of fire accurately and efficiently. The advantages of proposed inverse method are that no initial guesses and no iterations in the calculating process are required and the inverse problem can be solved in a linear domain [5].

2 Numerical model of Large Eddy Simulation

Fire Dynamics Simulator (FDS) is a 3D computational fluid dynamics model of fire-driven fluid flow. FDS solves numerically a form of the Navier-Stokes equations appropriate for low-speed, thermally-driven flow with an emphasis on smoke and heat transport from fires. FDS has been aimed at solving practical fire problems in fire protection engineering. The physical models included in FDS are described as follows:

2.1 Hydrodynamic model

An approximate form of the Navier-Stokes equations for low Mach number is used in this model. The approximation involves the filtering out of acoustic waves while allowing for large variations in temperature and density and small variation in pressure. This approximation proposed by Rehm and Baum [6] avoids the inefficient and inaccurate problems in most computational algorithms design for compressible flows at low Mach number. Turbulence is treated by means of the Smagorinsky form of Large Eddy Simulation (LES), or Direct Numerical Simulation (DNS) if the underlying numerical grid is fine enough. Details of the hydrodynamics and turbulence model are supplied in [5].

2.2 Combustion model

The mixture fraction-based combustion model is used in FDS code. The local heat release rate is based on Huggett's [7] relationship of oxygen consumption.

$$\dot{\mathbf{q}}''' = \Delta\mathbf{H}_0\dot{\mathbf{m}}_0''' \tag{1}$$

Here, $\Delta\mathbf{H}_0$ is the heat release rate per unit mass of oxygen consumed, $\dot{\mathbf{m}}_0'''$ is the oxygen mass burning rate (subscript o is denoted as oxygen, superscript ''' means production rate per unit volume) that can be calculated by transforming the oxygen mass conservation equation

$$\rho\frac{\mathbf{DY_0}}{\mathbf{Dt}} = \nabla\cdot\rho\mathbf{D}\nabla\mathbf{Y_0} + \dot{\mathbf{m}}_0''' \qquad (2)$$

into an expression for the local heat release rate using the conservation equation for the mixture fraction and the state relation for oxygen.

$$-\dot{\mathbf{m}}_0''' = \nabla\cdot(\rho\mathbf{D}\frac{\mathbf{dY_0}}{\mathbf{dZ}}\nabla\mathbf{Z}) - \frac{\mathbf{dY_0}}{\mathbf{dZ}}\nabla\cdot\rho\mathbf{D}\nabla\mathbf{Z} = \rho\mathbf{D}\frac{\mathbf{d^2Y_0}}{\mathbf{dZ^2}}\left|\nabla\mathbf{Z}\right|^2 \qquad (3)$$

the conservation law is showed as: $\rho\frac{DZ}{Dt} = \nabla\cdot\rho D\nabla Z$, and the ideal state relation for oxygen is introduced based on the assumption that the chemistry is fast so that fuel and oxidizer cannot co-exist.

$$\mathbf{Y_0(Z)} = \begin{cases}\mathbf{Y_0^\infty(1 - Z/Z_f)} & \mathbf{Z < Z_f} \\ 0 & \mathbf{Z > Z_f}\end{cases} \qquad (4)$$

where $\mathbf{Y_0}$ means mass fraction of oxygen, $\mathbf{Z}$ is mixture fraction, $\mathbf{Z_f}$ is flame surface, and $\mathbf{Y_0^\infty}$ is undepleted ambient value. In the numerical algorithm, the local heat release rate is computed by first locating the flame sheet, then computing the local heat release rate per unit area, and finally distributing this energy to the grid cells cut by the flame sheet.

2.3 Thermal Radiation model

The Radiation Transport Equation (RTE) for a non-scattering gray gas is

$$s\cdot\nabla I_\lambda(x,s) = \kappa_\lambda(x,\lambda)\left[I(x) - I(x,s)\right] \qquad (5)$$

where $I_\lambda(x,s)$ is the radiation intensity at wavelength λ, *I(x)* is overall radiation intensity, s is the direction vector of the intensity, $\kappa_\lambda(x,\lambda)$ is absorption coefficient. Due to the spectral dependence can not be solved easily, the radiation spectrum is divided into a relatively small number of bands in FDS code and the total intensity $I_\lambda(x,s)$ is calculated by summing over all the bands. The band specific RTE's are showed as follows:

$$s\cdot\nabla I_n(x,s) = \kappa_n(x)\left[I_{b,n}(x) - I(x,s)\right],\ n = 1\ \ldots N \qquad (6)$$

where $I_n(x,s)$ is the intensity integrated over the band n, and $\kappa_n(x)$ is the appropriate mean absorption coefficient inside the band. The radiant heat flux vector is defined as $q_r(x) = \int sI(x,s)\,d\Omega$, so that the radiative loss term in the energy equation could be expressed as:

$$-\nabla\cdot q_r(x) = \kappa(x)\left[U(x) - 4\pi I_b(x)\right];\ U(x) = \int_{4\pi} I(x,s)\,d\Omega \qquad (7)$$

The nct radiant energy gained by a grid cell is the difference between that which is absorbed and which is emitted. The source term κI_b is defined as

$$\kappa I_b = \begin{cases} \kappa \sigma T^4 / \pi & \textit{Outside flame zone} \\ \chi_r q''' / 4\pi & \textit{Inside flame zone} \end{cases} \tag{8}$$

Here, q''' is the chemical heat release rate per unit volume and χ_r is the local fraction of the energy emitted as thermal radiation.

2.4 Numerical method

In FDS, all spatial derivatives in the governing equations described in the above subsection are discretized by second order central differences on a rectilinear grid and the thermodynamic variables are updated in time using an explicit second order predictor-corrector scheme. The total pressure differential equation formed in Poisson equation is solved efficiently by a direct FFT (Fast Fourier Transform) solver. A more detailed description regarding numerical method could refer to other literature [8].

2.5 Grid test

It is important consideration to study the influence of the grid resolution on simulation results for any CFD analysis. The choice of grid size should reflect the impact of fire dimension and fire size. For a fire plume, the minimum length scale that must be resolved is the characteristic fire diameter D^*, that is $D^* = \left(\frac{\dot{Q}}{\rho_\infty c_p T_\infty \sqrt{g}} \right)^{2/5}$, where ρ_∞ is the ambient density, c_p is constant pressure specific heat, T_∞ is the ambient temperature, g is the acceleration of gravity. Here we adopt the resolution of fire plume simulation defined by Ma and Quintiere [9] as a dimensionless parameter, $R^* = \frac{\max(\delta x, \delta y, \delta z)}{D^*}$.

In this paper, a 20cm by 20cm square burner with 24kW steady heat release rate is used to verify the applicability of FDS in predicting the fire plume. The domain simulated is 1.5m on a side and 3.5m in height. All runs are done with uniform grid size in all three dimensions. The centerline temperature and velocity are averaged over 10s after the fire reached the approximate steady state. Figure 1 shows the time averaged velocity and temperature distribution above the fire source for different grid sizes. It indicates that the plume dynamics can only be accurately simulated and correlate well with McCaffery's correlations if the resolution limit of grid is about R*=0.1 or smaller. Therefore, the numerical experiments performed in this study obey this criterion in order to make the accurate results.

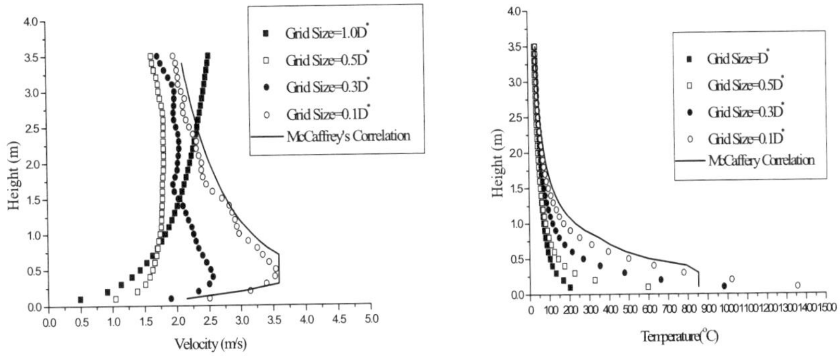

(a) (b)

Figure 1: (a) The time averaged velocity distribution for different grid sizes; (b) The time averaged temperature distribution for different grid sizes.

3 The inverse problem

The inverse problem in this paper is to identify the applied unknown heat release rate and location of fire from the temperature measurements. In general, the inverse method includes the process of analysis and the process of optimization. In this study, a generalized theory to predict gas velocities, gas temperatures of fire-driven ceiling jet flows developed by Alpert [10] and the implicit finite-difference method (FDM) are used to execute the analysis process. The discrete recursive equation can be expressed in the following matrix formulation:

$$[A]\{T_i\} = [B]\{Q_i\} + \{\eta_i\} \tag{9}$$

Matrix $[A]$ is the tridiagonal stiffness matrix of $\{T_i\}$, the components of [A] are composed of thermal properties, scale of spatial and temporal coordinate. The components of vector $\{T_i\}$ are the unknown temperature in the discretized points, matrix $[B]$ is the coefficient matrix of $\{Q_i\}$, and the components of matrix $[B]$ are determined by the boundary conditions and source terms. Vector $\{Q_i\}$ is unknown heat release rate, $\{\eta_i\}$ is a previous time temperature vector.

The inverse method is ill posed problem, that is, a slight measurement noise could induce the tremendous estimated error. In order to improve the instability property of the inverse method, the sequential method with the concept of future time [11] is utilized in this study. The sequential procedure is to assume temporarily that several future heat release rate of fire are constant. Thus, the unknown conditions in the future are assumed to be equal. The algebraic equations are reconstructed as the linear symbolic form so that this linear inverse model can lead to solving the estimated heat release rate of fire by using the linear least-squares error method. Therefore, the heat release rate of fire at a specific time, $t = t_i$, can be expressed as following equation:

$$\hat{\theta} = \left(\Phi^T \Phi\right)^{-1} \Phi^T \Gamma \tag{10}$$

where $\left(\Phi^T \Phi\right)^{-1} \Phi^T$ is the reverse matrix of the inverse problem and is denoted by $R_{m\times n}$. Matrix Γ is denoted as the measured temperature distribution, $\hat{\theta}$ is denoted as the estimated heat release rate, Φ is rearranged matrix from $[A]$, $[B]$, $\{\eta_i\}$.

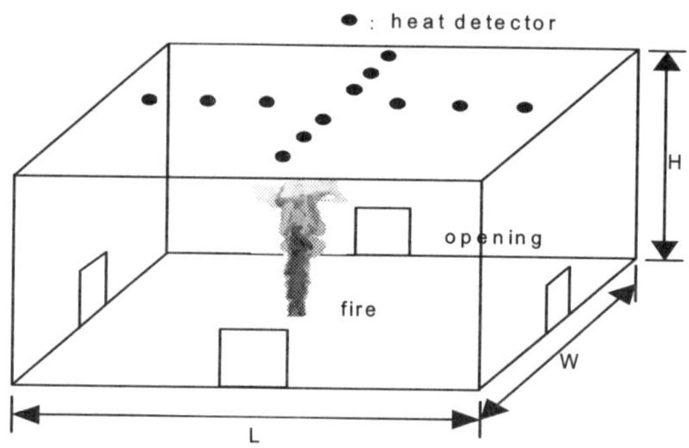

Figure 2: The schematic diagram of inverse fire detection problem.

4 Results and discussion

In this section, the authors conducted three examples to demonstrate the applicability of the proposed method in estimating the heat release rate and location of fire. The inverse fire detection problem is illustrated in fig. 2. The domain simulated is 10m on a side and 3m in height. Twelve temperature sensors placed at different locations near the ceiling are used to measure the ceiling jet temperature. In order to consider the effect of the sensor's noise, random errors of measurement are added to the exact temperature. It can be shown in the following equation:

$$Y_i = T_i(1 + \omega\sigma) \tag{11}$$

where T_i is the exact temperature and Y_i is the measured temperature at specific location respectively, σ is the measurement error, and ω is the random number between –1 and 1. In our studied cases the authors discuss the performance of the sequential method with future time as measurement errors are included. The detailed descriptions for three examples are shown as follows:

4.1 Case 1 the estimation of constant heat release rate of fire

In order to verify the performance of the proposed method, we assumed the unknown heat release rate of fire as a constant value 200kW in case 1. The calculated time domain is from 0 to 180 seconds with 1.0 second increment. The heat release rate of fire in the solution of forward problem is shown in fig. 3. Figure 4 shows the estimated heat release rate with future time r = 3 as the measurement error is 0.5%. Note that the algorithm of sequential method is to assume temporally that the several future conditions are constant with time. In the case of future time r =3, the undetermined conditions Q_i, Q_{i+1}, Q_{i+2} are assumed to be equal. As shown in fig. 4, the time lag phenomena between predicted value and exact heat release rate are revealed at the early stage of the simulation result. Furthermore, the simulation results also indicate that the oscillation of predicted value is tremendous in the first 25 seconds. But the estimation of HRR distribution becomes more stable after 25 seconds. The average relative error in this case is 1.5%. Therefore, the heat release rate of fire obtained by employing the sequential method with r = 3 is in good agreement with the exact value. It should be noted that in order to study the sensitivity of proposed inverse method on measurement error, the random errors of measurement are added to the temperature computed from the solution of direct problem, i.e. $Y_i = T_i(1+\omega\sigma)$. The measurement errors are caused by the factors due to calibration, fluctuation in sensor reading during measurement. The average relative error stated in discussion of Figure 4 means the average of heat release rate estimation error. The reason why we use the average relative error is that the estimated HRR value is very sensitive to the random error in the transient temperature measurements and systematic error in the forward problem solution, so that it is meaningless to judge the heat release rate and fire location by one set of measuring data only.

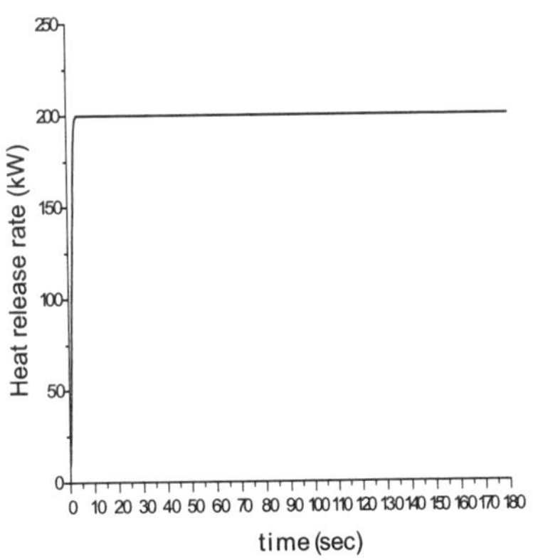

Figure 3: The heat release rate of fire in the solution of forward problem.

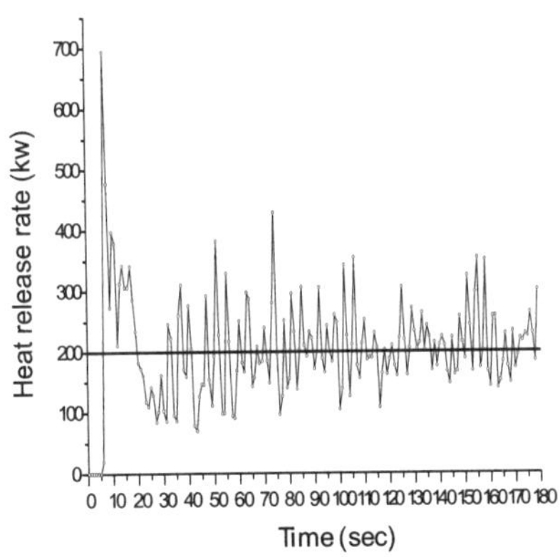

Figure 4: The estimation of HRR in Case 1 (measurement error=0.5%).

4.2 Case 2 the estimation of a sudden variation of heat release rate and location of fire by using two measurement points

The unknown heat release rate of fire is expressed in the following form:

$$HRR = 100kW \quad 0\,\text{sec} \le t < 120\,\text{sec}$$
$$HRR = 200kW \quad 120\,\text{sec} \le t < 240\,\text{sec}$$
$$HRR = 100kW \quad 240\,\text{sec} \le t < 360\,\text{sec}$$

The calculated time domain is from 0 to 360 seconds with 1.0 second increment. The location of fire source is predicted by using two measure points in case 2. Figure 5 shows the estimate of HRR for case 2. The simulation result indicated that the estimation of HRR gets worse at t=120 to 240 second due to the abrupt change of HRR. The average relative discrepancies are 11.3%, 0.3% and 31.5% at three periods, respectively. The estimations of the distance between the fire location and measuring point at two different sensors are shown in figs. 6 and 7, respectively. According to the estimations of the distance between the fire location and measuring point at every sensor, the location of fire could be determined by using the least square method. The estimation of x-coordinate of fire source in this case is shown in fig. 8. The average relative error of estimated location of fire is 1.27%.

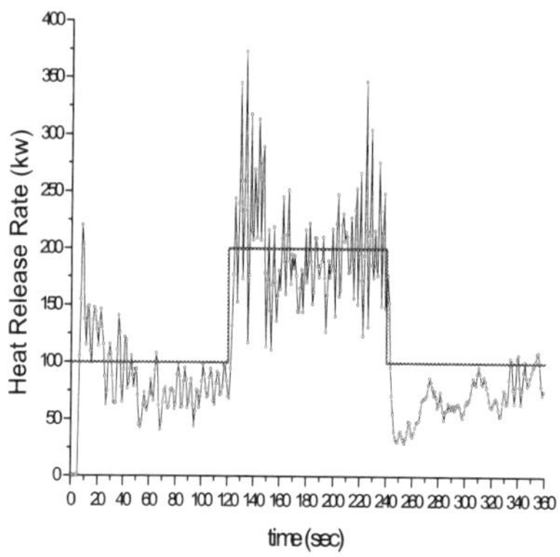

Figure 5: The estimation of HRR in Case 2 (measurement error=0.5%).

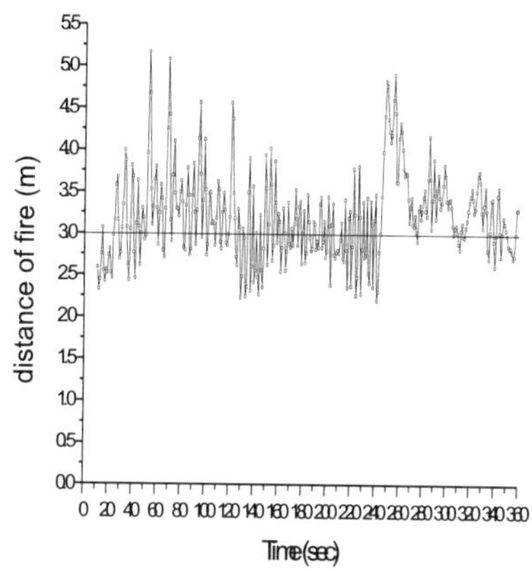

Figure 6: The estimation of distance between fire source and measuring point in Case 2 (measurement point located at X=2, Y=5).

4.3 Case 3 the estimation of a sudden variation of heat release rate and location of fire by using 12 measurement points

Case 3 is the same as above case, except that 12 measurement sensors are employed to estimate the location of fire. The estimation of fire location is shown in fig. 9. The simulation results indicated that the number of sensors is more than that of above case, resulting in a better estimated result. The estimated HRR value is very sensitive to the random error in the transient temperature measurements and systematic error in the forward problem solution. The sensitivity depends on the position of measuring points, the measurement error and the number of measuring sensors.

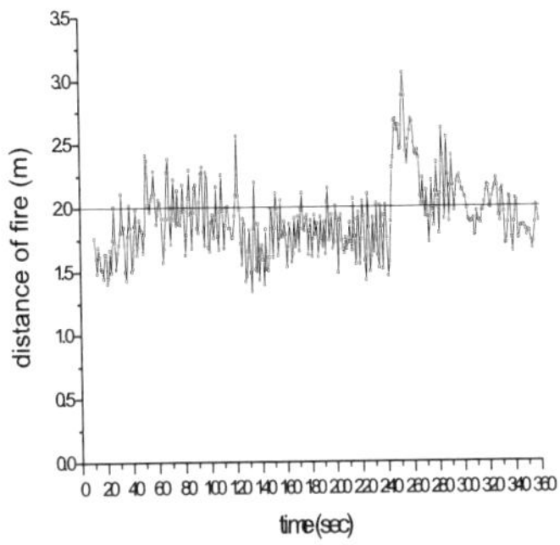

Figure 7: The estimation of distance between fire source and measuring point in Case 2 (measurement point located at X=3, Y=5).

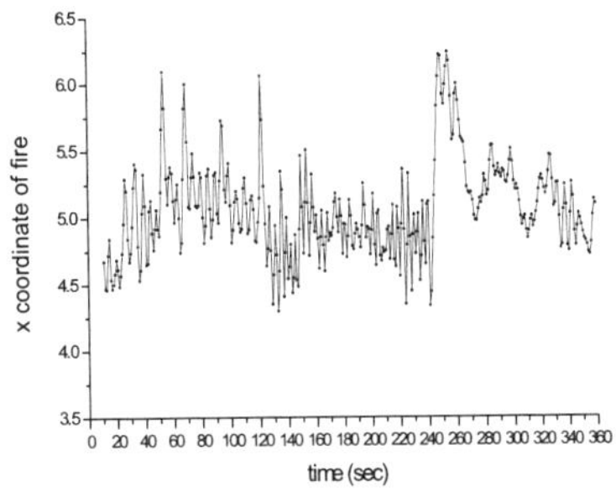

Figure 8: The estimation of x-coordinate of fire by using two measurement sensors.

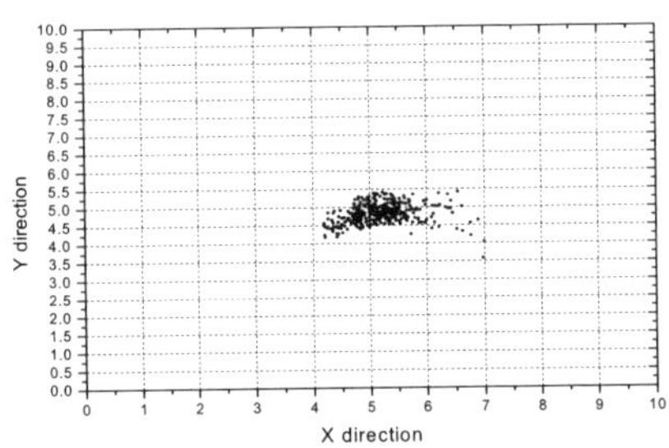

Figure 9: The estimation of fire location by using 12 measurement sensors.

5 Conclusions

This paper has presented the inverse method to estimate the heat release rate and location of fire. The advantage of using inverse method to predict the fire heat release rate is that only few temperature data measured at some locations are required. The simulation results in this paper show that the measurement random error and systematic error in the forward problem solution have significant effect on estimation of the heat release rate and location of fire in the inverse method. The sequential inverse method with future time concept proposed in this paper could predict the heat release rate and location of fire efficiently. Therefore, from a fire safety engineer's point of view, the inverse method can get more reliable and rapid results than the traditional approach.

Acknowledgement

The current authors gratefully acknowledge the financial support provided to this study by the National Science Council of the Republic of China under Grant No. NSC94-2212-E006-038.

References

[1] Chu Y. H., Hsu P. T. & Liu Y. H., A two-dimensional inverse problem in estimating the heat generation rate in an electronic device. *Journal of the Chinese Society of Mechanical Engineers*, **23(6)**, pp. 567-576, 2002.

[2] Yang C.Y., A sequential method to estimate the strength of the heat source based on symbolic computation. *Int. J. Heat Mass Transfer*, **41(14),** pp. 2245-2252, 1998.

[3] Niliot C. L., Rigollet F. & Petit D., An experimental identification of line heat sources in a diffusive system using the boundary element method. *Int. J. Heat Mass Transfer*, **43(12),** pp. 2205-2220, 2000.

[4] McGrattan, K. B., Baum, H. R., Rehm, R. G., Hamis A. & Forney, G. P., *Fire Dynamics Simulator – User's Manual*, NISTIR 6469, National Institute of Standards and Technology, Gaithesburg, MD, 2002.

[5] Lin J. H., Chen C. K. & Yang Y. T., An Inverse method for simultaneous estimation of the center and surface thermal behavior of a heated cylinder normal to a turbulent air stream. *ASME Journal of Heat Transfer*, **124**, pp. 601-608, 2002.

[6] Rehm, R. G. & Baum, H. R., The equations of motion for thermally driven. *Journal of Research of the NBS*, **83,** pp. 297-308, 1978.

[7] Huggett C., Estimation of the rate of heat release by means of oxygen consumption measurements. *Fire Material*, **4,** pp. 61-65, 1980.

[8] Zhang W., Hamer A., Klassen M., Carpenter D. & Roby R., Turbulence statistics in a fire room model by large eddy simulation. *Fire Safety Journal*, **37,** pp. 721-752, 2002.

[9] Ma T. G. & Quintiere J. G., Numerical simulation of axi-symmetric fire plumes: accuracy and limitations. *Fire Safety Journal*, **38**, pp. 467-492, 2003.

[10] Alpert R. L., Calculations of response time of ceiling mounted fire detectors, *Fire Technology*, **8(3)**, pp. 181-195, 1972.

[11] Yang C. Y., Inverse estimation of mix-typed boundary conditions in heat conduction problems. *Journal of thermophysics and heat transfer*, **12(4),** pp. 552-561, 1998.

Section 4
Electrical engineering and electromagnetics

Computation of maximal electric field value generated by a power substation

N. Kovač, D. Poljak, S. Kraljević & B. Jajac
Split University, Croatia

Abstract

A procedure for a computation of the maximal value of extremely low frequency (ELF) electric field from a power substation is proposed. The present technique is based on a multiquadric approximation of the electric field. The approximation is obtained using discrete field values calculated by the Source Element Method (SEM) representing a variant of Indirect Boundary Element Method (IBEM). The approximation sufficiently handles multidimensional multiextreme functions by interpolating their discrete values accurately. Subsequently, the maximal electric field value is evaluated by minimizing the negative multiquadric approximation via the stochastic optimization method – differential evolution. Therefore, the procedure provides the maximal field value assessment on the basis of a limited number of computed discrete values, thus reducing the computational effort.

1 Introduction

A computation of the maximal value of extremely low frequency (ELF) electric field from a power substation has been of great interest in last decades. The reason for this is the increasing public concern regarding the possible health risk for human exposure to ELF electric field and magnetic flux density. It is to be noticed that the present piece of work has been focused only toward the technical aspects of electric field assessment. The ELF electric field from substations has been determined via the computations or/and measurements in [1–6], presenting a number of three-dimensional as well as contour plots. Nevertheless, they provided a useful description of the field characteristics, no technique incorporating certain optimization method for assessing the maximal electric field value has been applied in these papers. The spatial electric field function

WIT Transactions on Modelling and Simulation, Vol 42, © 2006 WIT Press
www.witpress.com, ISSN 1743-355X (on-line)
doi:10.2495/BE06017

may have a number of local extremes arising from the fact that substation represents a concentration of electric field sources.

Consequently, an application of a stochastic optimization method – differential evolution [7–9], which is in the scope of this work, could be a useful tool for the maximal field value assessment. The stochastic optimization method shows the well-known advantages over the gradient-based methods, namely: (i) the parallel global minimum search technique, (ii) the simplified set-up of the optimization task and (iii) the ability to find the global minimum. The optimization procedure is carried out using a multiquadric approximation [10, 11] of electric field, since the computed values are distributed in a set of discrete points through the volume of interest. The approximation based on the radial basis functions is capable to handle multidimensional multiminima functions by interpolating their discrete data accurately.

2 Electric field computation in a set of discrete points

The spatial distribution of electric field can be assessed using the source integration procedure featuring the Charge Simulation Method (CSM) or the Source Element Method (SEM). The approach has been already utilized for the electric field computation in the power substation environment [1–3] as well as in the modeling of a metallic post protection zone [12]. It can be referred to as a variant of the Indirect Boundary Element Method (IBEM).

Electric potential of a point *P(x,z)* due to a charged straight wire considered as a segment of a conductor (electrode), Figure 1, is given by [12]:

$$\varphi(x,z) = \frac{1}{4\pi\varepsilon}\int_{-L}^{+L}\frac{\rho_l\,(x')dx'}{R}, \tag{1}$$

where: ρ_l denotes the line charge density,

2L stands for the wire length,

R is the distance between a point on the straight wire and an arbitrary point *P*.

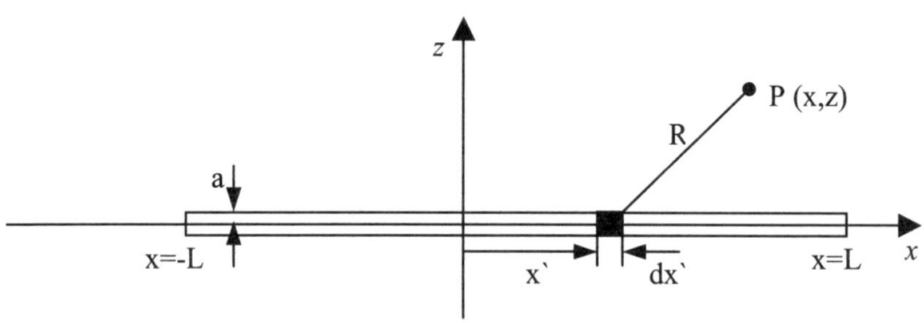

Figure 1: Straight wire geometry.

If the potential function φ_l along the straight wire is known, the integral equation (1) can be written as:

$$\varphi_l = \frac{1}{4\pi\varepsilon} \int_{-L}^{+L} \frac{\rho_l(x')dx'}{|x - x'|}. \tag{2}$$

Obviously, there is a singularity for $x=x'$. The problem can be overcome by assuming a finite wire radius a, i.e. it follows:

$$\varphi_l = \frac{1}{4\pi\varepsilon} \int_{-L}^{+L} \frac{\rho_l(x')dx'}{\sqrt{(x - x')^2 + a^2}}, \tag{3}$$

According to the standard Source Element Method (SEM) procedure, the conductors are divided into a number of segments, i.e. boundary elements [1]-[3]. Having performed a discretization of the substation conductors a system of equations for unknown charges along the each segment is obtained. Therefore, the integral equation (3) transforms into a corresponding matrix equation [1]:

$$\begin{bmatrix} P_{11} & P_{12} & \cdot & \cdot & \cdot & P_{1n} \\ P_{21} & P_{22} & \cdot & \cdot & \cdot & P_{2n} \\ \cdot & \cdot & & & & \\ \cdot & \cdot & & & & \\ \cdot & \cdot & & & & \\ P_{n1} & P_{n2} & \cdot & \cdot & \cdot & P_{nn} \end{bmatrix} \cdot \begin{bmatrix} q_1 \\ q_2 \\ \cdot \\ \cdot \\ \cdot \\ q_n \end{bmatrix} = \begin{bmatrix} \varphi_1 \\ \varphi_2 \\ \cdot \\ \cdot \\ \cdot \\ \varphi_n \end{bmatrix}, \tag{4}$$

where: $\varphi_1, \varphi_2, \dots, \varphi_n$ stand for the boundary element potentials,
$q_1, q_2, \dots, q_n$ denote the boundary element charges,
$P_{11}, P_{12}, \dots, P_{nn}$ are Maxwell coefficients.

The unknown charges can be calculated by inverting the Maxwell coefficient matrix. The details regarding the Maxwell coefficients can be found in [1].

The electric field components in Cartesian system at an arbitrary point *(x,y,z)*, produced by *i-th* boundary element, Figure 2, can be computed by means of the equations related to the finite length wire [1]:

$$E_x = -\frac{\partial \varphi}{\partial x} = \frac{q_i}{4\pi\varepsilon_0 L_i} \left[\frac{1}{\sqrt{(L_i - x)^2 + W^2}} - \frac{1}{\sqrt{x^2 + W^2}} \right], \tag{5a}$$

$$E_y = -\frac{\partial \varphi}{\partial y} = \frac{q_i}{4\pi\varepsilon_0 L_i} \frac{y}{W^2} \left[\frac{L_i - x}{\sqrt{(L_i - x)^2 + W^2}} + \frac{x}{\sqrt{x^2 + W^2}} \right], \tag{5b}$$

$$E_z = -\frac{\partial \varphi}{\partial z} = \frac{q_i}{4\pi\varepsilon_0 L_i}\frac{z}{W^2}\left[\frac{L_i - x}{\sqrt{(L_i - x)^2 + W^2}} + \frac{x}{\sqrt{x^2 + W^2}}\right], \tag{5c}$$

where: q_i denotes the charge of *i-th* segment,
L_i stands for the length of *i-th* segment, and

$$W^2 = y^2 + z^2. \tag{6}$$

The total field components are assembled from each boundary element.

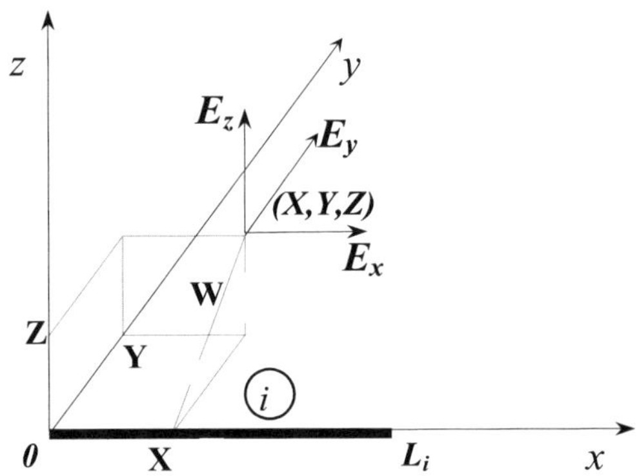

Figure 2: Electric field components produced by *i-th* segment.

3 Maximal field value assessment

The procedure for assessing the maximal electric field value generated by a power substation is outlined in a few steps: (A) multiquadric approximation of electric field, (B) minimization of objective function.

3.1 Multiquadric approximation

The vector of independent parameters, represented by the space co-ordinates, can be written by:

$$\mathbf{P} = \begin{bmatrix} x & y & z \end{bmatrix}^t. \tag{7}$$

The parameters are generally subjected to both inequality and equality constraints:

$$g_m(\mathbf{P}) \geq 0, m = 1,2,\dots,n_{t1}, \tag{8}$$

$$h_m(\mathbf{P}) = 0, m = n_{t1}+1, n_{t1}+2,\dots, n_{t1}+n_{t2}, \tag{9}$$

where n_{t1}, n_{t2} are the total numbers of inequality and equality constraints, respectively.

Applying the procedure, a vector related with the maximal electric field value is assessed, fulfilling the imposed constraints. The computation of discrete field values would require rather high effort, if applied to a large number of parameter vectors. Hence, the procedure of the maximum assessment is carried out using a multiquadric approximation [10, 11] of electric field.
The approximation is determined by an expression:

$$E_{MQ}(\mathbf{P}) = \sum_{j=1}^{M_p} c_j r(\mathbf{P})_j \, , \tag{10}$$

where: M_p is the total number of sample points $\mathbf{P}_j$ having the computed values $Ej(\mathbf{P}_j)$,
c_j stands for the approximation coefficients, while

$$r(\mathbf{P})_j = \sqrt{\left\|\mathbf{P} - \mathbf{P}_j\right\|^2 + h_z}\, , \; j = 1, 2, ..., M_p \tag{11}$$

denotes radial basis functions, where $\left\|\mathbf{P} - \mathbf{P}_j\right\|$ is the Euclidean norm and h_z is the shift parameter.
The value of M_p should be a compromise between opposite requirements: the higher accuracy of the multiquadric approximation and the lower computational effort. The shift parameter h_z can be determined by means of a few additional samples used for the error estimation.

The unknown coefficients c_j are calculated from the matrix equation:

$$\mathbf{c} = \mathbf{r}^{-1} \cdot \mathbf{E}\, , \tag{12}$$

where:

$$r_{ij} = \sqrt{\left\|\mathbf{P}_i - \mathbf{P}_j\right\|^2 + h_z}\, , \; i, j = 1, 2, ..., M_p\, , \tag{13}$$

$$\mathbf{E} = [\, E_1(\mathbf{P}_1) \;\; E_2(\mathbf{P}_2) \;\; \cdots \;\; E_{M_p}(\mathbf{P}_{M_p})\,]^t. \tag{14}$$

Hence, the discrete electric field data, obtained by the SEM computation, are related with M_p sample points only, thus reducing the required computational effort of the optimization procedure.

3.2 Minimization of objective function

Minimization of the objective function $-E_{MQ}(\mathbf{P})$, resulting in the maximal electric field value assessed, is formulated by:

$$min(-E_{MQ}(\mathbf{P}))\, , \tag{15}$$

In order to perform the minimization procedure, i. e. to find the global minimum, a stochastic optimization method – differential evolution (DE) is applied [7]-[9].

A constant number n_p of parameter vectors, representing members of a population, is used in each generation G:

$$\mathbf{P}_{i,G}, i = 0,1,\ldots,n_p - 1. \tag{16}$$

The population of the first generation is chosen randomly via *rand* function. There are several variants of DE algorithms. The one, using the best population vector to generate a new population member, is selected as a part of the proposed procedure. Hence, for each $\mathbf{P}_{i,G}$ there is the corresponding vector:

$$\mathbf{V}_{i,G+1} = \mathbf{P}_{best,G} + F(\mathbf{P}_{r_1,G} - \mathbf{P}_{r_2,G}), \tag{17}$$

where: $\mathbf{P}_{best,G}$ is the member having the lowest objective function value of the generation G,
r_1, r_2 denote randomly adopted integers from $[0,n_p-1]$,
F is a real factor controlling the amplification of a weighted difference.

To enhance a new population diversity, the crossover of the vectors $\mathbf{V}_{i,G+1}$ and $\mathbf{P}_{i,G}$ is introduced, by which the new parameter vector:

$$\mathbf{U}_{i,G+1} = [u_{0i,G+1} \; u_{1i,G+1} \; \ldots \; u_{(n-1)i,G+1}]^t \tag{18}$$

is obtained as:

$$\mathbf{U}_{i,G+1} = \begin{cases} v_{ji,G+1} & for \;\; j = \langle k_n \rangle_n, \langle k_n + 1 \rangle_n, \ldots, \langle k_n + L_n - 1 \rangle_n \\ p_{ji,G} & for\ all\ other\ j \in [0, n-1] \end{cases}, \tag{19}$$

where: n is the total number of independent parameters,
$\langle\, \rangle_n$ stands for the modulo function with modulus n,
k_n denotes the randomly chosen integer from $[0,n-1]$,
L_n is the number of exchanged parameters from $[1,n]$.

The evaluation algorithm of L_n is given by the following pseudo-code lines:

```
L_n = 0;
do {
      L_n = L_n + 1
} while (rand () < prob_C) and (L_n < n));
```

where $prob_C$ denotes the crossover probability used as a control variable.

If the resulting vector $\mathbf{U}_{i,G+1}$ gives the objective function value lower than the one corresponding to the vector $\mathbf{P}_{i,G}$, and if $\mathbf{U}_{i,G+1}$ fulfills the constraints, as well, it replaces $\mathbf{P}_{i,G}$ being a population member of the generation G+1: $\mathbf{P}_{i,G+1} = \mathbf{U}_{i,G+1}$. Otherwise, $\mathbf{P}_{i,G}$ is retained as a member of the generation G+1.

The vector, related to the maximal value, equals the best vector assessed upon the prescribed number of generations:

$$\mathbf{P}_{max} = \mathbf{P}_{best,G_l}, \tag{20}$$

where G_l is the number of the last generation.
The corresponding maximal electric field value from a power substation is then given by:

$$E_{MQ}^{\max}(\mathbf{P}_{\max}) = \sum_{j=1}^{M_p} c_j r(\mathbf{P}_{\max})_j \ . \tag{21}$$

4 Computational results

The procedure presented so far is illustrated by an example of *110/10 kV/kV* transmission substation of GIS (Gas-Insulated Substation) type in Split, Croatia. All the electric field levels ($f = 50\ Hz$) are computed at height $z = 1\ m$ above ground. There is no real possibility of a public exposure within the transmission substation, while a professional exposure is strictly limited to duration. Consequently, the computations are performed outside the fence of the substation, only. A simplified two-dimensional layout of the substation is shown in Figure 3.

It is to be underlined that the equipment having grounded shields (power cables), sheaths (GIS buses), or metallic casings (transformers, switchgears) can be neglected due to the shielding effect [5]. Namely, the metallic enclosures are connected to ground, thus producing negligible electric field strength. Consequently, the significant electric field sources regarding the substation considered are overhead transmission lines, as well as the unshielded conductors of both the high and middle voltage level. The corresponding domains, where the greater electric field intensities or magnetic flux density values are expected, can be seen in Figure 3. The areas assigned as 3 and 4 are the important ones for the magnetic flux density computation only, which is not within the scope of this work. The multiquadric approximations of the electric field distributions concerning the domains assigned as 1 and 2 are shown in Figures 4 and 5, respectively. The dependencies refer to *ABC-CBA* phase arrangement resulting in the greatest electric field values.

The field intensity increases by approaching the overhead line route, as shown in Figure 4. Moreover, there is a local intensity increase in the vicinity of the transformer due to the unshielded conductors connecting the transformer to the GIS buses. However, for the shortness of unshielded conductors, as well as their considerable distances from the substation fence, the main electric field sources are the overhead lines. This statement is confirmed by the field levels in domain 2, Figure 5. Minimizing the negative multiquadric approximation, the vector $\mathbf{P}_{opt} = [55.35 \quad -7.84 \quad 1]^t$, resulting in the maximal electric field intensity $E_{max} = 380.70\ V/m$, is obtained. The maximum point is located just below the overhead line route. The field distribution over the area No. 5 is not presented, since no field value exceeds *32 V/m*. The low field values are obtained as the main field sources (overhead lines) are located rather far away from the domain No. 5.

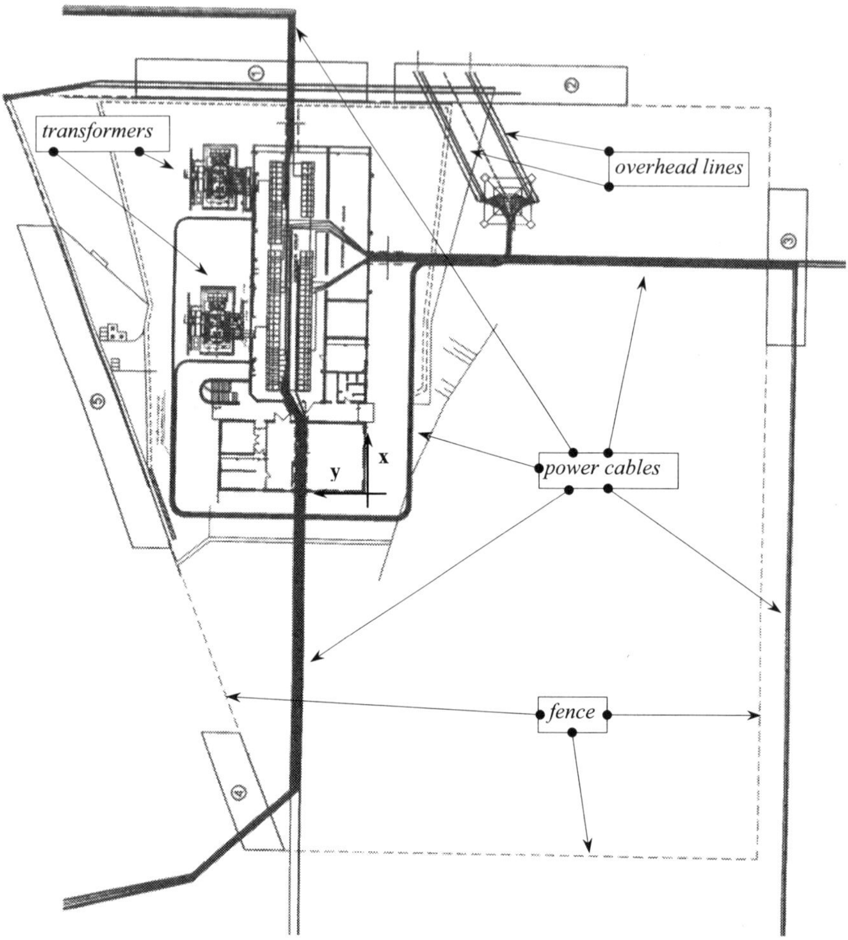

Figure 3: Substation layout.

5 Concluding remarks

An efficient technique for the assessment of maximal ELF electric field value from a power substation is presented in this work. The procedure is based on the multiquadric approximation of electric field which is obtained using discrete field values. The values are calculated by the Source Element Method (SEM) representing a variant of Indirect Boundary Element Method (IBEM). The approximation is capable to handle multidimensional multiextreme functions by accurate interpolating their discrete values. Subsequently, the maximal value is assessed by minimizing the negative multiquadric approximation via the stochastic optimization method – differential evolution. Therefore, the presented procedure provides the maximal field value assessment on the basis of a limited number of calculated discrete values, thus reducing the computational cost.

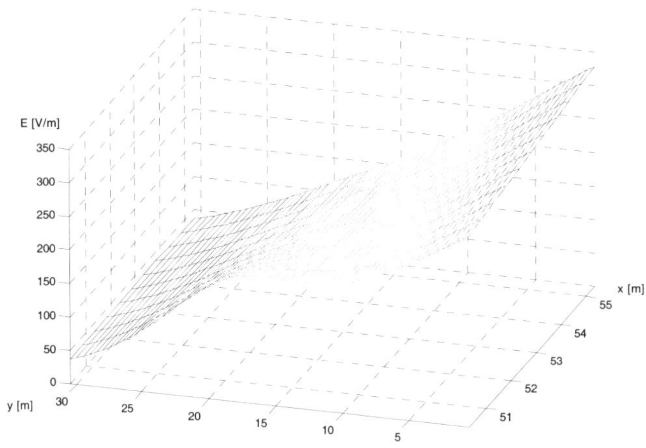

Figure 4: Multiquadric approximation of electric field in the domain No. 1.

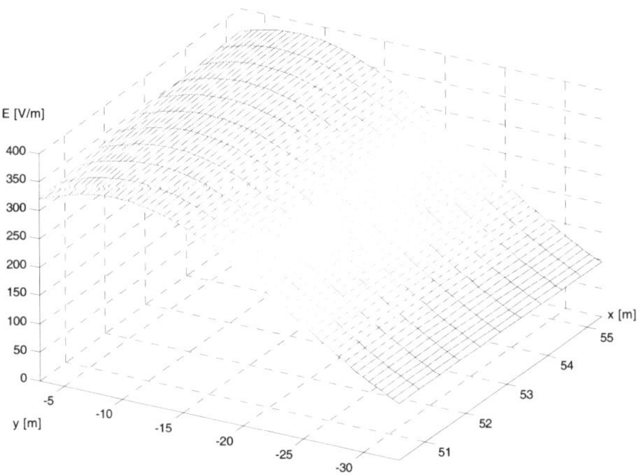

Figure 5: Multiquadric approximation of electric field in the domain No. 2.

References

[1] J.E.T. Villas, F.C. Maia, D. Mukhedkar, V.S. Da Costa: Computation of Electric Fields Using Ground Grid Performance Equations, *IEEE Trans. on Power Delivery*, **2(3)**, pp. 709-716, July 1987.

[2] S.H. Myung, B.Y. Lee, J.K. Park: Three Dimensional Electric Field Analysis of Substation Using Nonuniform Optimal Charge Simulation, *9th Int. Symposium on High Voltage Engineering,* Graz Austria, August 1995.

[3] B.Y. Lee, J.K. Park, S.H. Myung, S.W. Min, E.S. Kim: An Effective Modelling Method to Analyze Electric Field around Transmission Lines and Substations Using a Generalized Finite Line Charge, *IEEE Trans. on Power Delivery*, **12(3)**, pp. 1143-1150, July 1997.

[4] S. A. Sebo, R. Caldecott: Scale Model Studies Of AC Substation Electric Fields, *IEEE Trans. on Power Apparatus and Systems,* **98(3)**, pp. 926-939, May/June 1979.

[5] P. S. Wong, T. M. Rind, S. M. Harvey, R. R. Scheer: "Power Frequency Electric and Magnetic Fields from 230 kV Gas-insulated Substation", *IEEE Trans. on Power Delivery*, **9(3)**, pp. 1494-1501, July 1994.

[6] A. S. Safigianni, C. G. Tsompanidou: "Measurements of Electric and Magnetic Fields due to the Operation of Indoor Power Distribution Substations", *IEEE Trans. on Power Delivery*, **20(3)**, pp. 1800-1805, July 2005.

[7] R. Storn, K. Price, 'Differential Evolution – A Simple and Efficient Adaptive Scheme for Global Optimization over Continuous Spaces, Technical Report TR-95-012, ICSI, Berkeley, USA, 1995.

[8] R. Storn, 'On the Usage of Differential Evolution for Function Optimization', NAFIPS, Berkeley, pp. 519-523, 1996.

[9] R. Storn, K. Price, 'Minimising the Real Functions of the ICEC'96 Contest by Differential Evolution', *IEEE Conference on Evolutionary Computation*, Nagoya, pp. 842-844, 1996.

[10] P. Alotto, A. Caiti, G. Molinari, M. Repetto, 'A Multiquadrics-based Algorithm for the Acceleration of Simulated Annealing Optimization Procedures, *IEEE Trans. on Magnetics*, **32(3)**, pp. 1198-1201, 1996.

[11] P. Alotto, M. Gaggero, G. Molinari, M. Nervi, 'A Design of Experiment and Statistical Approach to Enhance the Generalised Response Surface Method in the Optimization of Multiminima Problems, *IEEE Trans. on Magnetics*, **33(2)**, pp. 1896-1899, 1997.

[12] D. Poljak, C. A. Brebbia: *Boundary Elements for Electrical Engineers,* WIT Press, Southampton-Boston, 2005.

Transient analysis of coated thin wire antennas in free space via the Galerkin-Bubnov indirect Boundary Element Method

D. Poljak[1] & C. A. Brebbia[2]
[1]*Department of Electronics, University of Split, Croatia*
[2]*Wessex Institute of Technology, Southampton, UK*

Abstract

This paper deals with a time domain analysis of dielectric coated wire antennas in free space. The formulation is based on the Hallen integral equation for loaded wires having a dielectric coating. The coating has been taken into account by means of an equivalent magnetic coating load term. The numerical solution has been carried out via the time domain variant of Galerkin-Bubnov indirect Boundary Element Method (GB-IBEM).

1 Introduction

Dielectric coated antennas are usually used for packaging or for the isolation of wire antennas [1]. Using such antennas the undesirable contact between them and the surrounding medium is avoided and the radiation properties of these antennas can be also modified appreciably [1, 2].

An important application of coated antennas is also the treatment of cancer by heating the malignant tissue. This tissue heating must be localized to maintain the temperature within the tumor tissue up to 43°C, for a given time period, while the neighbouring tissue temperature level must be kept far below 43°C [2]. Dielectric coated wire antennas have been analyzed in the frequency domain in a number of papers, e.g. [2–5].

The approximate formulas for the near field generated by an insulated dipole antenna immersed in a lossy medium have been proposed by King et al. in [3] while a more rigorous Method of Moments (MoM) approach to the near field analysis, based on the solution of the corresponding Pocklington equation by using piecewise sinusoids have been presented in [2].

WIT Transactions on Modelling and Simulation, Vol 42, © 2006 WIT Press
www.witpress.com, ISSN 1743-355X (on-line)
doi:10.2495/BE06018

A simplified bare loaded antenna concept for the treatment of wires having dielectric, magnetic or fixed coatings has been proposed in [4]. The indirect Galerkin Bubnov boundary element modelling of the dielectric coated antennas featuring the formulation presented in [4] has been proposed in [5]. More details on modelling with the frequency domain Galerkin Bubnov indirect Boundary Element method (GB-IBEM) can be found in [6] and [7].

To the best of our knowledge, one of the rather rare papers dealing with transient analysis of dielectric coated wire antennas is the one published by Bretones et al. in 1994 [1]. In that paper Pocklington integral equation approach is applied to the solution of this problem featuring the MoM and marching-on-in-time procedure. As the Pocklington integral equation approach generally suffers from numerical instability the Hallen integral equation is usually used to overcome this problem. Basically, the apparent advantage of the Hallén approach arises from the fact it contains neither space nor time derivatives which are proved to be the origin of the numerical instabilities. More details on Hallen integral equation modeling can be found in [8–10].

This paper deals with the transient analysis of dielectric coated wire antennas utilizing the Hallen integral equation approach. Dielectric coating is modelled via an equivalent magnetic coating in a similar manner as presented in [4] and [5], namely the influence of dielectric coating can be taken into account via an additional impedance term. The transient current along the coated antenna is obtained by solving the Hallen equation applying the time domain version of the Galerkin Bubnov indirect Boundary Element Method (GB-IBEM).

2 The space-time integral equation for a thin dielectric coated wire antenna

A straight thin wire antenna of length $2h$ and radius a having a dielectric coating of thickness $(b\text{-}a)$ with dielectric constant ε_r, shown in Figure 1 is considered.

The wire is excited by a time-dependent voltage source. The derivation of the corresponding Hallen integral equation starts with the continuity conditions for the tangential electric field components at the wire antenna surface in the Laplace domain:

$$E_z^{inc}(z,s)+E_z^{sct}(z,s)=I(z,s)Z_L(s) \tag{1}$$

where E_z^{inc} is the incident field, E_z^{sct} is the field scattered by the coated wire and Z_L is the general type loading per unit length of the wire, i.e. the corresponding impedance per unit length, and s denotes the Laplace variable.

Applying the straightforward convolution to the equation (1) results in its time domain counterpart:

$$E_z^{inc}(z,t)+E_z^{sct}(z,t)=\int_0^t Z(\tau)I(z,t-\tau)d\tau \tag{2}$$

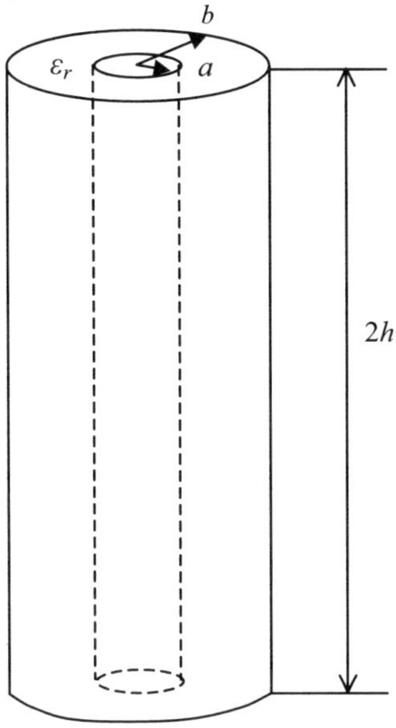

Figure 1: A straight thin wire antenna with dielectric coating of thickness (b-a).

Dielectric coating can be taken into account through the load impedance term, i.e. the dielectric coating can be expressed by means of a purely magnetic coating [4], [5]. Thus the load impedance has a magnetic character and it can be written as:

$$Z_L(s) = sL \tag{3}$$

where L is the equivalent inductance per unit length of the wire given by:

$$L = \frac{\mu_0}{2\pi}\frac{\varepsilon_r - 1}{\varepsilon_r}\ln\frac{b}{a} \tag{4}$$

The right-hand side term of equation (1) can be written as:

$$Z_L(s)I(z,s) = sLI(z,s) \tag{5}$$

and performing the Laplace transform to equation (5) yields:

$$\int_0^t Z(\tau)I(z,t-\tau)d\tau = L\frac{\partial I(z,t)}{\partial t} \tag{6}$$

The scattered electric field vectore can be expressed in terms of the vector and scalar potentials:

$$\vec{E}^{\text{sct}} = -\left(\frac{\partial \vec{A}}{\partial t} + \nabla \varphi\right) \tag{7}$$

The magnetic vector potential is determined by the relation:

$$\vec{A} = \frac{\mu}{4\pi} \iint_{S'} \frac{\vec{J}_s(\vec{r}\,', t - R/c)}{R} dS' \tag{8}$$

where $\vec{r}\,'$ denotes the distance co-ordinate of the source point.
The electric scalar potential is given by:

$$\varphi = \frac{1}{4\pi\varepsilon} \iint_{S'} \frac{\rho_s(\vec{r}\,', t - R/c)}{R} dS' \tag{9}$$

where ρ_s and $\vec{J}_s$ are the surface charge and surface current densities, respectively, satisfying the continuity equation:

$$\nabla_s \vec{J}_s = -\frac{\partial \rho_s}{\partial t} \tag{10}$$

$\nabla_s \vec{J}_s$ denotes the surface divergence of $\vec{J}_s$.

As only the axial component of current density J_z exists it can be expressed in terms of the unknown current along the wire, as follows:

$$I(z',t) = \int_0^{2\pi} J_z(z',t)\, a d\phi = J_z(z',t) \cdot 2\pi a \tag{11}$$

and the linear charge distribution can be expressed in terms of current [10], i.e.:

$$q = -\int_0^t \frac{\partial I(z',t)}{\partial t} dt \tag{12}$$

Combining equations (1) to (12) results in the time domain Pocklington integro-differential equation for a loaded wire in free space:

$$-\varepsilon\frac{\partial E_z^{inc}}{\partial t}=\left[\frac{\partial^2}{\partial z^2}-\frac{1}{c}\frac{\partial^2}{\partial t^2}\right]\int_0^{2h}\frac{I(z',t-R/c)}{4\pi R}dz'-L\frac{\partial^2 I(z,t)}{\partial t^2} \tag{13}$$

where, c is the velocity of light and R is the distance from the source point z' to the observation point z (both located on the thin wire antenna surface) given by:

$$R=\sqrt{(z-z')^2+a^2} \tag{14}$$

The Hallen integral equation counterpart of equation (13) is then obtained by performing a straightforward convolution, i.e. it follows:

$$\int_0^{2h}\frac{I(z',t-R/c)}{4\pi R}dz'=F_0(t-\frac{z}{c})+F_L(t-\frac{2h-z}{c})$$
$$+\frac{1}{2Z_0}\int_0^{2h}E_z^{inc}(z',t-\frac{|z-z'|}{c})dz'-\frac{1}{2Z_0}\int_0^{2h}L\frac{\partial}{\partial t}I(z',t-\frac{|z-z'|}{c})dz' \tag{15}$$

where Z_0 is the wave impedance of free space and the unknown functions $F_0(t)$ and $F_L(t)$ are related to the multiple reflections from the open wire ends.

Finally without any loss of generality it is assumed that the antenna is not excited before $t = 0$ [8]. Invoking the integral equation (15) for $z = 0$ and $z = L$, equation (15) can be rewritten in such a way that the current $I(z, t)$ remains the only unknown.

3 The boundary element solution of the Hallen equation

The time domain integral equation (15) can be solved applying the time domain scheme of Galerkin-Bubnov Indirect Boundary Element Method (GB-BIEM). The Galerkin-Bubnov formalism is used for space discretization, while the marching-on-in-time procedure is used for time sampling. Pulse, linear and parabolic shape functions were used for space discretization, and the linear functions were found to be the optimal choice regarding accuracy and the convergence rate [8].

These functions are Lagrange's polynomials and can be expressed, as follows:

$$f_i(x)=\prod_{i=1}^{n}\frac{x-x_j}{x_i-x_k},\ j\neq i \tag{16}$$

It is convenient to write the Hallen integral equation (15) symbolically:

$$L(I)\text{-}Y=0 \tag{17}$$

where L is the linear integral operator and I is the unknown function to be determined for a given excitation Y. The local approximation for current over a space segment can be written as:

$$I(z',t) = \{f\}^T \{I\} \tag{18}$$

where $\{f\}$ is the shape function vector and $\{I\}$ is the time-dependent solution vector.

According to the Galerkin-Bubnov procedure the integral equation (15) is multiplied by the test functions and integrated over the calculation domain. In order to satisfy the requirement that the interpolation error is minimazed over the wire segment Δl, the inner product of the functions $L(I)$-$Y=0$ and $\{f\}_j$ has to vanish, i.e.:

$$\sum_{i=1}^{M} \int_{\Delta l_j} \left[L\left(\{f\}_i^T \{I\}_i \right) - Y \right] \{f\}_j \, dz = 0, \quad j = 1,2,\ldots M \tag{19}$$

where M is the number of space segments over the entire length of the wire.

Assembling the contributions from all segments over the wire length the global matrix system is obtained.

The space discretization of equation (15) results in the following linear equation system for *i-th* source and *j-th* observation element:

$$\begin{aligned} &\sum_{i=1}^{M} \int_{\Delta l_j} \int_{\Delta l_i} \{f\}_j \{f\}_i^T \frac{1}{4\pi R} dz' dz \{I\} \Big|_{t-\frac{R}{c}} = \int_{\Delta l_j} F_0 (t - \frac{z}{c}) \{f\}_j dz + \int_{\Delta l_j} F_L (t - \frac{L-z}{c}) \{f\}_j dz + \\ &+ \frac{1}{2Z_0} \sum_{i=1}^{M} \left(\int_{\Delta l_j} \int_{\Delta l_i} E_z^{inc} (z', t - \frac{|z-z'|}{c}) \{f\}_j dz' dz + \int_{\Delta l_j} \int_{\Delta l_i} L \frac{\partial}{\partial t} I(z', t - \frac{|z-z'|}{c}) \{f\}_j dz' dz \right), j = 1,2,\ldots N_g \end{aligned} \tag{20}$$

It is worth noting that more accurate and more stable results are obtained by interpolating the known excitation over space segments with the same shape functions as those used in current expansions [8]-[10], i.e. it follows:

$$E_x^{inc}(x', t - \frac{|x - x'|}{c}) = \{f\}^{\mathrm{T}} \{E\} \tag{21}$$

where $\{E\}$ is the time dependent excitation vector.

Substituting the relation (21) into the equation (20), taking into account the thin wire approximation, the following global matrix system is obtained after some mathematical manipulation:

$$
\begin{aligned}
[A]\{I\}\Big|_{t-\frac{R}{c}} &= [B]\{E\}\Big|_{t-\frac{|z-z'|}{c}} - \frac{[Z]}{\Delta t}\left(\{I\}\Big|_{t-\frac{|z-z'|}{c}} - \{I\}\Big|_{t-\frac{|z-z'|}{c}-\Delta t}\right) \\
&+[C]\left\{\sum_{n=0}^{\infty} I^n\right\}\Bigg|_{t-\frac{z}{c}-\frac{2nL}{c}-\frac{R_0}{c}} - [B]\left\{\sum_{n=0}^{\infty} E^n\right\}\Bigg|_{t-\frac{z}{c}-\frac{2nL}{c}-\frac{z'}{c}} + \\
&+\frac{[Z]}{\Delta t}\left(\{I\}\Big|_{t-\frac{z}{c}-\frac{2nL}{c}-\frac{z'}{c}} - \{I\}\Big|_{t-\frac{z}{c}-\frac{2nL}{c}-\frac{z'}{c}-\Delta t}\right) \\
&-[D]\left\{\sum_{n=0}^{\infty} I^n\right\}\Bigg|_{t-\frac{z}{c}-\frac{2n+1}{c}L-\frac{R_L}{c}} + [B]\left\{\sum_{n=0}^{\infty} E^n\right\}\Bigg|_{t-\frac{z}{c}-\frac{2n+1}{c}L-\frac{L-z'}{c}} \\
&-\frac{[Z]}{\Delta t}\left(\{I\}\Big|_{t-\frac{z}{c}-\frac{2n+1}{c}L-\frac{L-z'}{c}} - \{I\}\Big|_{t-\frac{z}{c}-\frac{2n+1}{c}L-\frac{L-z'}{c}-\Delta t}\right) \\
&+[D]\left\{\sum_{n=0}^{\infty} I^n\right\}\Bigg|_{t-\frac{L-z}{c}-\frac{2nL}{c}-\frac{R_L}{c}} - [B]\left\{\sum_{n=0}^{\infty} E^n\right\}\Bigg|_{t-\frac{L-z}{c}-\frac{2nL}{c}-\frac{L-x'}{c}} \\
&+\frac{[Z]}{\Delta t}\left(\{I\}\Big|_{t-\frac{L-z}{c}-\frac{2nL}{c}-\frac{L-x'}{c}} - \{I\}\Big|_{t-\frac{L-z}{c}-\frac{2nL}{c}-\frac{L-x'}{c}-\Delta t}\right) \\
&-[C]\left\{\sum_{n=0}^{\infty} I^n\right\}\Bigg|_{t-\frac{L-z}{c}-\frac{2n+1}{c}L-\frac{R_0}{c}} + +[B]\left\{\sum_{n=0}^{\infty} E^n\right\}\Bigg|_{t-\frac{L-z}{c}-\frac{2n+1}{c}L-\frac{z'}{c}} \\
&-\frac{[Z]}{\Delta t}\left(\{I\}\Big|_{t-\frac{L-z}{c}-\frac{2n+1}{c}L-\frac{z'}{c}} - \{I\}\Big|_{t-\frac{L-z}{c}-\frac{2n+1}{c}L-\frac{z'}{c}-\Delta t}\right)
\end{aligned}
\tag{22}
$$

where [A], [B], [Z], [C] and [D] are the global space-dependent matrices that have been assembled from the local matrices in a following manner:

$$[\mathrm{A}] = \sum_{\mathrm{i=1}}^{\mathrm{M}} [A]_{ji}^{e}, j = 1,2,...M; [A]_{ji}^{e} = \int_{\Delta l_j}\int_{\Delta l_i} \{f\}_j \{f\}_i^T \frac{1}{4\pi R} dz' dz \tag{23}$$

$$[\mathrm{B}] = \sum_{\mathrm{i=1}}^{\mathrm{M}} [B]_{ji}^{e}, j = 1,2,...M; [B]_{ji}^{e} = \frac{1}{2Z_0} \int_{\Delta l_j}\int_{\Delta l_i} \{f\}_j \{f\}_i^T dz' dz \tag{24}$$

$$[\mathrm{C}] = \sum_{\mathrm{i=1}}^{\mathrm{M}} [C]_{ji}^{e}, j = 1,2,...M; [C]_{ji}^{e} = \int_{\Delta l_j}\int_{\Delta l_i} \{f\}_j \{f\}_i^T \frac{1}{4\pi R_0} dz' dz \tag{25}$$

$$[\mathrm{D}] = \sum_{\mathrm{i=1}}^{\mathrm{M}} [D]_{ji}^{e}, j = 1,2,...M; [D]_{ji}^{e} = \int_{\Delta l_j}\int_{\Delta l_i} \{f\}_j \{f\}_i^T \frac{1}{4\pi R_L} dz' dz \tag{26}$$

$$[Z]=\sum_{i=1}^{M}[Z]_{ji}^{e},\ j=1,2,...M;[Z]_{ji}^{e}=\int_{\Delta l_j}\int_{\Delta l_i}\{f\}_j\{f\}_i^T Ldz'dz \tag{27}$$

In each of the terms in equation (22) the space-dependent matrix multiplies the time dependent current or excitation vector.

The time discretization is also performed by using the weighted residual approach. The solution in time on the *i-th* space segment is assumed in the form:

$$I_i(t)=\{T\}^T\{I\} \tag{28}$$

where $\{T\}$ is the vector containing linear time domain shape functions.

Applying the weighted residual approach over a time increment leads to the integral:

$$\int_{t_k}^{t_k+\Delta t}([A]\{I\}_{|_{t-\frac{R}{c}}}-\{g\})\theta_k dt=0,\ k=1,2,...N_t \tag{29}$$

where θ_k is the set of time-domain weights and $\{g\}$ is the right-hand side of the corresponding matrix expression and N_t is the total number of time samples. Choosing the delta impulses for testing procedures results in the time sampling:

$$[A]\{I\}_{|_{t_k-\frac{R}{c}}}=\{g\}_{|all\ previous\ discrete\ instants} \tag{30}$$

i.e. one deals with the explicit scheme of the time domain stepping procedure. Satisfying the Courant discretization condition [8]-[10]:

$$\Delta t\leq\frac{\Delta z}{c} \tag{31}$$

the transient response of thin wire configuration can be obtained via the recurrence formula instead of performing a tedious matrix inversion procedure.

In addition, if the space discretization of the wire is sufficiently fine which ensures that the current does not vary significantly over the space segment within the time increment Δt then the recurrence formula for the calculation of the current at the *k-th* instant for *j-th* considered space node is given as follows:

$$A_{jj}^{'}I_{j|_{t_k}}+\left[\bar{A}'\right]\{I\}_{|_{t_k-\frac{R}{c}}}=\{g\}_{|alldiscreteins\tan ts} \tag{32}$$

where:

$$A_{jj}^{'}=A_{jj+}\frac{Z_{jj}}{\Delta t}\qquad\left[\bar{A}'\right]=\left[\bar{A}\right]+\frac{\left[\bar{Z}\right]}{\Delta t} \tag{33}$$

and overbar line denotes the absence of the diagonal terms. In relation (32) the transient current on a segment of observation, on which there is no time delay is expressed by means of all currents and excitations at retarded instants of time.

Consequently, the final recurrence relation for space-time dependent current is given by:

$$I_{j|_{t_k}} = \frac{-\sum_{i=1}^{N_g} \overline{A_{jj}^{'}}\, I_{i|_{t_k - \frac{R}{c}}} + g_{j}\Big|_{\text{all retarded times}}}{A_{jj}^{'}} \tag{34}$$

where N_g and N_T are the total numbers of space nodes and time samples, respectively.

The above presented time-marching procedure is stable for an arbitrary time interval and no smoothing procedure is required for averaging the solution in time.

4 A computational example

The numerical example under consideration is a monopole antenna of length $h = 0.125$m, and radius $a = 3.18$mm. The monopole antenna is mounted on the perfectly conducting (PEC) ground plane and driven at its base by a source voltage $v(t)$ given by:

$$v(t) = V_0 e^{-g^2(t-t_0)^2} \tag{35}$$

where: $V_0 = 1$V $g = 2.5*10^9 s^{-1}$, and $t_0 = 0.86$ns.

The monopole antenna is modeled as an equivalent dipole antenna driven at its center by twice the value of the impressed voltage v(t).

Fig 2 shows the transient current induced at the base of the bare and coated monopole, respectively, for various values of relative permeability.

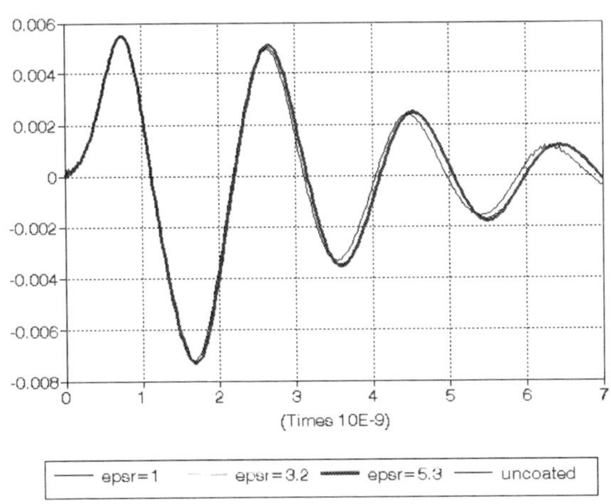

Figure 2: Transient current induced at the base of the monopole.

A slight shift can be noticed between the transient current curves related to the coated and bare wire, respectively. As there is a shift to the right of the dielectrically coated wire antenna it means that the transient current wave propagating along the wire is reflected later than in the bare wire case.

Therefore, the coating effect is related to the increase of the antenna effective length. The transient current curves presented in Fig 2 are in agreement with the ones available in [1], where this effect of increasing the effective length of the antenna has also been discussed.

5 Conclusion

The paper deals with the time domain analysis of dielectric coated wire antennas in free space via the Galerkin-Bubnov Boundary Element solution method and basically represents an extension of the previously reported frequency domain analysis.

The formulation of the problem is based on the Hallen integral equation for a loaded straight thin wire antenna. A load term with an equivalent magnetic coating accounts for the dielectric coating effects.

An illustrative computational example has been presented in the paper and the results demonstrate that in the time domain the coating effect is equivalent to the increase of the antenna effective length.

References

[1] A.R. Bretones, A. Salinas, R. Gomez Martin, I. Sanchez Garcia, Time Domain Analysis of Dielectric-Coated Wire Antennas and Scatterers, IEEE Trans. Antennas and Propagation, Vol. AP-42, pp. 815-819, June 1994.

[2] P.E. Atlamazoglu, N.K. Uzunoglu, A Galerkin Moment method for the Analysis of an Insulated Antenna in a Dissipative Dielectric Medium, IEEE Trans MTT, Vol 46, No7, pp 988-996, July1998.

[3] R.W.P. King et al., The electromagnetic Field of an Insulated Antenna in a Conducting or Dielectric Medium, IEEE Trans. Microwave Theory Tech., Vol. MTT-31, pp. 574-583, July 1983

[4] J. Moore, M.A. West, Simplified Analysis of Coated Wire Antennas and Scatterers, IEE Proc Microwaves Antennas Propag, Vol 142, No 1, pp 14-18, Feb 1995.

[5] D. Poljak, C.A. Brebbia, Indirect Galerkin-Bubnov Boundary Element Analysis of Coated Thin Wire Antenna in Free Space, Boundary Elements XXVII, pp 515-525, 2005

[6] D. Poljak, C.A. Brebbia, Indirect Galerkin-Bubnov Boundary Element method for Solving Integral Equations in Electromagnetics, *Engineering Analysis with Boundary Elements*, EABE 28, No 7, pp 771-778, July 2004.

[7] D. Poljak, C.A. Brebbia, *Boundary Elements for Electrical Engineers*, WIT Press, Southampton-Boston, 2005.

[8] D. Poljak, *Electromagnetic Modelling of Wire Antenna Structures*, WIT Press, Southampton-Boston, 2002.
[9] D. Poljak, C.Y. Tham, *Integral Equation Techniques in Transient Electromagneticsoundary Elements for Electrical Engineers*, WIT Press, Southampton-Boston, 2003.
[10] D. Poljak (Ed), *Time Domain Techniques in Computational Electromagnetics*, WIT Press, Southampton-Boston, 2004.

Synthesis method of the Cassegrain type unsymmetrical antennas

R. Dufrêne[1], W. Kołosowski[2], E. Sędek[1] & A. Jeziorski[2]
[1]*Telecommunications Research Institute, Poland*
[2]*Military University of Technology, Poland*

Abstract

A new method for the synthesis of the Cassegrain type unsymmetrical dual-reflector antennas is presented. The algorithm of synthesis is based on a scheme similar that which is used for designing reflector cosecant antennas. The "open Cassegrain" antenna is used as the basic prototype. The calculations solve differential equations by means of the Runge-Kutta method.
Keywords: Cassegrain antenna, offset reflector antenna.

1 Introduction

The synthesis of unsymmetrical dual-reflector antennas is one of the most important problems in the antenna theory. It is formulated in the following manner: having a horn with an axial symmetrical radiation pattern $G(\varphi)$ we must obtain a geometry of the sub-reflector and reflector of the unsymmetrical antenna that will give a desired aperture field distribution $E(\rho)$.

This problem has been studied by many authors, but the results obtained by them are very complicated [1, 2].

The proposed solution of synthesis problem for the unsymmetrical dual-reflector antennas consists of three main stages:

1. Calculation of the function $\rho(\varphi)$ for main cross-section (aperture is crossed at point ρ by sending the ray from the source under the angle φ, as in Fig.1);
2. Calculation of the main cross-section of the antenna;
3. Calculation of the sub-reflector and reflector surfaces.

It follows from above that the method of synthesis described here is based on the well-known idea of synthesis of the cosecant reflector antenna. From the point of view of the design considerations, this synthesis problem is solved by

WIT Transactions on Modelling and Simulation, Vol 42, © 2006 WIT Press
www.witpress.com, ISSN 1743-355X (on-line)
doi:10.2495/BE06019

means of changing the shapes of the reflector and sub-reflector surfaces of the antenna, which is referred to here as the "prototype-antenna". Geometric parameters of these "prototype-antennas" are similar to the parameters that are expected as the solution of our synthesis problem. The "antennas-prototype" is a special variant of the "open Cassegrain" antenna, having a symmetrical field distribution on the aperture [4, 5].

For describing the geometry of such antennas the knowledge of exactly 5 independent parameters is needed. In this paper, the following parameters are used: F – local length of the reflector, φ_t – width of the tube beam, φ_s – the angle between the tube axis and the reflector axis, e – sub-reflector eccentricity, R_k – distance between focuses of the paraboloid and the hiperboloid (points O and K in Fig.1). The calculation method used for determining the geometrical parameters of the "open Cassegrain" antenna is described in [5].

2 Synthesis of Cassegrain type unsymetrical dual-reflector antennas

2.1 Calculation of the function $\rho(\varphi)$

The equation describing conservation of energy in the main cross-section of the reflector antenna is used for obtaining the $\rho(\varphi)$ function. It describes the fact that the energy radiated from the source must be equal to the energy obtained on the aperture.

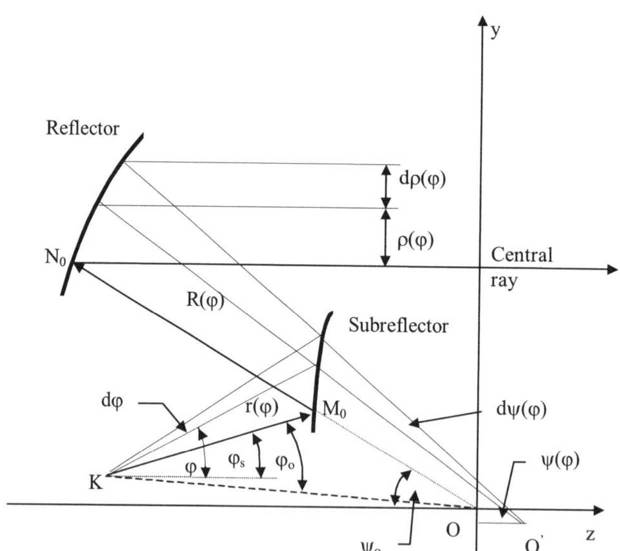

Figure 1: Main cross-section of unsymmetrical dual-reflector antenna.

This requirement may be expressed as the equality of energy in the elementary cross-sections of foreheads of the spherical wave radiated from the source dφ and of the plain wave appearing on the aperture of the reflector dρ [1–3].

$$G^2(\varphi)\sin(\varphi)d\varphi = k E^2(\rho)\rho d\rho \tag{1}$$

where: E(ρ) – field distribution on the reflector aperture;
G(φ) – radiation pattern of the source;
k – a real constant.

It is assumed that the whole energy radiated by the source is contained in the area between angles +/- φ_t, and at the aperture of the reflector in the region limited by the rays +/-ρ_0.

2.2 Calculation of the main cross-section

The main cross-section of a dual-reflector antenna should realize the function ρ(φ), which relates the radiation characteristic of the horn G(φ) with the field distribution on the aperture E(ρ). In other words, we should find such cross-section that the ray sent from the horn under angle j hits, after reflections, the sub-reflector and reflector, at the point of the aperture, defined by the coordinate r(j) (Fig.1) [5].

2.2.1 Formulas used for calculations of the main cross-section

The main cross-section of the synthesized unsymmetrical dual-reflector antenna is presented on Fig.1. The ray r(j) is sent from the phase center of the source (point K) in the direction of the sub-reflector. The ray R(j) is a current ray of the reflector, and the point O' can be called its "local" focus. The point O' is identical to the point O – the origin of the coordinate system, but only for three values of angle (-j_t, 0, +j_t).

In the present paper, the angle j was taken as the variable for the remaining parameters of the antenna, i.e. r = r(j), y = y(j), R = R(j) and r = r(j).
The following formulas are used for calculation of the main cross-section of the antenna [3, 5]:

- for the sub-reflector (Snellius rule in the differential form):

$$\frac{dr(\varphi)}{d\varphi} = r(\varphi)\mathrm{tg}\left(\frac{\varphi + \psi(\varphi)}{2}\right) \tag{2}$$

where: φ + ψ(φ) is the angle between rays illuminating the sub-reflector and reflected by this sub-reflector.

- for the reflector:

1. The Snellius rule in the differential form:

$$\frac{dR(\varphi)}{d\varphi} = R(\varphi)\frac{d\psi(\varphi)}{d\varphi}\mathrm{tg}\left(\frac{\psi(\varphi)}{2}\right) \tag{3}$$

2. The collinearity condition for the rays at the reflector aperture:

$$\frac{d\psi(\varphi)}{d\varphi} = \frac{1}{R(\varphi)}\frac{d\rho(\varphi)}{d\varphi} \tag{4}$$

2.2.2 Algorithm for determination of the main cross-section (using the differential geometry)

1. Input parameters for the calculation:
- function $\rho(\varphi)$;
- geometrical parameters of the "prototype-antenna";
2. Functions $\psi = \psi(\varphi)$ and $R = R(\varphi)$ are obtained by solving the system of two differential equations (3) and (4).
3. Function $r = r(\varphi)$ is obtained by solving the differential equation (2). The main cross-section of the sub-reflector is obtained in this manner.
4. The main reflector's cross-section is created by the points belonging to the intersection of the reflector's ray R(j) with aperture rays ρ = ρ(j) (Fig. 1).

Solution of the equations (2), (3) and (4) is obtained using the Runge-Kutta numerical method.

2.3 Calculation of the sub-reflector and reflector surfaces

Let us define the notion of the "local" hyperboloids and paraboloids. It is a family of unsymmetrical dual-reflector antennas satisfying the symmetry condition for the field contribution on the reflector's aperture [4]. The members of this family may to have different values of geometrical parameters: $e(\varphi)$ – eccentricity of hyperboles, $F(\varphi)$ – focal length of the parabolas, $\varphi_0(\varphi)$ – angle between tubes and sub-reflectors axes etc. In other words, to every value of the angle φ there exists an individual dual-reflector antenna called – "local" hyperboloid and paraboloid. All "local" antennas have the same location of the point K and the same axes of the horn and the aperture's axes of the reflector. The functions $e(\varphi)$, $\varphi_s(\varphi)$ and all the others are obtained by means of finding the main cross-section of the antenna (stage 2).

All results obtain for the test antenna in the design process are shown in Fig. 2 [6].

The surfaces of the sub-reflector and reflector of the antenna are calculated in the following manner:

- the rays, which are sent from the point K at the angles φ and φ+dφ, create the so called "cone in the cone", which "cuts out" the rings (strip) in the "local" hyperboloids;
- these rays, after reflection from the sub-reflector, cut out the rings in the "local" paraboloid. These rings have the widths dρ;
- "adding up" all rings formed in this manner (for the angles φ laying in the range from 0 to φ_t), we obtain the surfaces of the sub-reflector and reflector of the antenna.

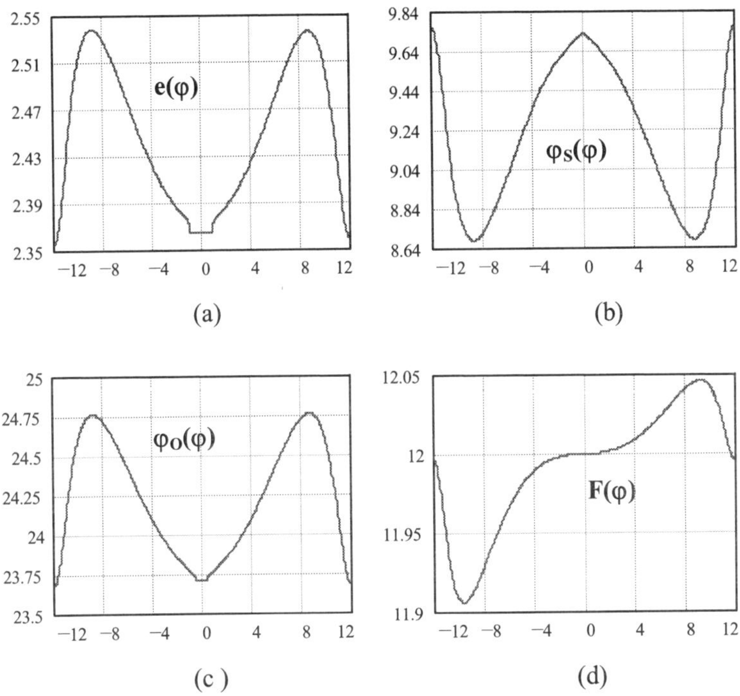

Figure 2: Functions e(φ), $\varphi_S(\varphi)$, $\varphi_O(\varphi)$, F(φ) for the antenna realizing the field distribution on the aperture $E(\rho) = 1/7 + 6(1-\rho^2)^2/7$ ($G(\varphi)=\cos^{88.1}(\varphi)$, for the parameters of the "antenna-prototype" – e=2.3646, $\varphi_S=9.80^0$, $\varphi_t=23.71^0$, F=12m, R_k=6.76m).

3 Conclusion

A new method for the synthesis of the unsymmetrical dual-reflector antennas was presented. This method is simple and accurate and may be used for the design of the large reflector antennas.

References

[1] V. Galindo-Israel, W.A. Imbriale, R. Mittra, On the Theory of the Synthesis of Single and Dual Offset Shaped Reflector Antennas. *IEEE Trans. Antennas Propagat*ion, vol. AP-35, No.8, pp.887-896, August 1987.

[2] P.S. Kildal, Synthesis of Multireflector Antennas by Kinematic and Dynamic Ray Tracing. *IEEE Trans. Antennas Propagat*ion, vol. AP-38, No.10, pp.1587-1599, October 1990.

[3] L.D. Bakchrach, G.K. Galimov, *Reflector scanning antennas,* Moscow, Nauka, 1981 (in Russian).

[4] C. Dragone, Offset Multireflector Antennas with Perfect Pattern Symmetry and Polarization Discrimination, *The Bell System Technical Journal*, vol. 57, No. 7, pp.2663-2685, September 1978.

[5] R. Dufrêne, A. Jeziorski, W. Kołosowski, E. Sędek, M. Wnuk, "Non-symmetrical dual-reflectors antennas", Prace PIT, No. 127, 2001 (in Polish)

[6] A. Jeziorski, Analysis and synthesis nonsymmetrical duo-reflector antennas, Ph.D. dissertation, Military University of Technology, Faculty of Electronics, Warsaw, 1999 (in Polish).

Numerical simulation of a 3D virtual cathode oscillator

F. Assous
Department of Computer Sciences and Mathematics,
College of Judea and Samaria, Israel

Abstract

The design of microwave devices requires the interaction of electromagnetic fields to be simulated with charged particle flow with high accuracy. The methods involved have to be, in particular, well adapted to the geometrical complexity of real devices, especially to take into account the 3-D effects. We have investigated a numerical method for solving the 3-D time dependent Vlasov-Maxwell equations in the relativistic case, on unstructured meshes. We present some numerical results obtained by simulating a three-dimensional virtual cathode system, and compare the resulting high power microwave radiation with a theoretical estimation.
Keywords: computer simulation, finite element, particle method; vircator.

1 Introduction

The numerical modelling in plasma physics as well as in microwave devices or accelerator technology, requires in some cases a full three-dimensional code for the resolution of the Vlasov-Maxwell equations, handling the time domain and well adapted to arbitrary complex geometries.

The Vlasov problem consists in determining the charged particle distribution function in the six-dimensional phase space $(\mathbf{x}, \mathbf{p})$, where $\mathbf{x}$ denotes the particle position, and $\mathbf{p}$ the momentum. This is achieved by a particle method. For computing the solution of Maxwell's equations, the first and probably most popular method was introduced by Yee [1], and is straightforward to implement in simple cases. However, despite its efficiency, as soon as the domain geometry becomes too complex, we have to use the flexibility of unstructured meshes to approximate complex geometries and to achieve local refinements.

WIT Transactions on Modelling and Simulation, Vol 42, © 2006 WIT Press
www.witpress.com, ISSN 1743-355X (on-line)
doi:10.2495/BE06020

We have developed a method based on a nodal finite element discretization, to ensure that the electromagnetic fields are continuous, (which is a stability condition for the Vlasov solver), using a space discretization that prevents the inversion of a linear system at each time-step. The control of the divergence of the fields is obtained by introducing Lagrange multipliers. We first recall the Vlasov-Maxwell model, the particle approximation of the Vlasov equation, and the nodal finite element discretization we used. We then derive the linear system to be solved. Finally, we describe the simulation of a three-dimensional virtual-cathode oscillator (*VIRCATOR*), which is a way to produce a high power microwave radiation from a relativistic electron beam. Numerical results and comparisons with a theoretical estimation are also given.

2 The Vlasov-Maxwell model

As it is well known, there exists a strong correlation between the Maxwell equations and models that describe the motion of particles. This correlation is at the origin of most of the coupled models, where the Maxwell equations appear in parallel with (and depending on) other models of equations.

Hence, the Maxwell equations are related to electric charged particles, the motion of which being a source for generating an electromagnetic field. Conversely, for a population of charged particles with a mass m and a charge q, the main force field is the electromagnetic Lorentz force

$$\mathbf{F} = q\left(\mathcal{E}(\mathbf{x},t) + \mathbf{v}(t) \times \mathcal{B}(\mathbf{x},t)\right), \tag{1}$$

that describes how the electromagnetic field $\mathcal{E}(\mathbf{x},t)$ and $\mathcal{B}(\mathbf{x},t)$ acts on a particle with a velocity $\mathbf{v}(t)$.

2.1 The Vlasov model

We consider a population of charged particles with a mass m and a charge q, submitted to a given force field $\mathbf{F}(\mathbf{x},\mathbf{v},t)$. The motion of these charged particles can be described in terms of particle distribution function $f(\mathbf{x},\mathbf{p},t)$, by the relativistic Vlasov equation

$$\frac{\partial f}{\partial t} + \mathbf{v}\cdot\nabla_{\mathbf{x}} f + \mathbf{F}\cdot\nabla_{\mathbf{p}} f = 0, \tag{2}$$

where $\mathbf{F}$ is the Lorentz force given in (1) and $\mathbf{p}$ denotes the momentum that verifies

$$\mathbf{p} = \gamma m\mathbf{v}, \qquad \text{with} \qquad \gamma m = \frac{\sqrt{|\mathbf{p}|^2 + m^2c^2}}{c}.$$

As we consider the Vlasov equation in a bounded domain Ω of the physical space $\mathbb{R}^3$ with a boundary Γ, we have to supplement this system with appropriate boundary conditions. We assume here that Γ consists of three disjoint parts Γ_e, Γ_a and Γ_r corresponding to emission, absorption and reflection of particles. Let us introduce $\mathbf{n} = (n_x, n_y, n_z)$ the unit outward normal of Γ. We let on the emissive boundary Γ_e

$$f(\mathbf{x}, \mathbf{p}, t) = g(\mathbf{x}, \mathbf{p}, t), \quad \mathbf{x} \in \Gamma_e, \quad \mathbf{p} \cdot \mathbf{n} < 0. \tag{3}$$

In practice, Γ_e coincides with the cathode and the given function g depends on the particle emission mechanism. It can be given points in the position-momentum phase space $(\mathbf{x}_k, \mathbf{p}_k)$ or a Child-Langmuir law of emission. On the absorbing boundary Γ_a

$$f(\mathbf{x}, \mathbf{p}, t) = 0, \quad \mathbf{x} \in \Gamma_a, \quad \mathbf{p} \cdot \mathbf{n} < 0. \tag{4}$$

On the reflecting boundary Γ_r

$$f(\mathbf{x}, \mathbf{p}, t) = f(\mathbf{x}, \mathbf{p} - 2(\mathbf{p} \cdot \mathbf{n})\mathbf{n}, t), \quad \mathbf{x} \in \Gamma_r, \quad \mathbf{p} \cdot \mathbf{n} < 0. \tag{5}$$

In addition we have to prescribe the initial distribution function of particles $f(\mathbf{x}, \mathbf{p}, t = 0) = f_0(\mathbf{x}, \mathbf{p})$.

2.2 The Maxwell equations

The coupling of the Maxwell and Vlasov equations appears first through the Lorentz force (1). It is also expressed by the charge and the current density induced by the motion of these particles with

$$\rho(\mathbf{x}, t) = q \int_{\mathbb{R}_v^3} f(\mathbf{x}, \mathbf{v}, t)\, d\mathbf{p}, \qquad \mathcal{J}(\mathbf{x}, t) = q \int_{\mathbb{R}_v^3} f(\mathbf{x}, \mathbf{v}, t)\, \mathbf{v}\, d\mathbf{p}, \tag{6}$$

that appear as (part of, if parts of ρ and $\mathcal{J}$ are due to external charge and current sources) the right-hand sides of the Maxwell equations

$$\frac{1}{c^2}\frac{\partial \mathcal{E}}{\partial t} - \nabla \times \mathcal{B} = -\mu_0 \mathcal{J}, \tag{7}$$

$$\frac{\partial \mathcal{B}}{\partial t} + \nabla \times \mathcal{E} = 0, \tag{8}$$

$$\nabla \cdot \mathcal{E} = \frac{\rho}{\varepsilon_0}, \tag{9}$$

$$\nabla \cdot \mathcal{B} = 0. \tag{10}$$

The constants ε_0, μ_0 are respectively the dielectric permittivity and the magnetic permeability in the vacuum, satisfy $\varepsilon_0 \mu_0 c^2 = 1$. As it is well known, the charge conservation equation

$$\frac{\partial \rho}{\partial t} + \nabla \cdot \mathcal{J} = 0, \tag{11}$$

can be viewed as a consequence of the Maxwell equations, obtained by taking the divergence of Ampere's law (7) combined with the time derivative of the divergence condition (9). Moreover, the property $\nabla \cdot (\nabla \times \mathcal{E}) = 0$ with (8) implies

$\frac{\partial}{\partial t}(\nabla \cdot \mathcal{B}) = 0$, so that (10) is satisfied for all time t if it is satisfied at the initial time $t = 0$. In the same way, $\nabla \cdot (\nabla \times \mathcal{B}) = 0$ with (7) and the charge conservation equation (11) imply that $\frac{\partial}{\partial t}(\nabla \cdot \mathcal{E} - \rho/\varepsilon_0) = 0$, and (9) is satisfied for all time t if it is satisfied at the initial time $t = 0$.

We supplement this system with appropriate boundary and initial conditions. On the part Γ_C of the boundary Γ that behaves as a perfect conductor, we prescribe $\mathcal{E} \times \mathbf{n} = 0$. On the other part Γ_A, we have $(\mathcal{E} - c\mathcal{B} \times \mathbf{n}) \times \mathbf{n} = (\mathbf{e} - c\mathbf{b} \times \mathbf{n}) \times \mathbf{n}$, or equivalently $(c\mathcal{B} + \mathcal{E} \times \mathbf{n}) \times \mathbf{n} = (c\mathbf{b} + \mathbf{e} \times \mathbf{n}) \times \mathbf{n}$. When $\mathbf{e}$ and $\mathbf{b}$ are prescribed electric and magnetic fields, this model the electromagnetic interactions between the domain Ω and the exterior. The special case when $\mathbf{e} = \mathbf{b} = 0$ corresponds to the Silver-Müller absorbing conditions.

3 Approximation methods

3.1 Particle approximation of the Vlasov equation

The Vlasov problem consists of determining the charged particle distribution function in the six-dimensional phase space $(\mathbf{x}, \mathbf{p})$. With a grid method (finite differences, finite volumes or finite element method) is almost impossible, since we rapidly reach the limit in memory available on a computer, leading then to an intractable cpu time. For this reason, a well suited method is the widely used particle method (see [2], [3]), in which the distribution function $f(\mathbf{x}, \mathbf{p}, t)$ is approximated at any time t, by a linear combination of delta distributions in the phase space:

$$f(\mathbf{x}, \mathbf{p}, t) \simeq \tilde{f}(\mathbf{x}, \mathbf{p}, t) = \sum_k w_k \, \delta(\mathbf{x} - \mathbf{x}_k(t))\delta(\mathbf{p} - \mathbf{p}_k(t)) \, , \tag{12}$$

where each term of the sum can be identified with a macro-particle, characterized by its weight w_k, its position $\mathbf{x}_k$ and its momentum $\mathbf{p}_k$. This distribution function is a solution of the Vlasov equation (2) if and only if $(\mathbf{x}_k, \mathbf{p}_k)$ is a solution of the differential system:

$$\frac{d\mathbf{x}_k}{dt} = \mathbf{v}_k = \frac{\mathbf{p}_k}{\gamma_k m} \, , \qquad \frac{d\mathbf{p}_k}{dt} = \mathbf{F}(\mathbf{x}_k, \mathbf{p}_k) \, , \tag{13}$$

which describes the time evolution of a particle k, submitted to the electromagnetic force $\mathbf{F}$.

To solve the system (13), we used an explicit time discretization algorithm - the leapfrog scheme - which is well-adapted in this case. Given a constant time step Δt, the particles positions are defined at time $t_n = n\Delta t$ and the particle momenta are computed at time $t_{n+1/2} = (n + 1/2)\Delta t$. To avoid the inversion of a system of non-linear equations for the computation of γ^n, we use a prediction of $\mathbf{p}^n$ following the Boris method [4]. We refer the reader to [5] for more details.

It remains now to deal with the initial and boundary conditions with the particle approximation. For the conditions (4) and (5), the situation is quite obvious: a particle k is cancelled if it reaches the absorbing boundary Γ_a, and is specularly reflected as it reaches the reflecting boundary Γ_r.

For the numerical simulation of the particle emission at the cathode Γ_e, we write the boundary condition (3) in the form $g(\mathbf{x}, \mathbf{p}, t) = n_0(\mathbf{x}, t)h(\mathbf{x}, \mathbf{p}, t)$, where $\int_{\mathbf{p}\cdot\mathbf{n}<0} h(\mathbf{x}, \mathbf{p}, t)\, d\mathbf{p} = 1$. We define the emission current density $\mathcal{I}_0(\mathbf{x}, t) = qn_0(\mathbf{x}, t) \int_{\mathbf{p}\cdot\mathbf{n}<0} \mathbf{v} \cdot \mathbf{n}\, h(\mathbf{x}, \mathbf{p}, t)\, d\mathbf{p}$. In most of the physical problems, one prescribes $\mathcal{I}_0$ and h instead of n_0 and h, which is equivalent. We consider a particle emission regime such that the space charge is not large enough for limiting the cathode current. Hence, all the particles emitted at the cathode contribute indeed to the formation of the beam. In this case, the emission current density $\mathcal{I}_0$ is known while the function h is prescribed. For instance in the frequent situation where the particles are emitted along the normal direction to the cathode, one can have for the monocinetic emission $h(\mathbf{x}, \mathbf{p}, t) = h(\mathbf{p}) = \delta(\mathbf{p} - m\mathbf{v}_0)$. Note that we have assumed that the velocities involved $\mathbf{v}_0$ is small so that $\mathbf{p}$ can be approximated by $m\mathbf{v}$. In practice, we have to create a certain number of particles at each time step Δt, that enter the domain from the cathode.

Finally, we have to approximate the initial condition, that is

$$f_0(\mathbf{x}, \mathbf{p}) \simeq \sum_k w_k\, \delta(\mathbf{x} - \mathbf{x}_k^0)\delta(\mathbf{p} - \mathbf{p}_k^0)\,,$$

that simply consists in imposing an initial distribution of particles in the domain Ω. The given initial positions $\mathbf{x}_k^0$ and velocities $\mathbf{v}_k^0$ are related to f_0, and are determined such that they preserve initial total charge and current Q_0 and $\mathbf{J}_0$. This requires to verify conditions such $q\sum_k w_k = Q_0$ or $q\sum_k w_k\mathbf{v}_k = \mathbf{J}_0$. Nevertheless such a distribution is not unique, and in general, a good choice is to select randomly the positions. Indeed, in most cases, a grid method is used to approximate the Maxwell equations requiring to a non-uniform mesh. In these conditions, criteria such the number of particles per cell or the positions of the particles into the cells influence the accuracy of the solution. Hence it is often useful to reduce the numerical noise by regularly reorganizing the particle distribution. This can be made by the aid of a coalescing process. We refer the reader to [6] and the references herein for a detailed study.

3.2 Finite element approximation of the Maxwell equations

As previously exposed, when Maxwell's equations are used in a Particle In Cell code, special care needs to be taken to the divergence conditions. To solve this problem, we have proposed (cf. [7]) a method based on a constrained wave equation formulation of Maxwell's equations. We derive here the magnetic field scheme, in the case of perfectly conducting boundary conditions. One can proceed similarly for the electric field $\mathcal{E}$.

We first introduce a formulation in terms of two second-order wave equations. By differentiating (8) with respect to t and using (7), we eliminate $\mathcal{E}$ in Equation (8). Now to take into account the constraint (10), we introduce a Lagrange multiplier p. Hence the magnetic field is computed using

$$\frac{\partial^2 \mathcal{B}}{\partial t^2} + c^2 \nabla \times (\nabla \times \mathcal{B}) - \nabla p = \frac{1}{\varepsilon_0} \nabla \times \mathcal{J}, \quad \nabla \cdot \mathcal{B} = 0,$$

along with boundary and initial conditions. We may impose homogeneous Neumann boundary condition on p ($\partial p / \partial \mathbf{n} = 0$) so that p is obviously identically zero. Thus, the system we solve is an equivalent formulation of the classical Maxwell equations.

We introduce now the variational formulation, which is the basis of the finite element method. Let us recall the definitions of the functional spaces $H_N(curl, \Omega) = \{\mathbf{C} \in L^2(\Omega)^3, \nabla \cdot \mathbf{C} \in L^2(\Omega), \mathbf{C} \cdot \mathbf{n} = 0 \text{ on } \Gamma\}$, and $H(div, \Omega) = \{\mathbf{C} \in L^2(\Omega), \nabla \cdot \mathbf{C} \in L^2(\Omega)^3\}$. Setting $X = H_N(curl, \Omega) \cap H(div, \Omega)$, the variational formulation for the constrained equation of the magnetic field on the whole domain reads

Find $(\mathcal{B}, p) \in X \times L^2(\Omega)$ *such that*

$$\frac{d^2}{dt^2} \int_\Omega \mathcal{B} \cdot \mathbf{C} \, d\mathbf{x} + c^2 \int_\Omega \nabla \times \mathcal{B} \cdot \nabla \times \mathbf{C} \, d\mathbf{x} + \int_\Omega p \nabla \cdot \mathcal{B} \, d\mathbf{x} = \frac{1}{\varepsilon_0} \int_\Omega \mathcal{J} \cdot \nabla \times \mathbf{C} \, d\mathbf{x}, \forall \mathbf{C} \in X, \tag{14}$$

$$\int_\Omega \nabla \cdot \mathcal{B} \, \psi \, d\mathbf{x} = 0 \qquad \forall \psi \in L^2(\Omega), \tag{15}$$

which can be seen as a standard mixed formulation for a constrained problem. Moreover, it can be easily proved that this variational formulation preserves the property $p = 0$.

To derive now a finite element approximation of the system (14-15), which leads to a well posed discrete problem, we used Taylor-Hood finite elements, following the method introduced in [7]. It requires to define two levels of meshes. A coarser tetrahedral one created by a mesh generator, and a finer one defined by dividing each tetrahedron into eight subtetrahedra. We then approximate the electromagnetic vector fields by P_1-conforming functions componentwise on the finer mesh, and the Lagrange multipliers by P_1-conforming functions on the coarser grid.

We introduce the matrices associated to the different terms in the variational formulation. We denote by $\mathbb{M}$ the lumped mass matrix for vectors on the finer mesh, by $\mathbb{K}$ the matrix corresponding to $\int_\Omega \nabla \times \mathcal{B} \cdot \nabla \times \mathbf{C} \, d\mathbf{x}$, by $\mathbb{L}$ the matrix corresponding to $\int_\Omega \nabla \cdot \mathcal{B} \, \psi \, d\mathbf{x}$, and by $\mathbb{R}$ the matrix corresponding to $\int_\Omega \mathcal{J} \cdot \nabla \times \mathbf{C} \, d\mathbf{x}$. Recall that The Taylor-Hood finite element was particularly interesting in our context, as it leads to a diagonal mass matrix $\mathbb{M}$ without any loss of precision. This property associated with a centered, second-order finite differences time discretization (a leap-frog scheme) leads to an explicit scheme without inverting any linear system at each time step. With these notations the problem after discretization becomes

$$\mathbb{M} B_h^{n+1} + \mathbb{L}^T p_{2h}^{n+1} = F^n,$$
$$\mathbb{L} B_h^{n+1} = 0,$$

where F^n contains all the terms being known at the time step $n + 1$. An efficient way to solve such a linear system is to use the Uzawa algorithm [8].

4 Simulation of a 3D vircator

We are concerned with the simulation of a three-dimensional vircator. Our purpose is to show the interest of three dimensional code computations, that can help to observe and analyze the virtual cathode formation and anticipate some experimental observations. Moreover, it can obviously be helpful for rearrangements of the Vircator geometry yielding better efficiency. For complete physical studies of this mechanism, see among others [9, 10]. Let us begin with a brief description of the vircator mechanism.

Consider an electron beam current injected into a drift tube. As the beam current approaches the critical current for pinching, electron trajectories are deflected by the magnetic field. If this current exceeds the space-charge limited current, the beam pinches along the anode plane and a virtual cathode forms at a distance where most of the electrons lose their kinetic energy. Roughly speaking, one part of these electrons is transmitted through the virtual cathode, another part is reflected from it, whereas the last part stagnates during a certain time in this area that is the virtual cathode. In general, the transmitted current is more or less close to the space-charge limited current. Virtual cathode systems are known as interesting tools for generating very high-power microwaves with variable frequencies. The source of these microwaves is explained by two mechanisms. First, this cathode oscillates in space and time generating thus a microwave emission. Second, the reflected electrons are trapped and their own oscillations contribute to a second microwave emission. The term vircator (VIRtual CAthode oscillaTOR) is commonly used to describe the whole mechanism for the emission of this radiation.

Now we describe the numerical experiment. The geometry of the vircator configuration we simulated consists mainly of a rectangular coax separated from a cylindrical waveguide by a metal sheet (the anode), which is transparent to the beam propagation, but is a perfect conductor for the electromagnetic fields. Due to the symmetry of the geometry and of the emission, the computational domain is divided by 4 by using well adapted symmetry boundary conditions for the EM fields and for the particles (see for example Figure 1). The unstructured mesh is finer in the region of interest, namely between the emission area and the virtual cathode. In this example, the unstructured mesh contains 134136 tetrahedra whereas the number of particles in the domain is approximatively 250000. The electrons are emitted from the cathode and accelerated by an exterior E_z field, $E_z = -10^8\ V/m$.

This simulation allows us to observe the appearance of the virtual cathode. For doing this, we can depict the positions of the particles (or eventually a sample of them) in the pattern of the device, the longitudinal velocity V_z being also indicated by the colour scale. At time $t = 2\ ns$ (Figure 1), the beam is pinched but the flow is still laminar. At time $t = 6\ ns$ (Figure 1), one can see an area where $V_z = 0$, which is a characterization of the presence of a virtual cathode. This accumulation of "slowed down" particles leads to an increasing potential and a part of the electrons are reflected (i.e. $V_z < 0$). Remark also that these reflected electrons are distributed following an annular shape, that is also characteristic of a vircator.

Another diagnostic is also useful: the phase space representation. Figure 2(a) shows an example of the phase space z/P_z obtained at time $t = 9\ ns$, with a characteristic diamond-shaped form, that confirms the presence of the virtual cathode and allows us to locate it.

Finally, one can compute the Poynting vector at the end of the device. Its spectrum, obtained from a Fourier transform, is depicted in Figure 2(b). One can observe a good agreement between theoretical estimations of the emission frequency of the virtual cathode, that is approximatively 8.5 GHz (see [11, 12, 13]), and the numerical value deduced from this spectrum. More complete analysis and detailed comparisons with experimental results can be done using such a numerical tool.

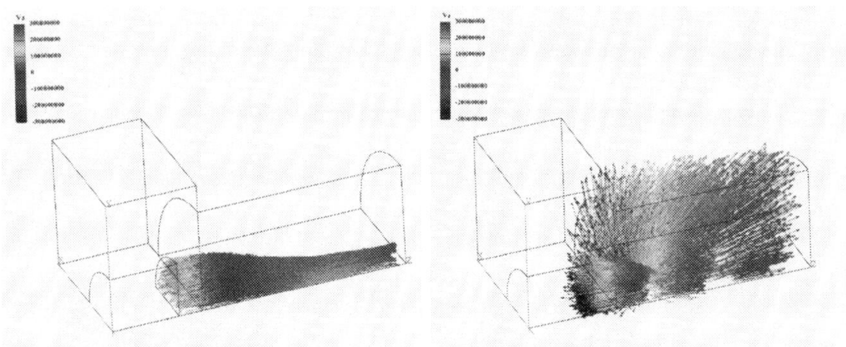

Figure 1: Particle positions at time $t = 2\ ns$ (left) and $t = 6\ ns$ (right).

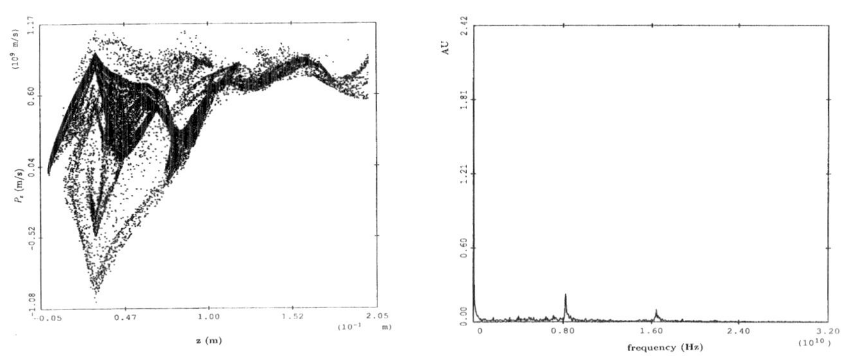

(a) (b)

Figure 2: (a) $z - \mathbf{p}_z$ at $t = 9\ ns$, (b) spectrum of the Poynting vector.

5 Conclusion

In this paper, we proposed a numerical simulation of a 3D virtual cathode oscillator, based on a time-dependent Vlasov-Maxwell solver, well-adapted to complex three-dimensional geometries. Comparisons with theoretical estimations confirm the efficiency of the method, and makes the study of such physical real configurations possible. The development of a parallel version of the method is an improvement in progress that can help to construct an efficient numerical tool for microwave devices simulations.

References

[1] Yee K.S., Numerical solution of initial boundary value problems involving Maxwell's equations in isotropic media, *IEEE Trans. Antennas and Propag.*, **14**, pp. 302–307, 1966.

[2] Birdsall C.K., Langdon A.B. & Okuda H., Finite sized particle Physics applied to Plasma simulation, *Methods in Computational Physics: Plasma Physics*,**9** (Plasma Physics), eds. B . Alder, S. Fernbach & N. Rotenberg, Academic Press, New York, pp. 241–258, 1970.

[3] Morse R.L., Multidimensional plasma simulation by the particle-in-cell method, *Methods in Computational Physics: Plasma Physics*, eds. B. Alder, S. Fernbach & N. Rotenberg, Academic Press, New York, **9**, pp. 213–239, 1970.

[4] Boris J.P., Relativistic Plasma Simulations - Optimization of a Hybrid Code, *Proc. of the* 4th. *Conf. Num. Sim. of Plamas*, Naval Res. Lab., Washington D.C., pp. 3–67, 1970.

[5] Assous F., Degond P. & Segré J., A particle-tracking method for the 3D electromagnetic PIC codes on unstructured meshes, *Comp. Phys. Com.*, **72**, pp. 105–114, 1992.

[6] Assous F., Pougeard Dulimbert T. & Segré J., A new method for coalescing particles in PIC codes, *J. Comput. Physics*, **187**, pp. 550–571, 2003.

[7] Assous F., Degond P., Heintzé E., Raviart P.A. & Segré J., On a finite element method for solving the three dimensional Maxwell equations, *J. Comput. Physics*, **109(2)**, pp. 222–237, 1993.

[8] Fortin M. & Glowinski R., *Augmented Lagrangian Methods*, Springer Series in Computational Mathematics 5,1986.

[9] Sze H., Benford J., Young T., Bromley D. & Harteneck B., A Radially And Axially Extracted Virtual-Cathode Oscillator (Vircator), *IEEE Trans. Plasma Sci.*, **PS-13**, pp. 492–497, 1985.

[10] Sze H., Benford J., Woo W. & Harteneck B., Dynamics of a virtual cathode oscillator driven by a pinched diode, *Phys. Fluids*, **29(11)**, pp. 3873–3880, 1986.

[11] Gouard P., Henry O. & Sellem F., Numerical Simulation of the Vircator, *Proc. of the Conf. Euro-electromagnetics*, ed. D. J. Serafin, p. 237, Bordeaux, France, 1994.

[12] Lin T., Chen W., Liu W., Hu Y. & Wu M., Computer Simulation of Virtual Cathode oscillations, *J. Appl. Phys.*, **68(5)**, pp. 2038–2061, 1990.

[13] Woo W., Two-dimensional features of virtual cathode and microwave emission, *Phys. Fluids*, **30(1)**, pp. 329–356, 1987.

Section 5
Fluid flow

Iterative coupling in fluid-structure interaction: a BEM-FEM based approach

D. Soares Jr[1,2], W. J. Mansur[1] & O. von Estorff[3]
[1]*Department of Civil Engineering,*
COPPE - Federal University of Rio de Janeiro, Rio de Janeiro, Brazil
[2]*Structural Engineering Department, Federal University of Juiz de Fora, MG, Brazil*
[3]*Modelling and Computation, Hamburg University of Technology, Hamburg, Germany*

Abstract

An iterative coupling of finite element and boundary element methods for the investigation of coupled fluid-solid systems is presented. While finite elements are used to model the solid, the adjacent fluid is represented by boundary elements. In order to perform the coupling of the two numerical methods, a successive renewal of the variables on the interface between the two subdomains is performed through an iterative procedure until the final convergence is achieved.
Keywords: iterative coupling, solid-fluid interaction, BEM, FEM, adjustable time-steps, nonlinear analysis.

1 Introduction

Most of the BEM/FEM coupling algorithms [1-3] are formulated in a way that, first, a coupled system of equation is established, which afterwards has to be solved using a standard direct solution scheme. Such a procedure leads to several problems with respect to accuracy and efficiency. First, the coupled system of equation has a banded structure only in the FE part, while in the BE part it is fully populated. Consequently, for its solution the optimized solvers usually used in the FEM cannot be employed anymore, which leads to rather expensive calculations with respect to computer time. Second, the duration of a time step needs to be the same in all subsystems. In general, however, the velocities of the

WIT Transactions on Modelling and Simulation, Vol 42, © 2006 WIT Press
www.witpress.com, ISSN 1743-355X (on-line)
doi:10.2495/BE06021

propagating waves in the solid and the fluid are quite different, such that a unified time step may cause serious problems in the numerical solution algorithms (instabilities, lack of accuracy etc.). Third, in the case of taking into account some nonlinearity within the FE sub-region, the rather big coupled system of equations needs to be solved in each step of the iteration process, i.e., a few times within each time step. This is very computer time consuming.

2 BEM/FEM coupling

Considering the coupling conditions at each node i of the BEM/FEM interface (superscript I), the following equations must hold ($\hat{\rho}$ is the fluid mass density):

$$T({}_F^I\mathbf{F})_i = 0 \tag{1}$$

$$N({}_F^I\mathbf{F})_i = -\ ({}_B^I\tilde{\mathbf{P}})_i \tag{2}$$

$$N({}_F^I\ddot{\mathbf{U}})_i = (1/\hat{\rho})\ ({}_B^I\mathbf{Q})_i \tag{3}$$

where - in order to obtain consistency between the FE (subscript F) and the BE formulation (subscript B) - $\tilde{\mathbf{P}}$ represents the resultant nodal hydrodynamic pressure force, which is obtained from the potential distributions $\mathbf{P}$. $\mathbf{F}$ denotes the FE nodal forces; $\mathbf{Q}$ is the fluid flux and $\ddot{\mathbf{U}}$ is the solid acceleration. The functions $N(.)_i$ and $T(.)_i$ lead to the normal and the tangential component of their arguments, respectively.

2.1 Iterative coupling

In the iterative BEM/FEM coupling, first the FE problem is solved and the accelerations ${}_F^I\ddot{\mathbf{U}}_{(k+\alpha)}^{F\,t}$ are obtained. Then a relaxation parameter α is introduced, according to equation (4), in order to ensure and/or to speed up convergence:

$${}_F^I\ddot{\mathbf{U}}_{(k+1)}^{F\,t} = \alpha\ {}_F^I\ddot{\mathbf{U}}_{(k+\alpha)}^{F\,t} + (1-\alpha)\,{}_F^I\ddot{\mathbf{U}}_{(k)}^{F\,t} \tag{4}$$

Once the FEM accelerations at the interface are computed, equation (3) can be used to obtain the BEM flux ${}_B^I\mathbf{Q}_{(k+1)}^{F\,t}$. In the present formulation different time-step durations in each subdomain can be taken into account by means of extrapolations and interpolations (with respect to time) of the variables at the interface. These interpolations and extrapolations are done according to the time interpolation functions adopted by the BE formulation, as indicated by equations (5-6) (piecewise constant interpolation for the flux and linear interpolation for the potential).

$$ {}^{I}_{B}\mathbf{Q}^{B\,t}_{(k+1)} = {}^{I}_{B}\mathbf{Q}^{F\,t}_{(k+1)} \tag{5} $$

$$ {}^{I}_{B}\mathbf{P}^{F\,t}_{(k+1)} = \left[\left({}_{B}t - {}_{F}t\right) / {}_{B}\Delta t\right] {}^{I}_{B}\mathbf{P}^{B\,t-{}_{B}\Delta t}_{(k+1)} + \left[1-\left({}_{B}t - {}_{F}t\right) / {}_{B}\Delta t\right] {}^{I}_{B}\mathbf{P}^{B\,t}_{(k+1)} \tag{6} $$

Once ${}^{I}_{B}\mathbf{Q}^{B\,t}_{(k+1)}$ is obtained (equation (5), the BE subdomain can be solved, having prescribed flux values at the interface. As the result, ${}^{I}_{B}\mathbf{P}^{B\,t}_{(k+1)}$ is obtained and it needs to be interpolated as well, in order to be used by the FEM. By means of equations (1-2) and ${}^{I}_{B}\mathbf{P}^{F\,t}_{(k+1)}$ (equation (6)) one finally obtains the new FEM nodal forces ${}^{I}_{F}\mathbf{F}^{F\,t}_{(k+1)}$, which are needed to solve the FE problem once more. The iterative loop goes on until convergence is achieved. A sketch of the iterative coupling is shown in Figure 1.

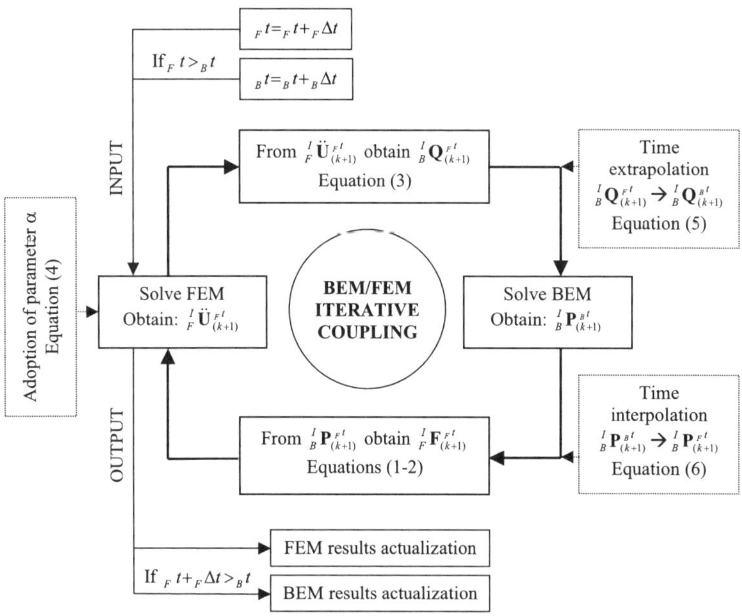

Figure 1: Iterative BEM/FEM coupling scheme.

2.2 Numerical example

In this example, a dam-reservoir system, as depicted in Figure 2, is analyzed. The structure is subjected to a sinusoidal, distributed vertical load on its crest, acting with an angular frequency ω = 18 *rad/s*. The material properties of the dam are: Poisson's ratio $\nu = 0.25$; Young's modulus $E = 3.437{\cdot}10^{6}$ N/m^{2}; mass density $\rho = 2.00$ Ns^{2}/m^{4}. A perfectly plastic material obeying the Drucker-Prager yield criterion is assumed: cohesion $c_{0} = 0.15$ N/m^{2}; internal friction angle ϕ =

20°. The adjacent water is characterized by a mass density $\hat{\rho}$ = 1.00 Ns^2/m^4 and a wave velocity c = 1436 m/s. The time-step duration adopted for the BEM and FEM are ${}_B\Delta t$ = 0.00350 s and ${}_F\Delta t$ = 0.00175 s, respectively. The transient behaviour of the vertical displacement at point A is shown in Figure 3(a). Results of linear as well as nonlinear analyses are given and two different water levels, namely h = $50\ m$ and h = $35\ m$, and their influence on the dam are investigated. In Figure 3(b) the transient hydrodynamic pressure at point B is depicted. A comparison of the results obtained with the iterative coupling procedure with those from the standard coupling scheme used in von Estorff and Antes [1] shows good agreement. The average number of iterations per time step (it was adopted α = 0.5), for the current linear model, was 2.4 (h = $50\ m$) and 2.2 (h = $35\ m$). For the nonlinear analysis, the average number of iterations per time step was 2.6 (h = $50\ m$) and 2.4 (h = $35\ m$). As one can observe, the convergence is quite fast: the iterative coupling can be regarded as a very attractive tool to deal with high scale linear or (especially) nonlinear models. For more details on the iterative BEM/FEM coupling taking into account solid-solid interactions, one is referred to Soares Jr et al. [4].

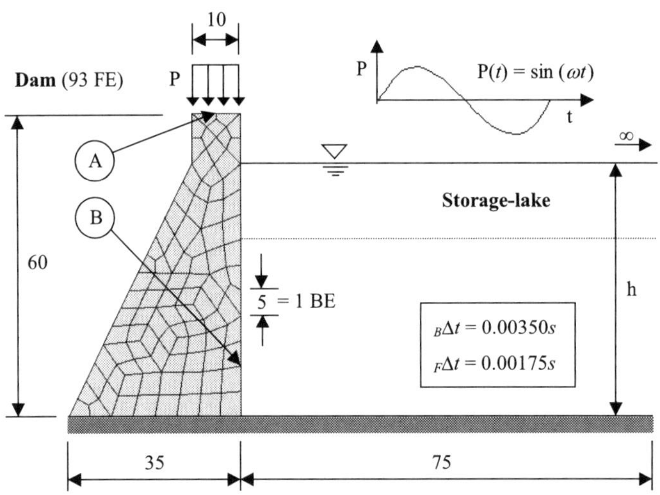

Figure 2: Dam with storage-lake.

3 Conclusions

In the present paper, an iterative coupling scheme for the investigation of a continuous solid coupled to an adjacent fluid was presented. The major advantage of such a procedure can be seen in the fact, that the FE and BE subsystems can be solved separately using optimised solution algorithms according to the special features of the respective system of equations, which are – a further advantage – much smaller than the coupled matrices resulting from a standard coupling approach. In addition, the iterative coupling offers two advantages: It is straightforward to use different time steps in each subdomain; and, moreover, to take into account nonlinearities (within the FE subdomain) in the same iteration loop that is needed for the coupling.

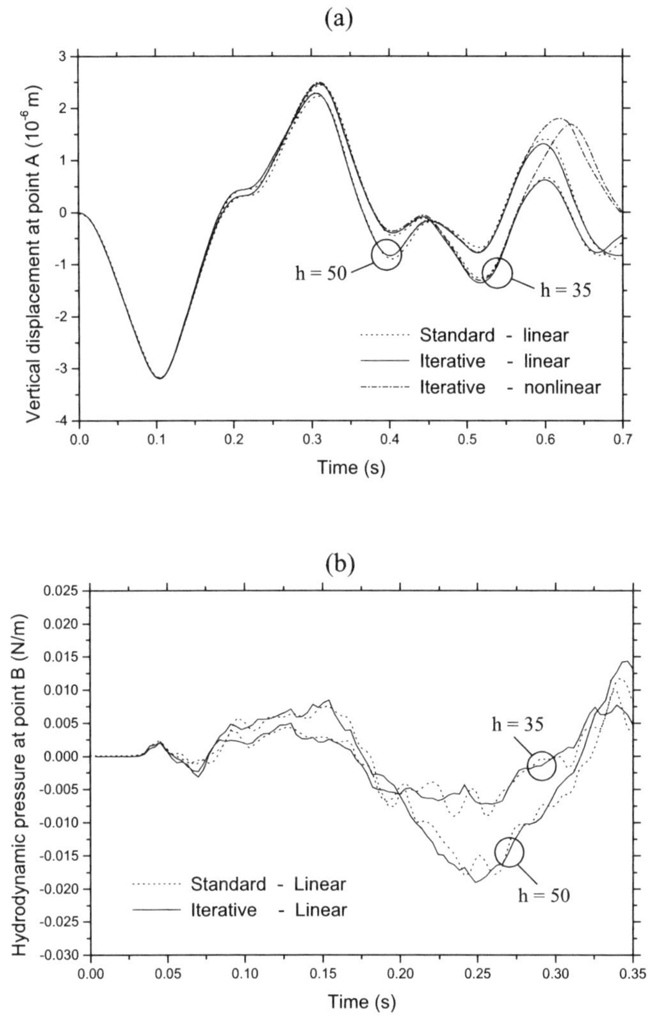

Figure 3: Results for the standard [1] and the iterative FEM/BEM coupling: (a) vertical displacements at point A; (b) hydrodynamic pressure at point B.

References

[1] von Estorff, O. & Antes, H., On FEM-BEM coupling for fluid-structure interaction analysis in the time domain. *International Journal of Numerical Methods in Engineering*, **31**, pp. 1151-1168, 1991.

[2] Czygan, O. & von Estorff, O., Fluid-structure interaction by coupling BEM and nonlinear FEM. *Engineering Analysis with Boundary Elements*, **26**, pp. 773-779, 2002.

[3] Yu, G.Y., Lie, S.T. & Fan, S.C., Stable boundary element method/finite element method procedure for dynamic fluid structure interactions. *Journal of Engineering Mechanics*, **128**, pp. 909-915, 2002.

[4] Soares Jr, D., von Estorff, O. & Mansur, W.J., Iterative coupling of BEM and FEM for nonlinear dynamic analyses. *Computational Mechanics*, **34**, pp. 67-73, 2004.

The Complex Variable Boundary Element Method for potential flow problems

M. Mokry
Institute for Aerospace Research, National Research Council, Canada

Abstract

The Cauchy type integral, used to represent the complex velocity, is converted to a line distribution of sources and vortices in the complex plane. The specification of the normal velocity on the bounding contour leads to a Riemann-Hilbert problem, which provides the theoretical foundation of the method. The boundary element discretization results in a simple algorithm for calculating potential flows in multiple-connected domains. Flow problems with periodic or homogeneous outer boundary conditions are treated using the concept of the Green's function in the complex plane.

1 Introduction

The Complex Variable Boundary Element Method (CVBEM) evolved as a numerical procedure for solving boundary value problems for analytic functions in terms of discretized Cauchy type integrals [1]. The method described in this paper is based on the same principle. However, instead of linking the integral to the complex potential, as is commonly done, it is linked to the complex velocity. The main advantages of this approach are: 1) the complex velocity is single-valued and hence no cuts in the computational domain are necessary and 2) the representing Cauchy type integral is equivalent to the contour distribution of source and vortex singularities. By selecting the Cauchy density as the boundary value of a function analytic in the external flow region, it is possible to specify the far field condition such that there is no flow in the complementary interior domain. In this particular case the normal and tangential velocities become decoupled, corresponding to the source and vortex densities respectively. The imposition of the normal-velocity boundary condition and the subsequent discretization by boundary elements leads to the vortex panel method in the complex plane, reported earlier [2].

WIT Transactions on Modelling and Simulation, Vol 42, © 2006 WIT Press
www.witpress.com, ISSN 1743-355X (on-line)
doi:10.2495/BE06022

2 Cauchy integral formulation

The complex disturbance velocity is represented by the Cauchy type integral

$$w(z) = \frac{1}{2\pi i}\int_C \frac{f(\zeta)}{\zeta - z} d\zeta, \tag{1}$$

where z and ζ are the complex coordinates of the observation and contour points respectively and C is the counterclockwise oriented, simple closed airfoil contour, Fig. 1a. The Cauchy density f is a continuous, complex-valued function defined on C. For multi-component airfoils, C is a union of nonintersecting simple-closed contours.

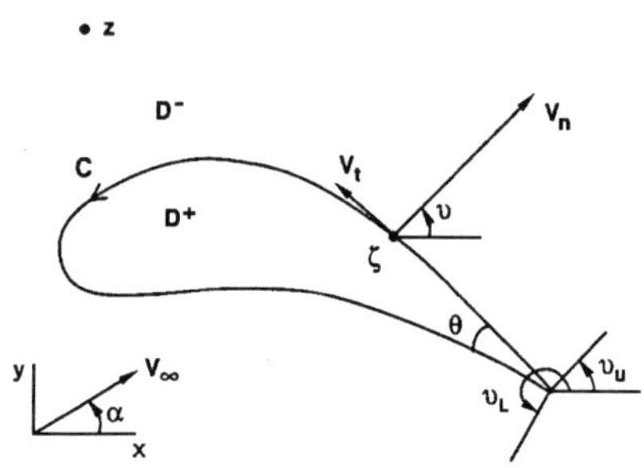

(a) geometrical description

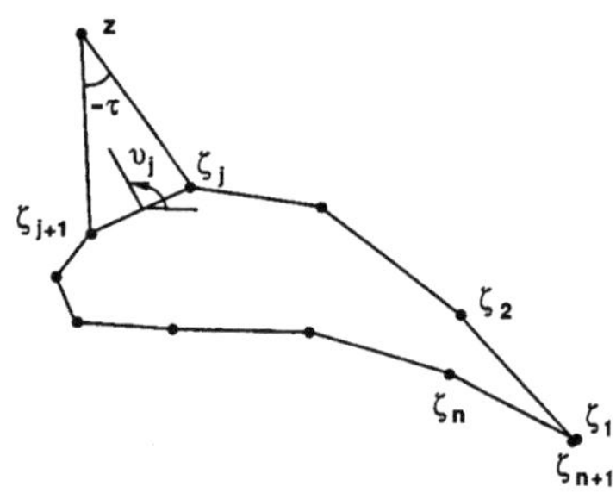

(b) approximation polygon

Figure 1: Single airfoil.

Introducing the angle ν between the outward normal to C and the real axis, the integral of Eq. (1) can be converted to the contour distribution of sources and vortices

$$w(z) = \int_C \left[\frac{\sigma(\zeta)}{2\pi(z-\zeta)} + \frac{i\gamma(\zeta)}{2\pi(z-\zeta)}\right]|d\zeta|, \tag{2}$$

using

$$d\zeta = [-\sin\nu(\zeta) + i\cos\nu(\zeta)]|d\zeta| = ie^{i\nu(\zeta)}|d\zeta| \tag{3}$$

and

$$f(\zeta) = -[\sigma(\zeta) + i\gamma(\zeta)]e^{-i\nu(\zeta)}. \tag{4}$$

The real-valued functions σ and γ are the source and vortex densities respectively.

There is a multiplicity of density functions f capable of representing a given analytic function w in either the internal domain D^+ (to the left of C) or the external domain D^- (to the right of C). For an external flow problem, it is natural to specify $f(\zeta)$ as the boundary value of a function $f(z)$ analytic in D^- and continuous in $D^- \cup C$. Applying the Cauchy integral formula to Eq. (1), we obtain

$$w(z) = \begin{cases} f(\infty) - f(z), & z \in D^- \cup C, \\ f(\infty), & z \in D^+, \end{cases} \tag{5}$$

where $f(\infty)$ is the value of $f(z)$ as $|z| \to \infty$.

If the free stream velocity is of unit magnitude and angle α to the real axis, the (total) complex velocity will be

$$W(z) = e^{-i\alpha} + w(z). \tag{6}$$

Choosing $f(\infty) = -e^{-i\alpha}$, it follows from Eqs. (5) and (6) that the fictitious interior flow vanishes whereas on the exterior face of the contour

$$W(\zeta) = -f(\zeta). \tag{7}$$

From Fig. 1a it can also be verified that in terms of the normal and tangential components

$$W(\zeta) = [V_n(\zeta) - iV_t(\zeta)]e^{-i\nu(\zeta)}. \tag{8}$$

Substituting Eq. (4) in (7) and comparing the latter with Eq. (8) shows that

$$V_n(\zeta) = \sigma(\zeta) \quad \text{and} \quad V_t(\zeta) = -\gamma(\zeta). \tag{9}$$

Accordingly, the discharge and circulation constants are given by

$$Q = \int_C V_n |d\zeta| = \int_C \sigma(\zeta)|d\zeta| \quad \text{and} \quad \Gamma = \int_C V_t |d\zeta| = -\int_C \gamma(\zeta)|d\zeta| \tag{10}$$

3 Airfoil boundary value problem

We assume that the normal component of velocity $V_n \equiv \sigma$ is prescribed (typically as zero) on the closed airfoil contour C. From Eqs. (6) and (8) the following boundary condition is obtained for the complex disturbance velocity:

$$\text{Re}\{[e^{-i\alpha} + w(\zeta)]e^{i\nu(\zeta)}\} = \sigma(\zeta). \tag{11}$$

Equation (11), written in the form

$$\text{Re}\left\{\frac{w(\zeta)}{q(\zeta)}\right\} = c(\zeta), \tag{12}$$

where

$$q(\zeta) = e^{-i\nu(\zeta)} \quad \text{and} \quad c(\zeta) = \sigma(\zeta) - \cos[\nu(\zeta) - \alpha], \tag{13}$$

are functions prescribed on the contour C, specifies the Riemann-Hilbert problem for the analytic function w. It resembles the Schwarz problem, except that the term in the curly brackets of Eq. (12) is not a boundary value of a function analytic in

D^-. Using Gakhov's regularization method [3], it can be shown that the solution can be made unique by specifying the circulation Γ if C is smooth [4] or by requiring that w be bounded if C possesses a single corner point [5], also known as the trailing edge. Since a discontinuity of the density function f gives rise to a logarithmic singularity in the integral of Eq. (1), this (Kutta-Joukowski) condition is synonymous with the requirement that f be equal on the upper and lower sides of the trailing edge: $f_U = f_L$. For the trailing edge angle

$$\theta = \nu_L - \nu_U - \pi, \qquad \pi > \theta > 0 \tag{14}$$

we obtain from Eq. (4) the conditions

$$\gamma_U = \frac{\sigma_L + \sigma_U \cos\theta}{\sin\theta} \quad \text{and} \quad \gamma_L = -\frac{\sigma_U + \sigma_L \cos\theta}{\sin\theta}. \tag{15}$$

Although an explicit solution of Eqs. (12)–(13) exists [4], its evaluation involves conformal mapping as an intermediary step and hence is less practical than a direct numerical solution obtained by the CVBEM.

4 The CVBEM algorithm

The airfoil contour is approximated by n straight-line boundary elements, as indicated in Fig. 1b. The nodal points, which are the vertices of the approximation polygon, are numbered counterclockwise, $1, 2, \ldots, n+1$, starting with the upper trailing edge point, ζ_1 and ending with the lower trailing edge point, ζ_{n+1}. The jth boundary element is the line segment between points ζ_j and ζ_{j+1}. For a closed trailing edge, $\zeta_{n+1} = \zeta_1$.

The Cauchy type integral of Eq. (1) is discretized as

$$w(z) = \sum_{j=1}^{n} \Delta_j w(z), \tag{16}$$

where

$$\Delta_j w(z) = \frac{1}{2\pi i} \int_{\zeta_j}^{\zeta_{j+1}} \frac{f(\zeta)}{\zeta - z} d\zeta \tag{17}$$

is the contribution of the jth boundary element to the complex disturbance velocity at the observation point z.

In the present method, the density function f between the boundary element end points ζ_j and ζ_{j+1} is represented by the linear trial function

$$\begin{aligned} f(\zeta) &= f_j + \frac{f_{j+1} - f_j}{\zeta_{j+1} - \zeta_j}(\zeta - \zeta_j) \\ &= \frac{f_{j+1} - f_j}{\zeta_{j+1} - \zeta_j}(\zeta - z) + f_{j+1}\frac{z - \zeta_j}{\zeta_{j+1} - \zeta_j} - f_j \frac{z - \zeta_{j+1}}{\zeta_{j+1} - \zeta_j}, \end{aligned} \tag{18}$$

where f_j and f_{j+1} are the values of $f(z)$ at the respective end points of the boundary element. Substituting Eq. (18) in Eq. (17), we find

$$\Delta_j w(z) = \frac{f_{j+1} - f_j}{2\pi i} + \frac{1}{2\pi i}\left[f_{j+1}\frac{z - \zeta_j}{\zeta_{j+1} - \zeta_j} - f_j \frac{z - \zeta_{j+1}}{\zeta_{j+1} - \zeta_j}\right] \ln \frac{\zeta_{j+1} - z}{\zeta_j - z}. \tag{19}$$

From Eq. (4) we have for the jth boundary element

$$\begin{aligned} f_j &= -(\sigma_j + i\gamma_j)e^{-i\nu_j} \\ f_{j+1} &= -(\sigma_{j+1} + i\gamma_{j+1})e^{-i\nu_j}, \end{aligned} \tag{20}$$

where

$$e^{i\nu_j} = \cos\nu_j + i\sin\nu_j = -i\frac{\zeta_{j+1} - \zeta_j}{|\zeta_{j+1} - \zeta_j|}. \tag{21}$$

The logarithmic term of Eq. (19) can be written

$$\ln\frac{\zeta_{j+1} - z}{\zeta_j - z} = \ln\left|\frac{\zeta_{j+1} - z}{\zeta_j - z}\right| + i\tau,$$

where $-\pi \leq \tau \leq \pi$ is the angle obtained by the rotation of the vector $\zeta_j - z$ into the direction of the vector $\zeta_{j+1} - z$, see Fig. 1b. In the limit as the point z approaches an interior point of the jth boundary element from the flow field side D^-, τ tends to the value $-\pi$. Accordingly, the complex disturbance velocity induced by the jth segment at its own midpoint

$$z_j = \frac{1}{2}(\zeta_j + \zeta_{j+1}) \tag{22}$$

is

$$\Delta_j w(z_j) = \frac{f_{j+1} - f_j}{2\pi i} - \frac{1}{4}(f_{j+1} + f_j). \tag{23}$$

From Eqs. (16),(19),(20) and (23), we obtain for the complex disturbance velocity at the midpoint z_k of the kth segment

$$w(z_k) = \sum_{j=1}^{n+1} C_{k,j}(\sigma_j + i\gamma_j), \quad k = 1, \ldots, n. \tag{24}$$

The complex matrix $C_{k,j}$ can be evaluated as

$$C_{k,j} = K_{k,j} + L_{k,j} \tag{25}$$

where

$$K_{k,j} = \begin{cases} \frac{e^{-i\nu_j}}{2\pi i}\left(1 + \frac{z_k - \zeta_{j+1}}{\zeta_{j+1} - \zeta_j}\ln\frac{\zeta_{j+1} - z_k}{\zeta_j - z_k}\right), & j \neq k,\, n+1 \\ \left(\frac{1}{4} + \frac{1}{2\pi i}\right)e^{-i\nu_k}, & j = k \\ 0, & j = n+1 \end{cases} \tag{26}$$

$$L_{k,j} = \begin{cases} \frac{-e^{-i\nu_{j-1}}}{2\pi i}\left(1 + \frac{z_k - \zeta_{j-1}}{\zeta_j - \zeta_{j-1}}\ln\frac{\zeta_j - z_k}{\zeta_{j-1} - z_k}\right), & j \neq 1,\, k+1 \\ \left(\frac{1}{4} - \frac{1}{2\pi i}\right)e^{-i\nu_k}, & j = k+1 \\ 0, & j = 1 \end{cases} \tag{27}$$

Equation (11) is to be satisfied at all boundary element midpoints:

$$\mathrm{Re}\big\{[e^{-i\alpha} + w(z_k)]e^{i\nu_k}\big\} = \frac{1}{2}(\sigma_k + \sigma_{k+1}), \quad k = 1, \ldots, n, \tag{28}$$

where σ_k and σ_{k+1} stand for the prescribed values of normal velocity at the end-points of the kth boundary element. Substituting from Eq. (24) and separating the terms containing the unknown values of vortex density from the given quantities, we obtain

$$\sum_{j=1}^{n+1} \mathrm{Im}\big\{e^{i\nu_k} C_{k,j}\big\}\gamma_j = \mathrm{Re}\Big\{e^{i\nu_k}(e^{-i\alpha} + \sum_{j=1}^{n+1} C_{k,j}\sigma_j)\Big\} - \frac{1}{2}(\sigma_k + \sigma_{k+1}) \qquad k = 1, \ldots, n. \tag{29}$$

This represents a system of n linear algebraic equations in $n+1$ unknown vortex densities γ_j. However, with the inclusion of the two trailing edge conditions of Eq. (23), we end up with $n+2$ linear equations in the $n+1$ unknown vortex densities.

Thus, adopting for Eq. (29) the matrix notation

$$A_{k,j}\gamma_j = b_k$$

and, inserting the trailing-edge conditions of Eq. (15) as the first and the last rows, we obtain

$$A_{1,j} = \begin{cases} 1, & j = 1 \\ 0, & j = 2, \ldots, n+1, \end{cases} \quad b_1 = \frac{\sigma_{n+1} + \sigma_1 \cos\theta}{\sin\theta}$$

$$A_{k+1,j} = \mathrm{Im}\big\{e^{i\nu_k} C_{k,j}\big\}, \quad k = 1, \ldots, n, \quad j = 1, \ldots, n+1$$

$$b_{k+1} = \mathrm{Re}\Big\{e^{i\nu_k}(e^{-i\alpha} + \sum_{j=1}^{n+1} C_{k,j}\sigma_j)\Big\} - \frac{1}{2}(\sigma_k + \sigma_{k+1}), \quad k = 1, \ldots, n$$

$$A_{n+2,j} = \begin{cases} 0, & j = 1, \ldots, n \\ 1, & j = n+1, \end{cases} \quad b_{n+2} = -\frac{\sigma_1 + \sigma_{n+1}\cos\theta}{\sin\theta}.$$

The cosine and sine of the trailing edge angle θ are obtained according to Eq. (14) as the real and imaginary parts of

$$e^{i\theta} = e^{i(\nu_n - \nu_1 - \pi)} = \frac{\zeta_n - \zeta_{n+1}}{|\zeta_n - \zeta_{n+1}|}\frac{|\zeta_2 - \zeta_1|}{\zeta_2 - \zeta_1}.$$

An extension of the algorithm to a multi-component airfoil is fairly straightforward [2]. Denoting by $m(1)$, $m(2)$, ..., the number of boundary elements on the airfoil components 1, 2, ..., the numbering is such that ζ_1, ζ_2, ..., $\zeta_{m(1)+1}$ are the corner points of the first component, $\zeta_{m(1)+2}$, $\zeta_{m(1)+3}$, ..., $\zeta_{m(1)+m(2)+2}$ the

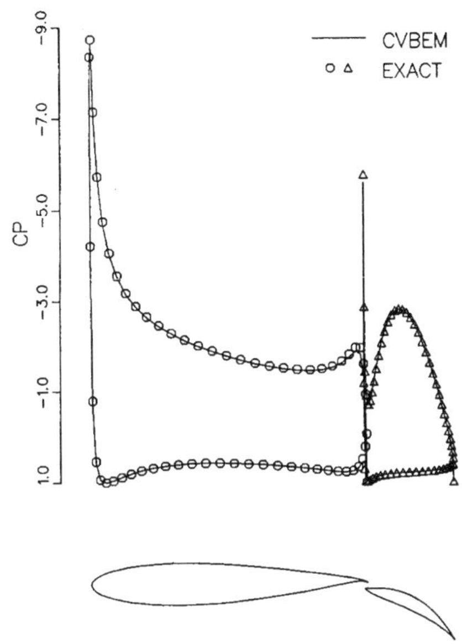

Figure 2: Williams airfoil.

corner points of the second component, and so on. The two trailing edge conditions (15) are satisfied on each airfoil component, leaving the system of linear equations overdetermined by the number of airfoil components. Several methods can be used to solve numerically such a system [2].

Once the values γ_j have been computed, the normal and tangential components of velocity at the corner points are obtained as equal to σ_j and $-\gamma_j$ respectively, see Eqs. (9). The corresponding pressure coefficients are obtained as

$$C_p(\zeta_j) = 1 - (\sigma_j^2 + \gamma_j^2). \tag{30}$$

A numerical example is given for Williams' two-component airfoil [6], configuration 'A', designed by conformal mapping. The number of boundary elements is $m(1) = m(2) = 62$, with the contour point coordinates listed in [6]. From Fig. 2 it is seen that the CVBEM reproduces the pressure distributions correctly, including the steep suction peaks.

5 Modified distributions of sources and vortices

If the external flow around a body is subject to periodic or outer constraints, which can be described by linear homogeneous boundary conditions, it is convenient to generalize the line distribution of sources and vortices as

$$w(z) = \int_C [\sigma(\zeta)G_\sigma(z,\zeta) + \gamma(\zeta)G_\gamma(z,\zeta)]|d\zeta|, \tag{31}$$

where C is the body contour and

$$G_\sigma(z,\zeta) = \frac{1}{2\pi(z-\zeta)} + H_\sigma(z,\zeta), \quad G_\gamma(z,\zeta) = \frac{i}{2\pi(z-\zeta)} + H_\gamma(z,\zeta) \quad (32)$$

are the Green's functions with the analytic parts H_σ and H_γ.

5.1 Cascades

Consider an infinite cascade of blades with the oncoming stream parallel to the x-axis as illustrated in Fig. 3a. The cascade is characterized by the spacing (pitch) t, stagger angle β, inlet angle β_1 and outlet angle β_2 (unknown).

The periodicity boundary condition is

$$w(z) = w(z \pm k\tau), \qquad k = 1, 2, 3, \ldots, \quad (33)$$

where τ is the complex spacing

$$\tau = te^{i\beta_1}. \quad (34)$$

Using the method of images [7]

$$\begin{aligned} H_\sigma(z,\zeta) &= \frac{1}{2\tau}\left[\cot\frac{\pi(z-\zeta)}{\tau} - \frac{\tau}{\pi(z-\zeta)}\right] + \frac{i}{2\tau} \\ H_\gamma(z,\zeta) &= iH_\sigma(z,\zeta). \end{aligned} \quad (35)$$

Important limiting values are

$$\begin{aligned} &\lim_{z\to\zeta} H_\sigma(z,\zeta) = \frac{i}{2\tau}, && \lim_{x\to-\infty} H_\sigma(z,\zeta) = 0, && \lim_{x\to\infty} H_\sigma(z,\zeta) = \frac{i}{\tau}, \\ &\lim_{z\to\zeta} H_\gamma(z,\zeta) = -\frac{1}{2\tau}, && \lim_{x\to-\infty} H_\gamma(z,\zeta) = 0, && \lim_{x\to\infty} H_\gamma(z,\zeta) = -\frac{1}{\tau}. \end{aligned}$$

From Eqs. (10) and (31)–(32),

$$\lim_{x\to-\infty} w(z) = 0, \qquad \lim_{x\to\infty} w(z) = \frac{1}{\tau}(\Gamma + iQ). \quad (36)$$

Using Eqs. (34) and (36), the deflection angle δ, measured positive in the clockwise direction, is obtained from

$$\tan\delta = \lim_{x\to\infty} \frac{\mathrm{Im}\{w(z)\}}{1 + \mathrm{Re}\{w(z)\}} = \frac{Q\cos\beta_1 - \Gamma\sin\beta_1}{t + \Gamma\cos\beta_1 + Q\sin\beta_1} \quad (37)$$

and the outlet angle from

$$\beta_2 = \beta_1 + \delta. \quad (38)$$

A numerical verification is given for the Gostelow compressor cascade [8]. The geometrical parameters of the cascade are: spacing to chord ratio $t/c = 0.99$, stagger angle $\beta = 52.5^o$ and inlet angle $\beta_1 = 36.5^o$, see Fig. 3a. The agreement of the CVBEM pressure distribution with the theoretical one is demonstrated in Fig. 3b. The calculated outlet angle $\beta_2 = 59.80^o$ compares reasonably well with Gostelow's exact value of 59.98^o.

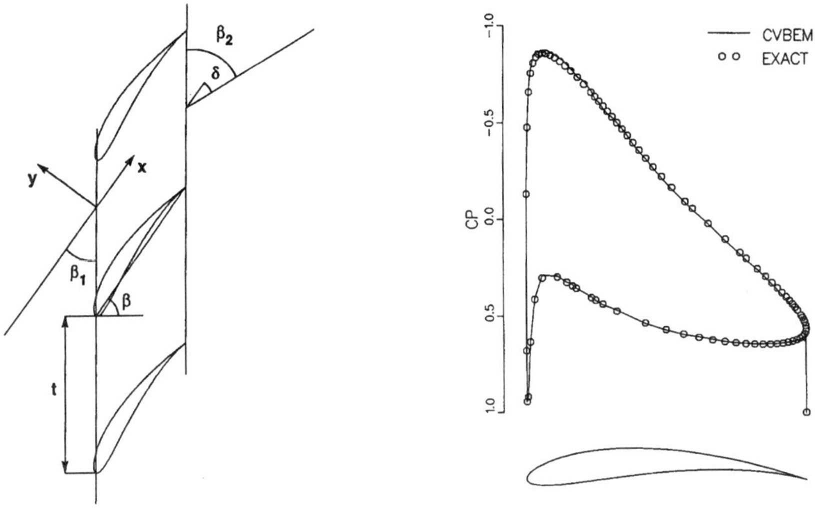

(a) geometry (b) pressure

Figure 3: Gostelow cascade.

5.2 Wind tunnel walls

Flow between two parallel walls, specified by the boundary condition

$$\mathrm{Im}\{w(z)\} = 0, \qquad -\infty < x < \infty, \quad y = \pm\frac{h}{2}, \tag{39}$$

can be handled similarly. By the method of images [2]

$$H_\sigma(z,\zeta) = B(z,\zeta) + E(z,\zeta) + \frac{1}{h}, \quad H_\gamma(z,\zeta) = i[B(z,\zeta) - E(z,\zeta)], \tag{40}$$

where, using the over bar for complex conjugation,

$$\begin{aligned} B(z,\zeta) &= \frac{1}{2h}\left[\exp\left(\pi\frac{z-\zeta}{h}\right) - 1\right]^{-1} - \frac{1}{2\pi(z-\zeta)} \\ E(z,\zeta) &= -\frac{1}{2h}\left[\exp\left(\pi\frac{z-\overline{\zeta}}{h}\right) + 1\right]^{-1} \end{aligned} \tag{41}$$

From Eqs. (31)–(32) and (40)–(41) it can be shown that

$$\lim_{x\to-\infty} w(z) = 0, \qquad \lim_{x\to\infty} w(z) = \frac{1}{h}Q, \tag{42}$$

which is consistent with the 'wake blockage' phenomenon.

6 Concluding remarks

This paper shows that the two-dimensional vortex panel method implemented in the complex plane can be regarded as a special CVBEM case. The method is particularly well suited for potential flow problems in multiple connected domains. Its accuracy and versatility has been demonstrated on two examples.

References

[1] Hromadka II, T.V., *The Complex Variable Boundary Element Method*, Springer-Verlag: Berlin, 1984.
[2] Mokry, M., "Calculation of the Potential Flow Past Multi-Component Airfoils Using a Vortex Panel Method in the Complex Plane," LR-596, National Aeronautical Establishment, National Research Council Canada, Nov. 1978.
[3] Gakhov, F.D., *Boundary Value Problems*, Pergamon Press: Oxford, 1966.
[4] Mokry, M., Potential Flow Past Airfoils as a Riemann-Hilbert Problem, AIAA-96-2161, 1st AIAA Theoretical Fluid Mechanics Meeting, New Orleans, June 1996.
[5] Murid, A., Nasser, M., and Amin, N., A Boundary Integral Equation for the 2D External Potential Flows, to be published in *J. Appl. Mech and Eng.*
[6] Williams, B.R., "An Exact Case for the Plane Potential Flow About Two Adjacent Lifting Aerofoils," TR 71197, Royal Aircraft Establishment, Sept. 1971.
[7] Giesing, J.P., "Extension of the Douglas Neumann Program to Problems of Lifting, Infinite Cascades," LB 31653, Douglas Aircraft Co., July 1964.
[8] Gostelow, J.P., "Potential Flow through Cascades - A Comparison between Exact and Approximate Solutions," C.P. No.807, Aeronautical Research Council, 1965.

BE DRM-MD for two-phase flow through porous media

T. Samardzioska[1] & V. Popov[2]
[1]*Faculty of Civil Engineering, Univ. St. Cyril & Methodius, Skopje, Macedonia*
[2]*Wessex Institute of Technology, Southampton, UK*

Abstract

The numerical model of two-phase flow through porous media is used in many branches of science and engineering, including groundwater hydrology, soil science, reservoir problems in petroleum engineering, storage of radioactive or chemical wastes, etc. Two-phase flow consists of two flowing fluids, which do not interchange mass and do not have reaction with the solid matrix. The equations governing the two-phase flow through porous media are special cases of general balance laws. From a numerical point of view, due to the high nonlinearity of the governing equations and their strong coupling, the simulation provides a challenging numerical problem, even when simplifying assumptions are invoked. A new numerical scheme based on the dual reciprocity method – multi domain (DRM-MD) for the numerical simulation is presented. The efficiency of the method is proved on one and two dimensional two-phase flow examples.

1 Introduction

Numerical simulation of multiphase flow through fractured rocks is important in many branches of science and engineering. Hydrologists, agricultural engineers, soil physicists and others have for many years been concerned with fluid movement in air-water porous media systems. In the most recent years, environmental concerns and search for secure atomic waste depositories and for remediation strategies for contaminated aquifers raised more interest in the movement of fluids in the so-called vadose or unsaturated zone interposed between the atmosphere and the groundwater (saturated zone). Many potential

WIT Transactions on Modelling and Simulation, Vol 42, © 2006 WIT Press
www.witpress.com, ISSN 1743-355X (on-line)
doi:10.2495/BE06023

groundwater contaminants are introduced at (or near) the soil surface via atmospheric deposition, spills, leakage from underground tanks, subsurface waste disposal, etc. Furthermore, a wide class of environmental contaminants consists of organic compounds of low water solubility which can occur as a separate liquid phase in the soil. Such liquids include many widely used industrial solvents and automobile and jet fuels which unfortunately often enter the ground via surface spills or leaks from underground storage tanks. Historically, the greatest stimulus for development of multiphase flow models was initiated by the commercial interest of the petroleum industry, induced by the lure of more efficient oil and gas recovery from reservoirs. Most of the existing flow and transport models are single-phase models that describe the groundwater flow and the convective-dispersive spreading of one or more components entirely dissolved in water. The unsaturated zone is a multiphase system, consisted of wetting phase (water), air and non-wetting phase (for example, oil). Two-phase flow consisted of two flowing fluids, which do not interchange mass and do not have reaction with the solid matrix, is the simplest multiphase flow. The elimination of the equation of the gas phase is often referred to as Richard's approximation and is the basis of conventional analyses of two-phase flow. The porous media itself is assumed to be incompressible.

2 Governing equations

The equations governing the two-phase flow of fluids are special cases of general balance laws. When two fluids coexist in a porous medium, one of them will generally have preferential wettability for the solid phase and will occupy the smaller voids (wetting phase, subscript W), while the less wetting fluid is consigned to larger voids (non-wetting phase, for example oil, subscript O). Darcy’s law holds for both the phases, see [1-3]:

$$\phi\frac{\partial}{\partial t}(S_W\rho_W)-\nabla[\rho_W\lambda_W K(\nabla p_W-\rho_W g)]-\rho_W q_W=0 \tag{1}$$

$$\phi\frac{\partial}{\partial t}(S_O\rho_O)-\nabla[\rho_O\lambda_O K(\nabla p_W+\nabla p_{COW}-\rho_O g)]-\rho_O q_O=0 \tag{2}$$

where:

S_α - saturation of the fluid phase α; ρ_α - density of the fluid phase α; p_α - pressure head in the fluid phase α; $\lambda_\alpha=k_{r\alpha}/\mu_\alpha$ - mobility of the fluid phase α; $k_{r\alpha}$ - relative permeability of the fluid phase α; μ_α - dynamic viscosity; K – permeability.

Equations (1) and (2) are coupled, since they have to satisfy the condition that the fluids fill up the pore volume, i.e. the additional relations:

$$S_w+S_O=1 \tag{3}$$

A pressure difference occurs along the fluid-fluid interface, whose magnitude depends on the interface curvature. This pressure diference is termed as capillary pressure p_{COW} and it must be positive by definition, since the non-wetting phase pressure must exceed that in the wetting phase:

$$p_{COW} = p_O - p_W \tag{4}$$

Darcy's law for the wetting and non-wetting phase can be formulated as follows:

$$v_W = -\lambda_W K(\nabla p_W - \rho_W g) \tag{5}$$

$$v_O = -\lambda_O K(\nabla p_W + \nabla p_{COW} - \rho_O g) \tag{6}$$

Introducing (5) and (6) into equations (1) and (2) one can obtain shortened form:

$$\phi \frac{\partial (S_W \rho_W)}{\partial t} + \nabla[\rho_W v_W] - \rho_W q_W = 0 \tag{7}$$

$$\phi \frac{\partial (S_O \rho_O)}{\partial t} + \nabla(\rho_O v_O) - \rho_O q_O = 0 \tag{8}$$

Summing up the equations (7) and (8), and introducing (3):

$$v_t = v_W + v_O$$

$$\nabla v_t = \nabla(v_W + v_O) = q_W + q_O - \frac{1}{\rho_O} v_O \cdot \nabla \rho_O - \frac{1}{\rho_W} v_W \cdot \nabla \rho_W$$
$$+ \phi \left[\frac{1 - S_W}{\rho_O} \cdot \frac{\partial \rho_O}{\partial t} + \frac{S_W}{\rho_W} \cdot \frac{\partial \rho_W}{\partial t} \right] = 0 \tag{9}$$

Neglecting the compressibility of both phases yields to:

$$\nabla v_t = q_W + q_O = q_t \tag{10}$$

The combination of equations (5) and (6) when the first one is multiplied by λ_O, and the second one by λ_W, leads to:

$$v_W \lambda_O - v_O \lambda_W = \lambda_O \lambda_W K(\nabla p_{COW} - \rho_O g + \rho_W g) \tag{11}$$

with $$v_O = v_t - v_W$$

$$v_O = \frac{\lambda_O}{\lambda_O + \lambda_W} [v_t - \lambda_W K(\nabla p_{COW} - \rho_O g + \rho_W g)] \tag{12}$$

After substituting eq. (12) in eq. (8), the resulting equation is:

$$\nabla [f_O v_t - \bar{\lambda}(\nabla p_C + \rho_W g - \rho_O g)] = \phi \frac{\partial S_W}{\partial t} - q_O \tag{13}$$

where: $$f_O = \frac{\lambda_O}{\lambda_O + \lambda_W} \quad \text{- fractional flow for the oil} \tag{14}$$

$$f_w = \frac{\lambda_W}{\lambda_O + \lambda_W} \quad \text{- fractional flow for the water} \tag{15}$$

and $$\bar{\lambda} = \frac{\lambda_W \lambda_O}{\lambda_O + \lambda_W} \tag{16}$$

All the unknowns in the equation (13) are functions of the saturation, thus one can use following expressions for the derivatives:

$$\frac{\partial p_c}{\partial x_i} = \frac{dp_c}{dS_W}\frac{\partial S_W}{\partial x_i} \tag{17}$$

$$\frac{\partial (f_O v_t)}{\partial x_i} = v_t \frac{\partial f_O}{\partial S_W}\frac{\partial S_W}{\partial x_i} - f_O q_t \tag{18}$$

$$\frac{\partial}{\partial x_i}\left[\bar{\lambda}(\rho_W g - \rho_O g)\right] = (\rho_W g - \rho_O g)\frac{d\bar{\lambda}}{dS_W}\frac{\partial S_W}{\partial x_i} \tag{19}$$

Including (17), (18) and (19) into equation (13) yields the following equation:

$$\begin{aligned}&\frac{\partial}{\partial x_i}\left[\bar{\lambda}\frac{dp_c}{dS_W}\frac{\partial S_W}{\partial x_i}\right] - \left[v_t \frac{df_W}{dS_W} + (\rho_W g - \rho_O g)\frac{d\bar{\lambda}}{dS_W}\right]\frac{\partial S_W}{\partial x_i} \\ &- \phi\frac{\partial S_W}{\partial t} - q_W + f_W q_t = 0\end{aligned} \tag{20}$$

The first term in the two-phase saturation equation (20) represents dispersive part, the second term is convective part, the third is mass storage term and last ones a source and sink terms. The equation is parabolic because of the presence of the term that includes capillary pressure.

Neglecting only the gravitational effects and the source and sink term, for simplicity, and taking the significance of the capillary pressure gradient into account, one arrives to the so-called Mc Whorter problem:

$$\phi \cdot \frac{\partial S_W}{\partial t} + q_t \cdot \frac{df_W}{dS_W} \cdot \frac{\partial S_W}{\partial x_i} = -\frac{\partial}{\partial x_i}\left(D \cdot \frac{\partial S_W}{\partial x_i}\right) \tag{21}$$

$$D = k_{ro} \cdot K \cdot \frac{f_W}{\mu_O} \cdot \frac{dp_c}{dS_W} \text{ - dispersion tensor} \tag{22}$$

Assuming that all the properties dependent on water saturation (the dispersion tensor) are constant in a particular time step, equation (21) can be written as:

$$D \cdot \nabla^2 S_W = -\phi \cdot \frac{\partial S_W}{\partial t} - q_t \cdot \frac{df_W}{dS_W} \cdot \frac{\partial S_W}{\partial x_i} \tag{23}$$

A critical component of predictive two-phase flow models is the mathematical description of the relative permeability, saturation and pressures relations (k-S-p), sometimes reffered to as constitutive models. Reliable experimental measurements are difficult to obtain, they are time consuming, expensive and require advanced laboratory skills. A summary of different constitutive models is given by Helmig, see [3].

3 Numerical implementation

The numerical technique that is used to solve the partial differential equation defined by the model is the Boundary Element Dual Reciprocity Method – Multi

Domain scheme (BE DRM-MD) [4-8]. It has been used for the first time to solve a two-phase flow model. The method belongs to the boundary element techniques that transform the original partial differential equation into an equivalent integral equation by means of the corresponding Green's theorem and its fundamental solution. Simplicity, elegance of the formulation and its efficient solution extended the application of the BEM to a wide variety of time dependent and non-linear problems.

The governing equation that describes a linear time-dependent flow in porous media, in the general form can be written as,

$$\nabla^2 u = b(x, y, u, t) \text{ in the domain } D \tag{24}$$

with the 'essential' boundary conditions of the type $u = \bar{u}$ on Γ_1 and 'natural' boundary conditions $q = \partial u / \partial n = \bar{q}$ on Γ_2. Here $\Gamma=\Gamma_1+\Gamma_2$ is the exterior boundary that encloses the domain D and n is its outward normal.

For the two-phase flow model, u represents $S_w(x_i,t)$ in the equation (23) for the flow and $b = -\frac{\phi}{D} \cdot \frac{\partial S_W}{\partial t} - \frac{q_t}{D} \cdot \frac{df_W}{dS_W} \cdot \frac{\partial S_W}{\partial x_i}$.

Applying the DRM (Dual Reciprocity Method) approach to the equations (23), according the detailed explanation in references [6] and [7], yields:

$$HS_W - GQ = -\frac{S}{D}\left[\phi \cdot \frac{\partial S_W}{\partial t} + q_t \cdot \frac{df_W}{dS_W} \cdot \frac{\partial S_W}{\partial x_i}\right] \tag{25}$$

where:

$$S = \left(H\hat{S}_W - G\hat{Q}\right) \cdot F^{-1}, \tag{26}$$

The two boundary element characteristic matrices H and G on both sides of the equation (25) consist of coefficients, which are calculated assuming the fundamental solution is applied at each node successively, and depending only on geometrical data, see [4–7].

Applying BE DRM derivation:

$$\frac{\partial S_W}{\partial x_i} = \frac{\partial F}{\partial x_i} F^{-1} \cdot S_W \tag{27}$$

Including that:

$$T = S \cdot \frac{1}{D} \cdot q_t \cdot \frac{df_W}{dS_W} \cdot \frac{\partial F}{\partial x_i} F^{-1} \tag{28}$$

and

$$koef = \frac{\phi}{D}, \tag{29}$$

the main equation takes the following form:

$$HS_W - GQ = -S \cdot koef \cdot \frac{\partial S_W}{\partial t} - T \cdot S_W \tag{30}$$

$$(H + T)S_W - GQ = -S \cdot koef \cdot \frac{\partial S_W}{\partial t} \tag{31}$$

For the time marching scheme:

$$S_W = (1-\theta_s)\cdot S_W{}^{n} + \theta_s \cdot S_W{}^{n+1} \tag{32a}$$

$$q = (1-\theta_q)\cdot q^{n} + \theta_q \cdot q^{n+1} \tag{32b}$$

where θ_s and θ_q take values between 0 and 1. Time derivative is approximated using a finite-difference approximation:

$$\frac{\partial S_W}{\partial t} = \frac{1}{\Delta t}\cdot\left(S_W{}^{n+1} - S_W{}^{n}\right) \tag{33}$$

Finally the equation for two-phase Mc Whorter flow is:

$$\begin{aligned}&\left[\theta_S(H+T)+\frac{koef\cdot S}{\Delta t}\right]\cdot S_W{}^{n+1} - G\cdot\theta_q\cdot Q^{n+1} = \\ &=\left[\frac{koef\cdot S}{\Delta t} - (1-\theta_S)(H+T)\right]\cdot S_W{}^{n} + G\cdot(1-\theta_q)\cdot Q^{n}\end{aligned} \tag{34}$$

At each interface between the adjacent sub-domains, the corresponding full matching conditions are imposed. At the mth interface of two adjacent sub-domains, i and i+1, the saturation needs to be continuous and the flux leaving one sub-domain has to be equal to the flux entering the other sub-domain:

$$S_{wi}\big|_m = S_{wi+1}\big|_m \qquad \text{for saturation} \tag{35}$$

$$\left(\frac{\partial S_{wi}}{\partial n}\right)\Bigg|_m = -\left(\frac{\partial S_{wi+1}}{\partial n}\right)\Bigg|_m \qquad \text{for saturation flux} \tag{36}$$

where subscripts i and i+1 indicate the sub-domain, n_i and n_{i+1} represent the unit outward vectors normal to the interface looking from the sub-domain i and i+1 respectively.

DRM-MD scheme has been implemented in a code which offers high flexibility in mesh generation. The mesh can be built of sub-domains of various sizes, geometries and numbers of boundary elements per sub-domain. Coupling of large homogeneous domains that do not need to be discretized with very small domains is implemented in the code, possibility that had been previously obtained only with BEM-FEM formulations. The present formulation removes problems with the matching compatibility of the two methods.

The sub-domain formulation leads to block banded matrix systems with one block for each sub-domain and overlaps between blocks when sub-domains have a common interface. In the limit of a very large number of sub-domains the resulting internal mesh pattern looks like a finite element grid. One obvious advantage of the DRM-MD is that the matrix multiplications are done on the level of each sub-domain, accelerating this way the numerical scheme.

By using the DRM-MD one is able to solve the system for the field variables, in this case S_w, and their derivatives. The DRM-MD approach requires no major restrictions in the size and geometry of sub-domains.

4 Model verification

Two examples are given for model verification in this paper. For both of them, the gravitational effects are neglected. One of the most popular models, proposed by Brooks and Corey is employed for description of the relative permeabilities and capillary pressure constitutive relationships. Expression for the wetting phase, the power-law function, is:

$$K_{rw} = \left(S_w\right)^{\frac{2+3\lambda}{\lambda}} \tag{37}$$

Similarly for the non-wetting phase:

$$K_{row} = \left(1-S_w\right)^2\left(1-S_w^{\frac{2+\lambda}{\lambda}}\right) \tag{38}$$

$$p_{cow} = p_d \cdot S_w^{-1/\lambda} \tag{39}$$

where p_d is the bubbling pressure, which apparently represents the smallest capillary pressure at which a continuous non-wetting phase starts to enter the system.

The Brooks-Corey function is constrained by $P_c \geq P_d$, see [9], neglecting capillary pressure variation with saturation below the entry pressure.

For the most often used values of λ=2, the above equations obtain the following form:

$$k_{rw} = S_w^{\ 4};\quad k_{ro} = \left(1-S_w\right)^2\left(1-S_w^{\ 2}\right);\quad p_{cow} = p_d\big/\sqrt{S_w} \tag{40}$$

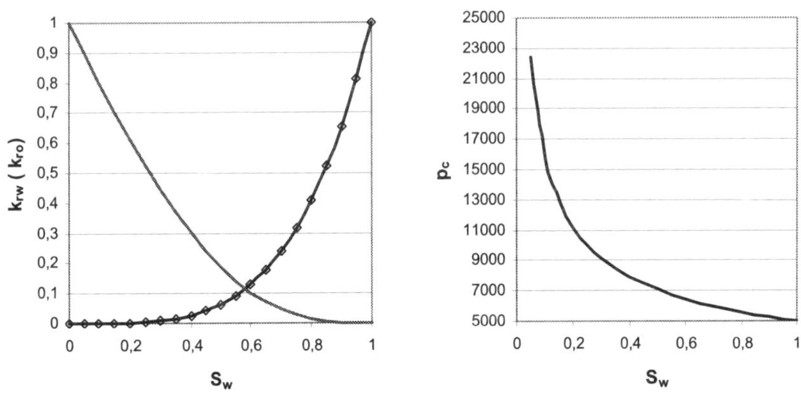

Figure 1: Brooks – Corey model; a) relative permeabilities; b) capillary pressure.

According the definition, the water saturation cannot be greater than 1 or negative. In this work both phases are assumed to be mobile and their ranges of saturations [0,1] cover the two-phase flow region.

Because there are not available analytical solutions for complex nonlinear processes, the results of these test cases can only be checked for plausibility and compared with other numerical results in the literature.

4.1 One-dimensional two-phase flow example

In a one-dimensional set-up the initial oil filling is extracted on one side and water is entering the system on the other side. The numerical example that corresponds to the example in the book of R. Helmig, see [3] is analyzed. Domain is long L=2.6m. Dirichlet boundary conditions are imposed: at x=0.0m, S_w=1.0 and at x=2.6m, S_w=0.01. These boundary conditions are chosen for sake of having a nonconductive right boundary. Initial saturation through the whole domain is S_w=0.01. Residual saturations for the water and the oil are $S_{wr}=S_{or}$=0.0. Densities are $\rho_w=\rho_o$=1000 kg/m^3 and dynamic viscosities are $\mu_w=\mu_o$=0.001 kg/(ms) for both fluids. Properties of the rock are: absolute permeability $K=10^{-10}$ m^2, porosity ϕ=0.3. The influence of the space discretization is analyzed with three meshes; the first one with 13 sub-domains of length 0.2m, the second mesh with 26 sub-domains of length 0.1m, and the third one is composed of 52 sub-domains x 0.05m. The results of the simulation with the grid with 52 sub-domains and time step dt=0.5sec are presented in Figure 2.

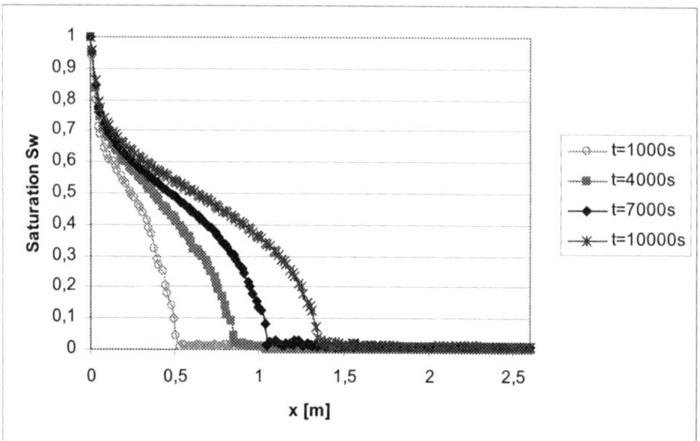

Figure 2: Saturation profiles at four time steps (52sub-domains, dt=0.5sec).

Even the coarse grid simulation produces useable and accurate solution, though the results on the finer grids are considerably better. It is oscilation free and shows a good mass balance. Testing simulation with different time steps on each of these meshes showed that one should be extremely careful with the choice of the time step when rather coarse mesh is used for the domain.

4.2 Two-dimensional two-phase flow example

The so-called "five spot example" discretized with square mesh 300m x 300m is analysed. The initial saturation of oil is assumed to be So=1.0 and initial pressure is Po=2·10^5 Pa. Initial saturation through the whole domain is S_w=0.0. Residual saturations for the water and the oil are assumed to be S_{wr}=S_{or}=0.0. Densities are ρ_w=ρ_o=1000 kg/m^3 and dynamic viscosities are μ_w=μ_o=0.001 kg/(ms) for both fluids. The properties of the rock are: absolute permeability K=10^{-7} m^2, porosity ϕ=0.2. Two cases are considered: one homogeneous with permeability K_1= 10^{-7}m^2 all over the domain and the second one is a problem with low permeability zone (in the center of the domain a zone of 112.5mx112.5m with low permeability of K_2=10^{-10}m^2, three orders of magnitude less than in zone 1).

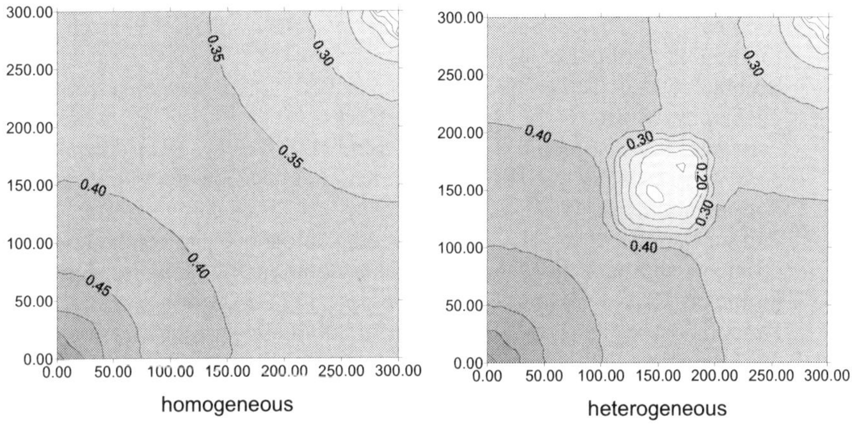

Figure 3: Homogeneous and heterogeneous five spot example.

5 Conclusions

In this work two-phase flow model was presented for the flow through fractured porous rocks. Even the coarse grid simulation produces useable and accurate solution, though the results on the finer grids are considerably better. It is oscilation free and shows a good mass balance. Testing simulation with different time steps on each of these meshes showed that one should be extremely careful with the choice of the time step when rather coarse mesh is used for the domain. Both examples showed good agreement for the representation of the saturations. They perform very well for non-homogeneous domains, even when there are great differences between the permeabilities within the domain. The models were implemented using the DRM-MD scheme, which showed flexibility in the modelling that previously has been obtained using the FEM-BEM coupling.

It is important to point out that to assemble a coherent set of boundary and initial conditions, appropriate material relationships and to use the two-phase model appropriately, indepth understanding of the physical processes is required, because of the strong nonlinear coupling.

References

[1] Parker J.C., Multiphase Flow and Transport in Porous Media, *Review of Geophysics*, **27**/3, pp. 311-328, 1989.

[2] Marle Ch.M., *Multiphase Flow in Porous Media*, Editions Technip, Paris, 1981.

[3] Helmig R., *Multiphase Flow and Transport Processes in the Subsurface: A Contribution to the Modeling of Hydrosystems*, Springer- Verlag Berlin Heidelberg New York, 1997.

[4] Brebbia C.A., *The boundary element method for engineers*, Pentech Press Limited, Plymouth London UK, 1978.

[5] Brebbia C.A., Telles J.C.F. and Wrobel L.C., *Boundary Element Techniques*, Springer-Verlag, Berlin, 1984.

[6] Partridge P.W., Brebbia C.A. and Wrobel L.C., *The dual reciprocity boundary element method*, Computational Mechanics Publications, Southampton UK, 1992.

[7] Popov V. and Power H., The DRM-MD integral equation method: An efficient approach for the numerical solution of domain dominant problems, *Int. J. Num. Meth. Engrg*, **44**, pp. 327-353, 1999.

[8] Samardzioska, T., Popov, V. (2005), 'Numerical comparison of the equivalent continuum, non-homogeneous and dual porosity models for flow and transport in fractured porous media', *Advances in Water Resources*, 28, 235-255.

[9] Brooks R.H. and Corey A.T., 'Hydraulic properties of porous media', *Hydrology papers*, **3**, Colorado State University, Fort Collins, Colorado, March 1964.

Boundary element method for the analysis of flow and concentration in a water reservoir

M. Kanoh[1], N. Nakamura[1], T. Kuroki[2] & K. Sakamoto[3]
[1]*Department of Civil Engineering, Kyushu Sangyo University, Japan*
[2]*Department of Civil Engineering, Fukuoka University, Japan*
[3]*Environment Division, Matsue Doken Co., Ltd. Japan*

Abstract

In an earlier study, a boundary element method (BEM) and a weighted finite difference method (WFDM) were combined to analyse the flow in a water reservoir. In this paper, the BEM is developed to obtain numerically stable and convergent results for the water flow and concentration distribution in the water reservoir. In order to apply the BEM to the problem, three techniques were used. (1) The first is the penalty method, in which the pressure terms are eliminated in the Navier-Storks equations for the BEM. (2) The second involves taking the uniform velocity into the basic solution in both the flow and concentration distribution analyses of our BEM. (3) The third is to identify the kinetic and dynamic conditions for pressure against the vertical wall or the bottom and on the free surface of the reservoir because it is necessary to ascertain the pressure conditions in case that the penalty method is not used. Referring to the velocity vectors of the water flow that are calculated by the WFDM and observed in the model simulation of a water reservoir constructed in our laboratory, the effect and accuracy of the alternative BEM are estimated.
Keywords: flow and concentration in a water reservoir, weighted finite difference method, observed velocity in a model simulation of a water reservoir, kinetic and dynamic conditions for pressure.

1 Introduction

A poor-oxygen layer, which is short of or lacking in dissolved oxygen (DO), often appears in the lower regions of water reservoirs in summer. The poor-oxygen condition results in the solving of heavy metals and phosphorous from

WIT Transactions on Modelling and Simulation, Vol 42, © 2006 WIT Press
www.witpress.com, ISSN 1743-355X (on-line)
doi:10.2495/BE06024

the bottom of water reservoirs. These heavy metals (e.g., manganese (Mn) and arsenic (As)) are so harmful to human health that the water in the lower layer cannot be used for drinking because the heavy metals may exceed the acceptable standards for drinking water. On the other hand, phosphorous leads to eutrophication or to the occurrence of large amounts of phytoplankton in water reservoirs. Thus, the poor-oxygen layer sometimes results in the pollution of the water in a reservoir. An attempt was made to ameliorate the concentration of oxygen in the lower layer of the reservoir by using a machine that supplies DO (Kanoh *et al.* [1]). In order to numerically confirm the efficiency of the improvement in DO, we developed a BEM to represent and calculate the slow but very delicate flow caused by the machine that supplies DO. The flow domain is divided into several sub-regions so that the flow velocity is uniform in every region, since the velocity of the flow changes only slightly and may be regarded as steady in the divided sub-region of the flow domain of the water reservoir. If the flow velocity is determined not to be uniform in a region, it is divided into a uniform and a deviation part; in our BEM, the deviation part is calculated as the volume integration. We anticipate that this refined BEM can be used to calculate the flow and concentration in higher Reynolds numbers than those obtained with the ordinary BEM, since the uniform velocity may be larger than the deviation part. The WFDM is used to confirm the precision of the kinetic and dynamic conditions for pressure against the vertical wall or the bottom and along the free surface of the water reservoir. The Poisson equation is introduced to compensate for the continuity of the Navier-Storks equations since the continuity equation excludes the pressure term.

2 Governing equations for the density flow

2.1 The equations for the problem

Three equations, namely, continuous, Navier-Stokes (N-S), and convective - diffusion equations, govern the water flow and concentration distribution in a water reservoir. In the vertical (x_1, x_2) plane, as illustrated in Figure 1, these equations are shown as follows:

$$\frac{\partial u_1}{\partial x_1}+\frac{\partial u_2}{\partial x_2}=0 \tag{1}$$

$$\rho\left(\frac{\partial u_1}{\partial t}+u_1\frac{\partial u_1}{\partial x_1}+u_2\frac{\partial u_1}{\partial x_2}\right)=\frac{\partial}{\partial x_1}\left(-P+\mu\frac{\partial u_1}{\partial x_1}\right)+\frac{\partial}{\partial x_2}\left(\mu\frac{\partial u_1}{\partial x_2}\right) \tag{2$_1$}$$

$$\rho\left(\frac{\partial u_2}{\partial t}+u_1\frac{\partial u_2}{\partial x_1}+u_2\frac{\partial u_2}{\partial x_2}\right)=\frac{\partial}{\partial x_1}\left(-P+\mu\frac{\partial u_2}{\partial x_1}\right)+\frac{\partial}{\partial x_2}\left(\mu\frac{\partial u_2}{\partial x_2}\right)+g \tag{2$_2$}$$

$$\frac{\partial T}{\partial t}+u_1\frac{\partial T}{\partial x_1}+u_2\frac{\partial T}{\partial x_2}-D\left(\frac{\partial^2 T}{\partial {x_1}^2}+\frac{\partial^2 T}{\partial {x_2}^2}\right)=0 \tag{3}$$

$$\rho = \alpha T + \beta \ , \tag{4}$$

$$\mu = \mu_m + \mu_t \tag{5}$$

where x_1 and x_2 are the horizontal and vertical directions, respectively, u_1 and u_2 describe the velocities of the x_1 and x_2 directions, respectively, P is the pressure, g is the gravity acceleration, μ is the viscosity, T is the water temperature, and D is the diffusion coefficient. Here, the density ρ is connected to the water temperature T, as written in Equation (4), with the coefficients α and β. The water temperature T is compatible to the concentration of DO in case it is necessary to calculate the DO, heavy metals, or other values. The viscosity μ is described as Equation (5) by using μ_m and μ_t, which are the molecular and time-dependent viscosity coefficients, respectively.

2.2 Viscosity coefficient for the flow in a water reservoir

The expression of the viscosity was as reported earlier in Kanoh *et al.* [1]:

$$\mu = k\,\mathbf{V}^2 \quad \text{or} \quad \mu = \lambda \mathbf{V}^2 \Delta t \ , \tag{6}$$

where **V** is the velocity vector at the position and the time under consideration, k or λ is a coefficient, Δt is the time increment. The value of the coefficient k or λ is estimated so that the WFDM solutions show good agreement with the velocity vectors observed in the model simulation of a water reservoir since we consider that k or α may change its value if the flow fields and conditions vary. The velocity **V** is given as the square root of the sum of the squares of u_1 and u_2, where u_1 and u_2 are the solutions of BEM calculated earlier [1].

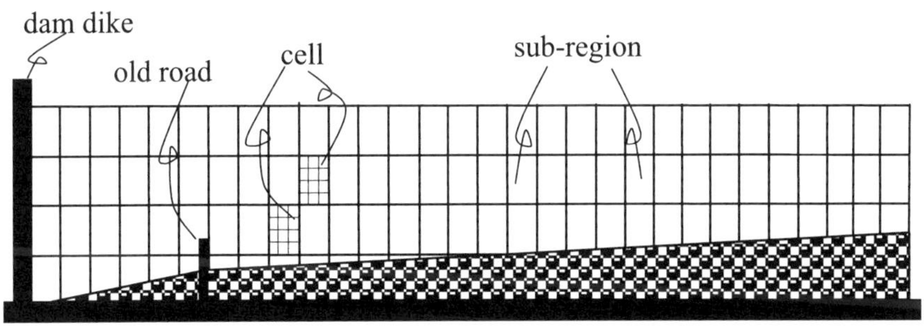

Figure 1: Analytical domain of a water reservoir, sub-regions, and cells.

3 Development of our methodology

In order to numerically confirm the efficiency of the improvement of DO, we attempted to refine our BEM to calculate the slow but very delicate flow caused by the machine that supplies DO in a case in which the concentration of DO is enhanced by the use of the DO-supplying machine improved by our research group.

3.1 BEM sub-regions and cells

In our boundary element method the flow domain is divided into several sub-regions so that the flow velocities can be regarded as nearly constant in every region, since the velocity of the flow changes only slightly and may be regarded as steady during the time increment (Δt) under consideration in the divided sub-region of the flow domain of the water reservoir. The sub-regions were divided into several inner cells, as illustrated in Figure 1, in order to calculate the volume integration of the homogenous terms that are described below.

3.2 BEM formulations

3.2.1 Penalty method

The flow-velocity solutions of ordinary BEM are heavily influenced by the boundary conditions for the pressure at the inflow or outflow of the water reservoir. It is difficult to identify appropriately the boundary conditions of the pressure in order to obtain convergent and comparative results with the observed value in our model simulation. The penalty method is introduced so that the pressure terms are eliminated in the N-S equations and the difficulty of the pressure boundary conditions can be avoided in the BEM.

3.2.2 Poisson equation for pressure

If the penalty method is not adopted, it is necessary, but rather difficult, to compensate for the continuity of the N-S equations since the continuity equation excludes the pressure term. Differentiating Equations (2_1) and (2_2) with respect to x_1 and x_2, respectively, summing both equations, and substituting Equation (1) for the sum, we obtained the Poisson equation regarding the pressure as described below

$$P_{,11} + P_{,22} = 2\rho(u_{1,1} {*}\, u_{2,2} - u_{1,2} {*}\, u_{2,1}) \,. \tag{7}$$

Solving the Poisson equation, we can compensate for the continuity of the N-S equations.

3.2.3 Kinetic and dynamic conditions for pressure

In order to obtain a convergent and accurate solution for the Poisson equation, it is important to identify the kinetic and dynamic conditions for pressure against the vertical wall or the bottom of the reservoir. Considering that the velocities along the vertical wall or on the bottom equal zero and substituting the zero values of the velocities to the N-S equations, we obtained the kinetic and dynamic conditions for pressure as described below:

$$P_{,n} = \rho \nu_e (u_{n,nn}) . \tag{8}$$

Applying the finite difference scheme to the above equation, we finally have:

$$P_{,n} = 2\rho\nu_e \cdot u_{n-1}/(\Delta n)^2 , \tag{9}$$

where u_{n-1} is the velocity value on the point that exists in the length of Δn to the boundary. Here, Δn describes the half-length in the normal direction between the bottom or wall boundary and the nearest element.

3.2.4 Taking velocity into a fundamental solution

Taking the uniform part of the velocities into the fundamental solution for the BEM, we attempted to reduce the influence of the volume integration to the calculated BEM values. As described above, the flow domain is divided into several sub-regions so that the flow velocity is uniform in every region. If the flow velocity cannot be regarded as uniform in a region, the flow velocity is divided into a uniform and a deviation part; in our BEM, the deviation part is calculated as the volume integration. We anticipate that the modified BEM can be used to calculate the flow and concentration in higher Reynolds numbers than those obtained using the ordinary BEM, since the deviation may be smaller than the uniform velocity and the volume integration of the deviation may also be smaller than that obtained with the ordinary BEM. For the analysis of the velocities u_1 and u_2, the methodology is described below. First, velocities u_1 and u_2 are divided into the uniform parts $\bar{u}_{1m}$ and $\bar{u}_{2m}$ and the deviations Δu_1 and Δu_2, respectively, as:

$$u_1 = \bar{u}_{1m} + \Delta u_1 \quad , \quad u_2 = \bar{u}_{2m} + \Delta u_2 . \tag{10}$$

Second the fundamental solution for analyzing u_1 and u_2 must be determined as:

$$w^* = -\frac{\ln r}{2\pi\nu_e} \exp\left\{ -\frac{\bar{u}_{1m}(x_1 - x_{1p}) + \bar{u}_{2m}(x_2 - x_{2p})}{2\nu_e} \right\} \tag{11}$$

where ν_e denotes kinetic viscosity and r is described as:

$$r = \{(x_1 - x_{1p})^2 + (x_2 - x_{2p})^2\}^{1/2.} \tag{12}$$

Here, the charging point has the coordinates (x_{1p}, x_{2p}). Finally, using this fundamental solution, the integral equation for analysing the flow (velocities u_1 and u_2) and the concentration distribution can be obtained.

3.3 Boundary conditions

3.3.1 Boundary conditions for flow analysis by BEM

We have previously defined the free surface and other boundary conditions for flow analysis by BEM in the flow region (Kanoh *et al.* [1, 2]). Earlier, we identified the kinetic and dynamic conditions for pressure against the vertical wall or the bottom of the water reservoir as described in Equation (10).

3.3.2 Boundary conditions for the concentration analysis by BEM

The boundary conditions for analysing the concentration of DO are simple since only one is unknown. On the free surface or along the cross sections of the inlet and outlet flows, the Dirichlet (value-given) condition is applied. The adiabatic condition ($\partial C/\partial n = 0$) can be used along the vertical wall or on the bottom, where C describes the relative concentration of DO.

3.3.3 Boundary conditions for flow analysis by the WFDM

The free surface, wall, and other boundary conditions for flow analysis by the WFDM have been previously defined by Nakamura *et al.* [3]. The outline is as follows: (1) the velocities that exist in normal and tangential directions at the wall or on the bottom are zero. (2) The pressure on the free surface is defined as zero. The pressure that is defined on the inside mesh neighbouring the wall or bottom is calculated as shown below:

$$P_{inside} = P_{outside} - \mu * u_{n-1} / (\Delta n), \tag{13}$$

where P_{inside} and $P_{outside}$ are the pressures on the inside and outside meshes, and u_{n-1} is the velocity value on the point that exists in the length of Δn to the boundary. Equation (13) is introduced from the kinetic equation of the water flow.

4 Machine that supplies DO

We have proposed a new DO-supplying machine that produces an outward flow of DO-rich water (a hundred milligram per litre: mg/L). The DO-rich water improves the overall content of oxygen in the lower layer of the reservoir. Thus, the DO-rich water furnished by our DO-supplying machine can help prevent water pollution, since the poor-oxygen layer is sometimes responsible for water pollution in the water reservoir. The DO-supplying machine has been demonstrated to improve the oxygen concentration in the lower layers of water reservoirs, lakes, and seas. Since field research is expensive and time consuming, we have used laboratory simulations to demonstrate the efficiency of the DO-supplying machine by using the modified BEM or WFDM and a laboratory model reservoir.

5 Results and discussion

5.1 Observed velocities in a model of a water reservoir

Figures 2 and 3 illustrate the observed velocity vectors in laboratory models of a water reservoir. The reservoir model (the so-called distorted model) is scaled down to 1/62 vertically and 1/775 horizontally. By using a digital video recorder (Sony VX-2000) and a stroboscope (Uchida TO-5D), we were able to take photographs of the very slow flow in the models. Aluminium flakes were suspended in the flow domain of the models in a darkened laboratory, and the stroboscope lit up the flakes four times per second. The digital VTR was then used to record four photographs of the moving flakes per second. Figures 2 and 3 demonstrate that the comparatively small old road has a significant influence on the flow vectors in the domain between the road and the dam dike. This old road was built on the bottom of the reservoir for use when transporting the materials for the construction of the dam dike and was not removed after the work was completed. The size (620000 m^2) and depth (22 m) of the B reservoir make field research prohibitive in terms of both money and time. To overcome this challenge, we also used the reservoir models and the BEM or WFDM to simulate the flow in the water reservoir.

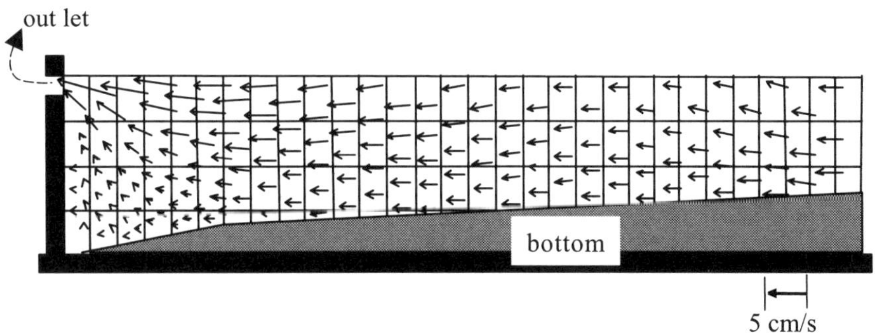

Figure 2: Observed flow vectors in a reservoir model (without an old road).

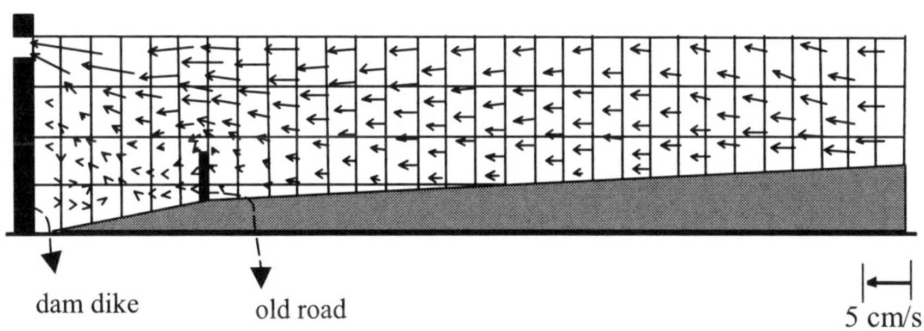

Figure 3: Observed flow vectors in a reservoir model (with an old road).

5.2 Numerical results and discussion

5.2.1 Flow velocities calculated by the BEM in the water reservoir

Figures 4 and 5 illustrate the velocity vectors calculated using the BEM in the B hydro-model, where the penalty method was used with the BEM. The BEM solutions can accurately replicate the very delicate flow in the hydro-model with and without an old road; therefore, the modified BEM with the penalty method can provide accurate solutions for flow analysis problems in a water reservoir.

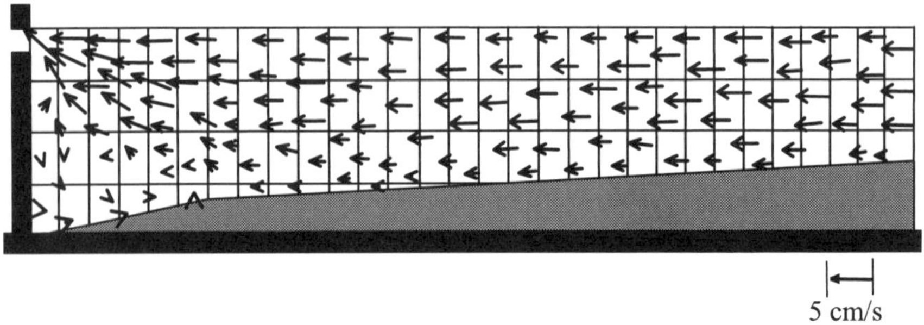

Figure 4: Flow vectors of the BEM in the domain (without an old road).

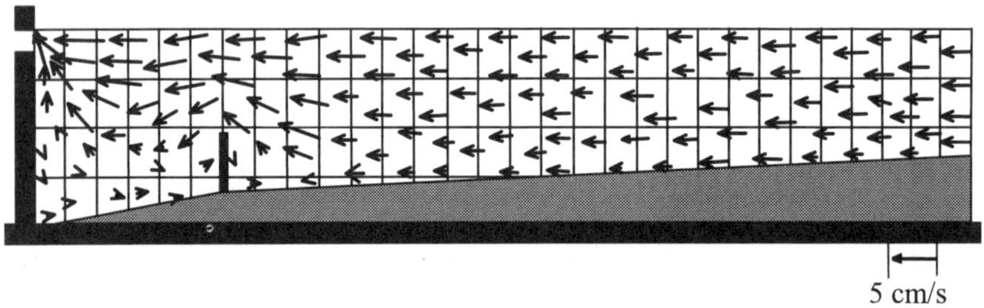

Figure 5: Flow vectors of the BEM in the domain (with an old road).

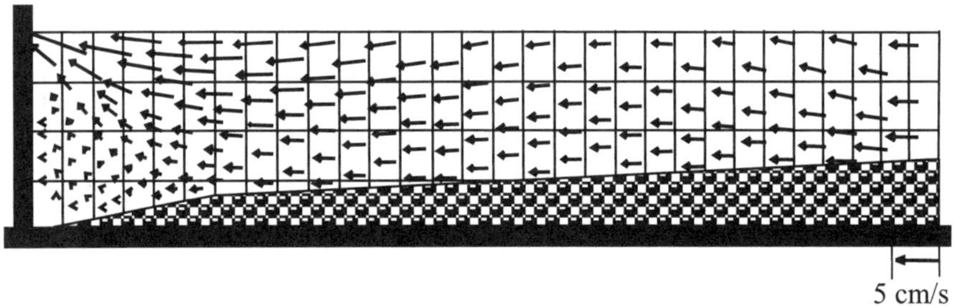

Figure 6: Flow vectors of the WFDM in the domain (without an old road).

5.2.2 Flow velocities calculated by the WFDM in the reservoir

Figures 6 and 7 show the velocity vectors calculated using the WFDM in the hydro-model. The WFDM solutions can also accurately reproduce the very delicate flow in the hydro-model in either case. We can then conclude that new techniques added to the WFDM for the analysis of the flow in a water reservoir provide satisfactory results.

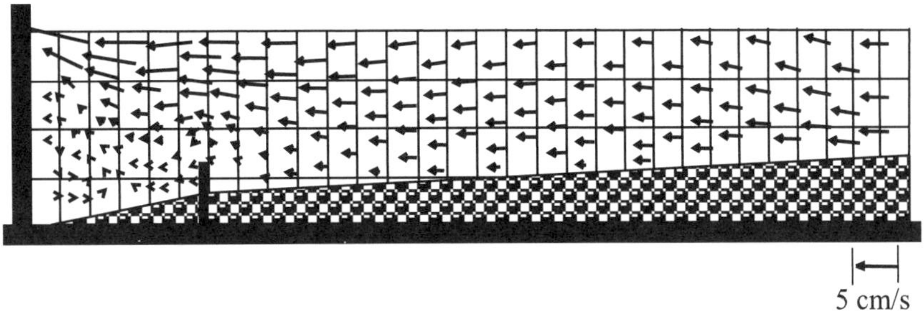

Figure 7: Flow vectors of the WFDM in the domain (with an old road).

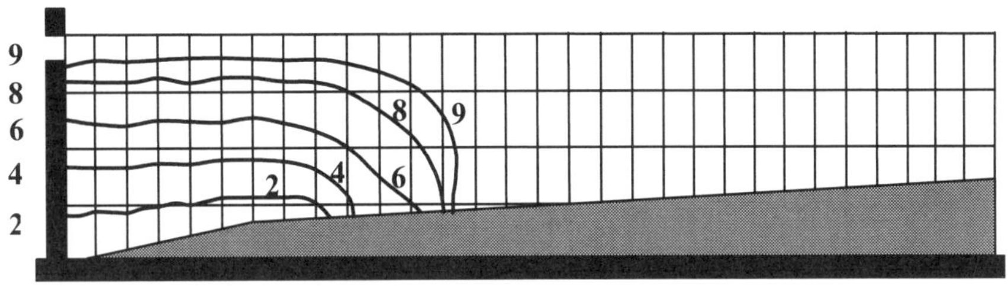

Figure 8: DO concentration (mg/L) of the BEM in a water reservoir.

5.2.3 Concentration distribution calculated by the BEM in the reservoir

Figure 8 is an illustration of the distribution of DO as calculated with the BEM in the hydro-model by considering the uniform part of the velocity in the fundamental solution. The concentration of DO calculated by the modified BEM appears to be reasonable; therefore, we anticipate that the new technique can be used to analyse the concentration-distribution of DO, heavy metals or other values in the reservoir in a satisfactory manner.

6 Conclusion

We refined the BEM to analyse the flow and concentration of DO in a water reservoir. To confirm the precision of the results obtained with the modified BEM, comparisons were made with results obtained using the WFDM and the observed flow vectors in the hydro-model. The concentration of DO calculated by the BEM appeared to be reasonable, and the modified BEM yielded accurate replications of the flow vectors that were observed in the hydro-model or calculated with the WFDM; therefore, it is anticipated that the alternative BEM will produce accurate calculations of the flow and the concentration-distribution of DO, heavy metals or other values in water reservoirs.

References

[1] Kanoh, M., Nakamura, N. and Kuroki T., Flow analysis in a water reservoir using a combined boundary element and weighted finite

difference method, *Proc. of the 27th World Conf. on Boundary Elements and Other Mesh Reduction Methods*, ed. C.A. Brebbia, WIT PRESS, Orlando, USA, pp. 429-438, 2005.

[2] Kanoh, M., Nakamura, N., Okuzono, H. and Kuroki T., Wave and water motion analysis using combined boundary element and weighted finite difference method, *Proc. of the 26th World Conf. on Boundary Elements and Other Mesh Reduction Methods*, ed. C.A. Brebbia, WIT PRESS, Bologna, Italy, pp. 35-44, 2004.

[3] Nakamura, N., Kanoh, M. and Kuga, Y., Application of two-dimensional weighted finite difference method to shallow water flow, *Hydraulic, Coastal and Environmental Engineering,* No.747/II-65, pp.125-134, 2003 (in Japanese).

[4] Tosaka, N. and Yagawa, G., Numerical simulation of free and moving boundary problem, *Yokendo, Tokyo, Japan*, 1995 (in Japanese).

Section 6
Computational techniques

A Laplace transform boundary element solution for the biharmonic diffusion equation

A. J. Davies & D. Crann
School of Physics, Astronomy and Mathematics, University of Hertfordshire, UK

Abstract

The most common diffusion problems involve the description of the diffusive term in terms of the Laplacian operator. Such problems have been solved successfully in a boundary element context using the Laplace transform in time together with a dual reciprocity approach. Some diffusion problems, *e.g.* heat transfer in certain oceanographic models and slow flow in oil films, have the diffusive term described by the biharmonic operator. Such problems can be written, on the introduction of a secondary dependent variable, as a pair of coupled equations, one of Poisson-type and the other of diffusion-type. The Laplace transform together with the dual reciprocity method can be used to solve the resulting pair of coupled equations.
Keywords: Laplace transform, boundary elements, dual reciprocity, biharmonic diffusion.

1 Introduction

Diffusion problems in which the diffusive operator is Laplacian are well-documented [1]. For such problems the most common numerical approach to the solution is to use a finite difference time-stepping process. The Laplace transform in time provides an alternative approach. In both cases the parabolic problem is reduced to an elliptic problem in the space variables and any suitable solver may be used.

Rizzo and Shippy [2] first used the Laplace transform in conjunction with the boundary integral equation method using an inversion process in terms of a prony series of negative exponentials in time. Stehfest's method [3, 4], which is much simpler to apply, was used by Moridis and Reddell [5]. The solution is developed directly at one specific time value without the necessity of intermediate values.

WIT Transactions on Modelling and Simulation, Vol 42, © 2006 WIT Press
www.witpress.com, ISSN 1743-355X (on-line)
doi:10.2495/BE06025

Once the elliptic problem has been solved it remains to invert the Laplace transform. The Stehfest method, as used by Moridis and Reddell [5], is recommended by Davies and Martin [6] in their study of a variety of numerical Laplace transform inversion methods as being simple to use and provides accurate results. Subsequently it has been used in a variety of circumstances by the current authors [7–9]. In particular the Laplace transform technique has been shown to provide a suitable approach for the solution of coupled diffusion-type problems [10].

A very good account of the Laplace transform technique in a boundary element context is given by Zhu [11].

A less well-known model for certain diffusion problems involves the description of the diffusion in terms of the biharmonic operator. Such problems occur in the slow flow of oil films [12] and ocean mixing models [13].

2 Biharmonic diffusion

The sophisticated ocean mixing problem described by Hunke *et al.* [13] includes both convection and diffusion, the equation being solved by a finite difference process. We wish to consider the applicability of the Laplace transform in time, so for the sake of simplicity we shall ignore convective terms. Crann *et al.* [10] show that in terms of Laplacian diffusion an additional convective term is easily incorporated and it is expected that this would be the case for biharmonic diffusion.

We consider problems in a two-dimensional region, Ω, bounded by the closed curve Γ.

The problem to be solved in Ω is

$$\nabla^4 u = \frac{1}{\alpha}\frac{\partial u}{\partial t} \tag{1}$$

together with suitable boundary and initial conditions.

We shall assume Dirichlet and Neumann conditions of the form

$$u = u(s,t) \text{ and } q \equiv \frac{\partial u}{\partial n} = q(s,t) \text{ on } \Gamma \tag{2}$$

together with the initial condition

$$u(x,y,0) = u_0(x,y) \tag{3}$$

Following the approach of Toutip *et al.* [14] we write the biharmonic operator in equation (1) in terms of a pair of Laplacian operators as follows:

$$\nabla^2 u = v \tag{4}$$

$$\nabla^2 v = \frac{1}{\alpha}\frac{\partial u}{\partial t} \tag{5}$$

To apply the boundary conditions we note that for properly-posed Laplacian problems we require either u or q, but not both, to be specified at each point on Γ.

In order to apply the boundary conditions in an appropriate manner we consider Γ to comprise two sections Γ_1 and Γ_2 such that $\Gamma = \Gamma_1 + \Gamma_2$. We then consider the boundary values as follows:
Write

$$u(s,t) = \begin{cases} u_1(s,t) & s \in \Gamma_1 \\ u_2(s,t) & s \in \Gamma_2 \end{cases}$$

$$q(s,t) = \begin{cases} q_1(s,t) & s \in \Gamma_1 \\ q_2(s,t) & s \in \Gamma_2 \end{cases}$$

then choose

$$u = u_1 \text{ on } \Gamma_1 \text{ and } q = q_2 \text{ on } \Gamma_2 \tag{6}$$

$$p \equiv \frac{\partial v}{\partial n} = p_1 \equiv \nabla^2 q_1 \text{ on } \Gamma_1 \text{ and } v = v_2 = \nabla^2 u_2 \text{ on } \Gamma_2 \tag{7}$$

3 The Laplace transform method

We use the Laplace transform with regard to the time variable only in equations (4) and (5). Define

$$\bar{u}(x,y;\lambda) = \int_0^\infty u(x,y,t)e^{-\lambda t}dt$$

$$\bar{v}(x,y;\lambda) = \int_0^\infty v(x,y,t)e^{-\lambda t}dt$$

so that

$$\nabla^2 \bar{u} = \bar{v} \tag{8}$$

$$\text{and} \quad \nabla^2 \bar{v} = \frac{1}{\alpha}(\lambda \bar{u} - u_0) \tag{9}$$

We now develop an iterative scheme for the solution of the coupled equations (8) and (9)

$$\nabla^2 \bar{v}^{(n+1)} = \frac{1}{\alpha}(\lambda \bar{u}^{(n)} - u_0) \tag{10}$$

$$\nabla^2 \bar{u}^{(n+1)} = \bar{v}^{(n+1)} \tag{11}$$

$$\text{with} \quad \bar{u}^{(0)} = \bar{u}_0$$

Equations (10) and (11) are both elliptic equations and may be solved by any suitable elliptic equation solver. We shall use the boundary element method together with the dual reciprocity method using the usual Laplacian fundamental solution

$$u^* = -\frac{1}{2\pi} \ln R \tag{12}$$

Equation (9) is solved subject to the boundary conditions

$$\bar{v} = \bar{v}_2 \text{ on } \Gamma_2, \quad \bar{p} = \bar{p}_1 \text{ on } \Gamma_1 \tag{13}$$

and equation (8) is solved subject to the boundary conditions

$$\bar{u} = \bar{u}_1 \text{ on } \Gamma_1, \quad \bar{q} = \bar{q}_2 \text{ on } \Gamma_2 \tag{14}$$

4 The dual reciprocity boundary element method

Equations (10) and (11) may be written

$$\nabla^2 \bar{v}^{(n+1)} = b_1(\bar{u}^{(n)}) \tag{15}$$

$$\nabla^2 \bar{u}^{(n+1)} = b_2(\bar{v}^{(n+1)}) \tag{16}$$

i.e. of the form

$$\nabla^2 \bar{u} = b \tag{17}$$

and we expand the domain functions b_1 and b_2 in the form

$$b \approx \sum_{j=1}^{N+L} \alpha_j f_j(R)$$

The integral equation equivalent to equation (17) is given by

$$c_\Gamma \bar{u}_\Gamma + \int_\Gamma q^* \bar{u}\, d\Gamma - \int_\Gamma u^* \bar{q}\, d\Gamma = \int_\Omega b u^*\, d\Omega \tag{18}$$

where u^* is the fundamental solution given by equation (12).

We apply the boundary element method in the usual manner, choosing N linear elements, and we approximate the domain function, b, by

$$b \approx \sum_{j=1}^{N+L} \alpha_j f_j(R) \tag{19}$$

where we choose L internal points, together with the N nodes on the boundary, and the approximating functions, f_j, are the usual linear radial basis functions [15]. With such functions, $f_j(R)$, we have

$$\nabla^2 \hat{u}_j = f_j(R)$$

for some $\hat{u}_j$.

Applying Green's theorem, the boundary element approximation to equation (18) may be written in the form

$$c_i\bar{u}_i + \sum_{k=1}^{N}\int_{\Gamma_k} q^*\bar{u}\,d\Gamma - \sum_{k=1}^{N}\int_{\Gamma_k} u^*\bar{q}\,d\Gamma =$$

$$\sum_{j=1}^{N+L}\alpha_j\left(c_j\hat{u}_{ij} + \sum_{k=1}^{N}\int_{\Gamma_k} q^*\hat{u}_j\,d\Gamma - \sum_{k=1}^{N}\int_{\Gamma_k} u^*\hat{q}_j\,d\Gamma\right)$$

for $i = 1, \ldots, N$

We write this system of equations in matrix form using the subscript B to denote that the coefficient is associated with a boundary node:

$$\mathbf{H}_B\bar{\mathbf{U}}_B - \mathbf{G}_B\bar{\mathbf{Q}}_B = \left[\mathbf{H}_B\hat{\mathbf{U}} - \mathbf{G}_B\hat{\mathbf{Q}}\right]\boldsymbol{\alpha} \tag{20}$$

We collocate at the $N + L$ nodes in equation (19) to obtain the vector $\boldsymbol{\alpha}$ by solving the system of equations

$$\mathbf{b} = \mathbf{F}\boldsymbol{\alpha} \tag{21}$$

Now, internal values are given by

$$\bar{u}_i = -\sum_{k=1}^{N}\int_{\Gamma_k} q^*\bar{u}\,d\Gamma + \sum_{k=1}^{N}\int_{\Gamma_k} u^*\bar{q}\,d\Gamma +$$

$$\sum_{j=1}^{N+L}\alpha_j\left(c_j\hat{u}_{ij} + \sum_{k=1}^{N}\int_{\Gamma_k} q^*\hat{u}_j\,d\Gamma - \sum_{k=1}^{N}\int_{\Gamma_k} u^*\hat{q}_j\,d\Gamma\right)$$

and we write in matrix form, using the subscript I to denote that the coefficient is associated with an internal node,

$$\mathbf{I}\bar{\mathbf{U}}_I = \mathbf{G}_I\bar{\mathbf{Q}}_B - \mathbf{H}_I\bar{\mathbf{U}}_B + \left[\mathbf{H}_I\hat{\mathbf{U}} - \mathbf{G}_I\hat{\mathbf{Q}}\right]\boldsymbol{\alpha} + \mathbf{I}\hat{\mathbf{U}}\boldsymbol{\alpha} \tag{22}$$

Since $b = b(x, y, \bar{u})$, we cannot find $\boldsymbol{\alpha}$ from equation (21). We can, however, write equation (21) as

$$\boldsymbol{\alpha} = \mathbf{F}^{-1}\mathbf{b} \tag{23}$$

Equations (20) and (22) may then be combined, using equation (23) in the form

$$\mathbf{H}\bar{\mathbf{U}} - \mathbf{G}\bar{\mathbf{Q}} = \left[\mathbf{H}\hat{\mathbf{U}} - \mathbf{G}\hat{\mathbf{Q}}\right]\mathbf{F}^{-1}\mathbf{b} \tag{24}$$

where

$$\mathbf{H} = \begin{bmatrix} \mathbf{H}_B & \mathbf{0} \\ \mathbf{H}_I & \mathbf{I} \end{bmatrix}, \mathbf{G} = \begin{bmatrix} \mathbf{G}_B & \mathbf{0} \\ \mathbf{G}_I & \mathbf{0} \end{bmatrix}, \bar{\mathbf{U}} = \begin{bmatrix} \bar{\mathbf{U}}_B \\ \bar{\mathbf{U}}_I \end{bmatrix}, \bar{\mathbf{Q}} = \begin{bmatrix} \bar{\mathbf{Q}}_B \\ \mathbf{0} \end{bmatrix}$$

We define the matrix $\mathbf{S}$, which depends only on the geometry, by

$$\mathbf{S} = \left[\mathbf{H}\hat{\mathbf{U}} - \mathbf{G}\hat{\mathbf{Q}}\right] \mathbf{F}^{-1}$$

then equation (24) provides our system of equations for u as

$$\mathbf{H}\bar{\mathbf{U}} - \mathbf{G}\bar{\mathbf{Q}} = \mathbf{S}\mathbf{b}_1 \tag{25}$$

and from equation (25) we find the boundary solutions, $\bar{\mathbf{U}}_B$ and $\bar{\mathbf{Q}}_B$, and the internal solution, $\bar{\mathbf{U}}_I$ simultaneously. Similarly we can set up the system of equations for v as

$$\mathbf{H}\bar{\mathbf{V}} - \mathbf{G}\bar{\mathbf{P}} = \mathbf{S}\mathbf{b}_2 \tag{26}$$

The discrete systems of equations associated with equations (15) and (16) are

$$\begin{aligned} \mathbf{H}\bar{\mathbf{V}}^{(n+1)} - \mathbf{G}\bar{\mathbf{P}}^{(n+1)} &= \mathbf{b}_1\left(\bar{\mathbf{U}}^{(n)}\right) \\ \mathbf{H}\bar{\mathbf{U}}^{(n+1)} - \mathbf{G}\bar{\mathbf{Q}}^{(n+1)} &= \mathbf{b}_2\left(\bar{\mathbf{V}}^{(n+1)}\right) \end{aligned} \tag{27}$$

with $\bar{\mathbf{U}}^{(0)} = \bar{\mathbf{U}}_0$

Equations (27), after the application of the boundary conditions, may be written in the form

$$\mathbf{A}_1\mathbf{x}^{(n+1)} = \mathbf{F}_1\left(\bar{\mathbf{U}}^{(n)}\right) \text{ and } \mathbf{A}_2\mathbf{y}^{(n+1)} = \mathbf{F}_2\left(\bar{\mathbf{V}}^{(n+1)}\right)$$

where $\mathbf{x}^{(n+1)} = \left[\bar{\mathbf{V}}^{(n+1)} \;\; \bar{\mathbf{P}}^{(n+1)}\right]^T$ and $\mathbf{y}^{(n+1)} = \left[\bar{\mathbf{U}}^{(n+1)} \;\; \bar{\mathbf{Q}}^{(n+1)}\right]^T$

The iteration is terminated by the stopping condition

$$\frac{\max\left|x_i^{(n+1)} - x_i^{(n)}\right|}{\max\left(|x_i^{(n+1)}| + |x_i^{(n)}|\right)} < \epsilon \text{ and } \frac{\max\left|y_i^{(n+1)} - y_i^{(n)}\right|}{\max\left(|y_i^{(n+1)}| + |y_i^{(n)}|\right)} < \epsilon$$

for some suitable tolerance ϵ.

Finally the approximate transform $\bar{\mathbf{U}}$ can be inverted to obtain the approximate solution $\mathbf{U}$.

5 Numerical inversion of the Laplace transform

The Stehfest numerical process [3, 4] is implemented as follows:

Choose a specific time value, τ, at which we seek the solution and define a discrete set of transform parameters given by

$$\left\{\lambda_j = j\frac{\ln 2}{\tau}: \quad j = 1, 2, \ldots, m;\ m \text{ even}\right\}$$

The dual reciprocity boundary element method is used for each λ_j to obtain a set of approximate boundary values

$$\bar{U}_{B,\,ij} \quad i = 1, \ldots, N; \quad j = 1, \ldots, m$$

and a set of approximate internal values

$$\bar{U}_{I,\,kj} \quad k = 1, \ldots, L; \quad j = 1, \ldots, m$$

The inverse transforms are then given as follows:

$$U_{B,\,r} = \frac{\ln 2}{\tau}\sum_{j=1}^{m} w_j \bar{U}_{B,rj}$$

$$\text{and} \qquad U_{I,\,r} = \frac{\ln 2}{\tau}\sum_{j=1}^{m} w_j \bar{U}_{I,rj}$$

where $r = 1, \ldots, N$ for boundary points and $r = 1, \ldots, L$ for internal points.

The weights, w_j, are given by Stehfest [3,4] as

$$w_j = (-1)^{\frac{m}{2}+j} \sum_{k=\left[\frac{1}{2}(1+j)\right]}^{\min\left(j,\frac{m}{2}\right)} \frac{k^{\frac{m}{2}}\,(2k)!}{\left(\frac{m}{2}-k\right)!k!\,(k-1)!\,(j-k)!\,(2k-j)!}$$

6 Results

In order to illustrate the process we consider the following example

$$\nabla^4 u = \frac{1}{\alpha}\frac{\partial u}{\partial t} + h(x, y, t)$$

in the unit square $\{(x, y);\ 0 < x < 1,\ 0 < y < 1\}$ subject to Dirichlet and Neumann boundary conditions appropriate to the exact solution in the case $\alpha = 1$

$$u(x, y, t) = (1 + x^4 + y^4)e^{-t}$$

The inclusion of the non-homogeneous term $h(x, y, t)$ makes only a trivial difference to the vector $\mathbf{b}_1$ in equation (25). We use 36 linear boundary elements and 9 internal nodes with $f(R) = 1 + R$ for the dual reciprocity method.

In Figure 1 we show the time development at three points along the diagonal line of symmetry and in Figure 2 we show the space variation along the diagonal for four specific times. We see that the approximation compares very well

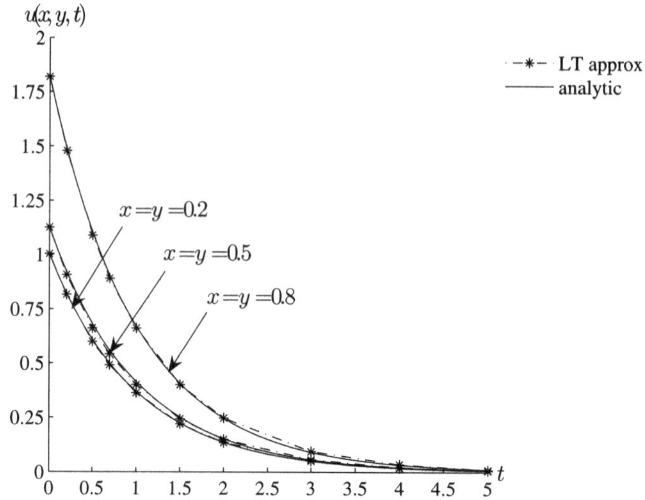

Figure 1: Time development at three points.

with the analytic solution. Although we haven't shown the results in Figures 1 and 2 we note here that solutions for small values of t are not in good agreement with the analytical values. This observation is consistent with conclusions given by Crann [16] when considering the Laplace transform for a variety of time dependent diffusion-type problems: The solution for small time values is often significantly less accurate than that for other times, almost certainly due to the ill-conditioning of the Laplace transform inversion.

We note here that in terms of the notation of Section 3 we have defined Γ_1 and Γ_2 as follows:

$$\Gamma_1 = \{(x, y) : \; y = 1; \; 0 \leq x \leq 1\} \cup \{(x, y) : y = 0; \; 0 \leq x \leq 1\}$$

$$\Gamma_2 = \{(x, y) : \; x = 1; \; 0 \leq y \leq 1\} \cup \{(x, y) : x = 0; \; 0 \leq y \leq 1\}$$

The choice of Γ_1 and Γ_2 is somewhat arbitrary. In order to investigate any effect that the choice may make we have considered a further two cases in which Γ_1 and Γ_2 comprise different combinations of the sides of the square.

We find that the choice of Γ_1 and Γ_2 does not affect the level of accuracy of approximate solutions.

7 Conclusions

The Laplace transform dual reciprocity method has been shown to provide a suitable approach for the solution of a model biharmonic diffusion problem of the form described by Hunke *et al.* [13]. Their model is significantly more complicated but

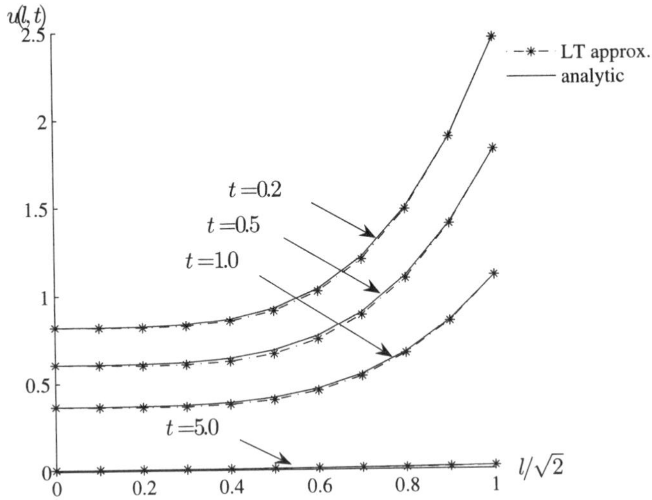

Figure 2: Space variation along the line of symmetry, l is the distance from the origin.

the results presented here suggest that the Laplace transform approach should be expected to offer a suitable alternative to the finite difference approach.

We conclude with a comment on the model described by Tanner and Berry [12] where the partial differential equation is $\nabla^4 u = -\frac{1}{\alpha}\frac{\partial u}{\partial t}$ *i.e.* a backward time problem. The negative sign associated with the time derivative leads to ill-conditioning problems in a finite difference approach. The model problem considered using the Laplace transform does not suffer from such difficulties. There are similar difficulties for small time values as with the forward time problem but in general the Laplace transform offers a suitable approach for the backward problem as well, details are given by Crann and Davies [17].

References

[1] Crank J, *The mathematics of diffusion*, Oxford University Press, 1975.

[2] Rizzo FJ and Shippy DJ, A method of solution of certain problems of transient heat conduction, *AIAA Journal*, **8**, 2004-2009, 1970.

[3] Stehfest H, Numerical inversion of Laplace transforms, *Comm. ACM.*, **13**, 47-49, 1970.

[4] Stehfest H, Remarks on Algorithm 368 [D5] Numerical inversion of Laplace transforms, *Comm. ACM.*, **13**, 624, 1970.

[5] Moridis GJ and Reddell DL, The Laplace transform boundary element (LTBE) numerical method for the solution of diffusion-type problems, *Boundary Elements XIII*, 83-97, 1991.

[6] Davies B and Martin B, Numerical inversion of Laplace transforms: A survey and comparison of methods, *J. Comput. Phys.*, **33**, 1-32, 1979.

[7] Davies AJ, Mushtaq J, Radford LE and Crann D, The numerical Laplace transform solution method on a distributed memory architecture, *Applications of High Performance Computing in Engineering V*, 245-254, 1997.

[8] Crann D, Davies AJ and Mushtaq J, Parallel Laplace transform boundary element methods for diffusion problems, *Boundary Elements XX*, 259-268, 1998.

[9] Honnor ME and Davies AJ, The Laplace transform dual reciprocity boundary element method for nonlinear transient field problems, *Boundary Elements XXIV*, 363-372, 2002.

[10] Crann D, Davies AJ and Christianson B, The Laplace transform dual reciprocity boundary element method for electromagnetic heating problems, *Advances in Boundary Element Techniques VI*, 229-234, 2005.

[11] Zhu S-P, Time-dependent reaction-diffusion problems and the LTDRM approach, *Boundary Integral Methods, Numerical and Mathematical Aspects*, ed. Goldberg M, 1-35, Computational Mechanics Publications, 1999.

[12] Tanner LH and Berry MV, Dynamics and optics of oil hills and oilscapes, *J. Phys. D; Appl. Phys.*, **18**, 1037-1061, 1985.

[13] Hunke EC, Lysne JA, Hecht MW and Maltrud ME, GM vs biharmonic ocean mixing in the Arctic, *J. Phys. Conference Series*, **16**, 348-352, 2005.

[14] Toutip W, Davies AJ and Kane SJ, The dual reciprocity method for solving biharmonic problems, *Boundary Elements XXIV*, 373-380, 2004.

[15] Partridge PW, Brebbia CA and Wrobel LC, *The dual reciprocity method*, Computational Mechanics Publications, 1992.

[16] Crann D, *The Laplace transform boundary element method for diffusion-type problems*, PhD Thesis, University of Hertfordshire, 2005.

[17] Crann D and Davies AJ. A Laplace transform solution of the biharmonic diffusion equation, *University of Hertfordshire Department of Physics, Astronomy and Mathematics Technical Report*, **97**, 2006.

Transformative models in reliability assessment of structures

P. Brož
Department of Concrete Structures and Bridges, Czech Technical University, Czech Republic

Abstract

In this paper, the relationship "load-loading effects" of engineering structures from the standpoint of applied calculation models is investigated. These simulations, both theoretical and experimental, replace the actual construction and are expected to express its existing characteristics under the given load. They are used for the specification of the response of the structures from the angle of the reliability conditions in the groups of limit states of load-carrying capacity and serviceability.

As an example of transformative simulations, the comprehensive methodology for determining the probability of fatigue failure for mixed-mode fatigue is presented. The loading is mixed-mode with randomness in the initial crack length, final crack length, initial crack angle, initial crack location, fatigue crack growth parameters, and the applied stress. The approach consists of calculating the reliability index that is applied to determine the first-order probability of collapse, by solving a constrained optimization problem.

The helpfulness of the probabilistic approach for structural analysis is indicated, namely by employing perturbation procedures.

Keywords: best guess, built-up beam, carrying resistance, crack path discretization, dynamic load factor, reference value, reliability index.

1 Introduction

Transformation models change the loading into the response of an individual construction to the application of load. These structural versions have to represent, if applicable, the real construction. A choice of the model and the study of relevant random variables incorporated must reflect fabrication and

WIT Transactions on Modelling and Simulation, Vol 42, © 2006 WIT Press
www.witpress.com, ISSN 1743-355X (on-line)
doi:10.2495/BE06026

installation tolerances, and also the influences of both environment and non-negligible imperfections. Transformation models fit in the reliability assessment method according to [1], in compliance with fig. 1.

Selection of a transformation model reflects the progress of structural mechanics and of the experimental approaches.

Structural analysis is the process of determining the response of the structure due to specific loadings. Data related to the determination and application of structural analysis models may be generally considered to be random variables. Examples of these random variables are geometrical and physical properties of the structure, imperfections, damping due to non-structural components, environmental effects, and soil-structure interaction. The accuracy of the selected structural analysis model depends on the evaluation and application of these variables and on their interaction. The structural analysis model chosen must be appropriate for the response and should represent the intended behaviour of a structure as closely as possible up to the limit state under consideration. We can launch simplifications to make the problem workable.

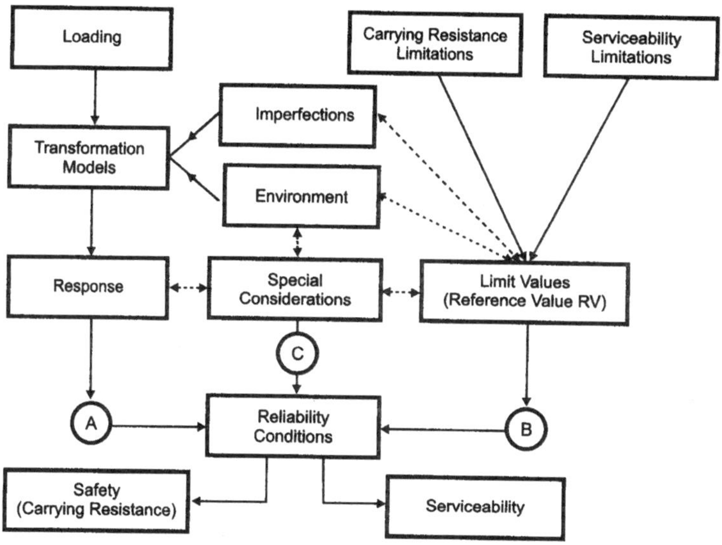

Figure 1: Reliability assessment process.

2 Overview of transformative simulations

To model the real construction and its characteristics when subject to loading and environmental influences, the version of structural analysis may be theoretical, empirical, or half-empirical. A half-empirical model is based partly on experimental results and partly on theoretical hypothesis. With respect to the expected response of the structure to the loading, and in conformity with the details of the actual problem, the model may be static or dynamic, elastic or elasto-plastic.

The structural mechanics model can consider first order or second order analysis, and also time dependence.

The leading mathematical and physical models accessible to the design engineer are illustrated in fig. 2.

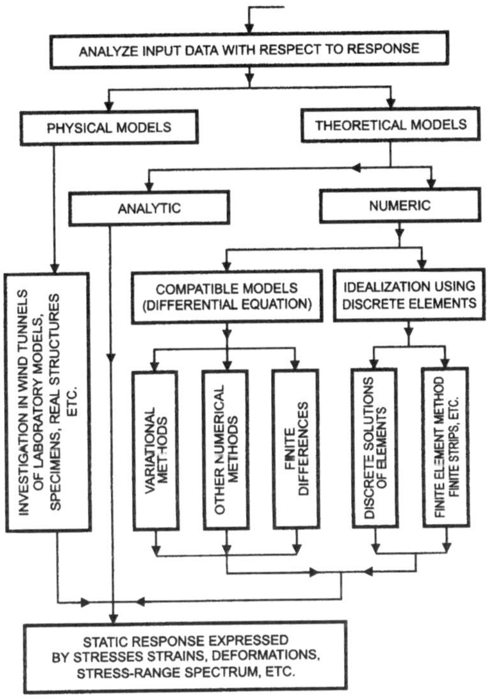

Figure 2: List of contents of the leading transformation version.

Physical versions serve as a guideline to examine experimentally the resistance and the serviceability. Tests may replace an analytical or numerical approach especially if adequate calculation models are not available, if a large number of similar components will be used, or if the real behaviour of a structure is of special interest. The following are examples of the types of tests that may be performed: tests to establish directly the ultimate resistance or serviceability properties of structural parts, tests to obtain specific material properties, tests to reduce uncertainties in transformation models (e.g. full size prototypes), control conformity tests to check the quality of mass-produced structures, and verification tests to check the behaviour of the actual structure after completion.

The issues may be applied to a definite structure or can function as a basis for the design of a broad range of construction, inclusive of the progress of rules in structural codes.

Mathematical models are of cardinal importance for constructional design usage. Such versions can be either analytical or numerical. Analytical models can be applied in everyday work such as in codes for structural design. In complex circumstances, numerical models can yield more suitable and detailed outcomes.

The principle of analytical modelling is to express a problem as a system of equation and to tackle it in closed form. Analytical models and their solution methods involve a very wide range of structural problems including plates, beams or columns elements, frames, and local structural details. The solutions of more complex problems using analytical models, however, are limited to those problems with special boundary conditions such as , types of supports, loading etc. Analytical methods are advantageous in that they provide a good physical understanding of the interaction of the factors and effects involved. They allow for simplifications to be introduced in relation to the significance of the problems (i.e., simplified formulas in codes for practical applications). In addition, they are simple enough for relatively inexpensive calculating tools like calculators. Analytical models can accelerate probabilistic simulation-based reliability assessment.

For problems that cannot be solved analytically, or in other adverse occurrences, one can use various numerical methods which are continually being improved. In so far as a given problem can be expressed by differential equations but cannot be solved in closed form, possibly because of the given boundary conditions, numerical solutions can be used. For these continuous models variational methods, finite difference methods, and other numerical procedures are available. These models were popular some time ago, when computers were already capable of resolving quite extensive systems of linear algebraic equations, but both the finite element method and boundary element technique had not yet been developed to their present-day perfection. Fair experience has been mustered in a similar way with other analogous algorithms, for example the collocation approach.

The present-day methods based on idealizing the construction by discrete elements are represented primarily by the FEM and BEM. Some other methods like finite strips and the folded plate method are available, however not extensively used nowadays. Today, the concept of the finite element method is frequently used. Major general aimed analysis programs have been worked out employing displacement-based finite element method.

2.1 Static model

(a) Time-independent loading.

In the case of time-independent loading, the equilibrium condition may be written down by a set of equations, in the form

$$K \times U = F \tag{1}$$

where the stiffness matrix K defines the load carrying system, the vector F defines the loading acting on the structure, and U is the function representing the resulting responses of the structure to the loading. Response can be expressed by deformations and corresponding stresses, moments, axial forces or variety of other quantities such as reactions at the supports. The solution to eqn. (1) means the static response of the structure to the loading.

(b) Time-dependent loading

As stated in the previous section, static models fit well with time-independent loading. Often, time-dependent loading causes dynamic response. If the forces are time-dependent, the designer must consider problems in which the inertia of accelerating masses must be taken into consideration. However, in some situations static response may be time-dependent. Such a situation can occur when the dynamic component of the response of a structure is negligible but loading is time-dependent. Basically, this consideration of inertia distinguishes the dynamic response from the static response. This is the case in many common situations in structural design where only "static" analysis is conducted and the response history need not be considered at all, or the response history is based on static analysis at selected points in time. Employing a point-in-time analysis, for special circumstances and if demanded, a static response description may be gained and applied such as in assessing fatigue damage accumulation and risk of fatigue collapse.

2.2 Dynamic models

In general, dynamic models applied to the analysis of dynamic response, may be rendered by means of the dynamic equilibrium equations

$$M \times U'' + C \times U' + K \times U = F(t) \tag{2}$$

where M stands for the mass matrix, C the damping matrix, K the stiffness matrix of the structural system, F(t) is the vector of the time-dependent loading function; U is function representing the resulting response of the structure to the loading (for example deformations), and U' and U" are the first and second derivatives of the function U. In this universal instance, the response of the construction is dynamic, and equation (2) is named the equation of motion.

In the majority of cases, dynamic response of a structure is created by a fast alteration of the magnitude, position or direction of the loads. However, a sudden change of stiffness or failure of a structural element such as the fracture of cables of a guyed mast or of a cable stayed bridge can gives rise to dynamic response as well.

Dynamic models may be formulated in the time domain or in the frequency one. When the load history is described in statistical terms, a statistical description of the response is also desirable. Based on such models, one can determine the probability of exceeding some limit state in a given reference period. Structural properties may be random as well as deterministic and time dependent as well as time independent. In a full probabilistic analysis, the variation of structural properties is taken into account. The models for dynamic analysis consist of stiffness, inertia and damping models. Applying to the stated models, for dynamic response in severe earthquakes, a ductility control and corresponding hysteretic energy dissipation may be essential to clarify cyclic degradation.

2.3 Quasi-dynamic models

Frequently, dynamic models are replaced by quasi-dynamic versions. At the same time, the time dependent loading F(t) (see eqn. 2) is replaced by an "adjusted" time-independent load F resulting in approximately the same response S (i.e., moments, stresses, deflections and others). We may formulate this approach by the following relationships, with respect to the response, S, of a construction to the loading:

$$\begin{aligned} &\text{dynamic model} && S_{dyn}(t) = H_{dyn} \times F(t) \\ &\text{quasi-dynamic model} && S_{q\text{-}dyn} = H_{stat} \times F \times DF \end{aligned} \tag{3}$$

where $S_{dyn}(t)$ – the dynamic response, and $S_{q\text{-}dyn}$ – the quasi-dynamic response of the structure to the loading, H are the transformation operators (static or dynamic) containing parameters of the load carrying system. F(t) is the time-dependent loading, F is the time-independent loading, and DF is the "Dynamic Load Factor". The DF is defined (usually) as the ratio of the dynamic deflection at any time to the deflection which would have resulted from the static application of some selected arbitrarily value of the load which is used in specifying the load-time variation.

DF depends on the natural frequency and the relative damping. The dynamic factor approach may be used to determine the extreme values of the response, although the response history cannot be obtained. The application of deterministic values of DF as given in specifications (which reflect both, the loading history as well as the properties of the structure), when not regarded conveniently, can result in considerable variances from the real dynamic response to the loading.

3 Fatigue fracture in the light of reliability

Advances in fracture mechanics have facilitated to qualify a cardinal problem in lifetime prediction: the reliability, or probable life, of a flawed construction. By a flawed construction, we understand a structure weakened by a crack below the threshold of elucidation by inspection or a crack of a known size.

For life expectance determination of flawed construction, the influence of the geometry of the component or structure and its interaction with the propagating crack should be contemplated. Alternatively, the effect of the state of stress that is really present in the body and its interaction with the singular stress state at the crack-tip are not precisely considered in the reliability appraisal.

There are two main difficulties to calculate the reliability of flawed construction by truly simulating the propagating crack in the structure: (i) a lack of data on the distributions of uncertain variables; and (ii) the great computational onus that is associated with the resultant probabilistic analysis.

The probabilistic finite element method (PFEM), in conformity with Besterfield et al. [3], provides a tool for estimating the effects of uncertainties in loading, material properties, and geometry on the uncertainties in the response variables. Via a fusion of the probabilistic finite element method with fracture mechanics, it has become possible to estimate the statistics of the stress intensity factor at the crack-tip.

Fatigue crack growth is sensitive to many characteristics and these parameters can rarely be stipulated accurately.

Uncertainties in the crack geometry, material properties, crack direction, crack growth, component geometry, and load time history, all play an important part. In this way, the prediction of fatigue collapse must be being an interpreted probabilistic problem. Several authors have recently studied probabilistic fatigue crack growth from a theoretical and statistical viewpoint using classical problems with known deterministic solutions.

Generally, for probabilistic fatigue crack growth, randomness in crack geometry, material properties, crack direction law, crack growth law, component geometry, and load time history are notable.

The reliability problem for fatigue crack growth is formulated by dint of an optimization procedure. In the optimization problem, the equilibrium equation and crack direction law at each discretized crack point are asserted by Lagrange multipliers. The solution methods employing PFEM and adjoint methods are then discussed. Finally, some detailed computational aspects are given.

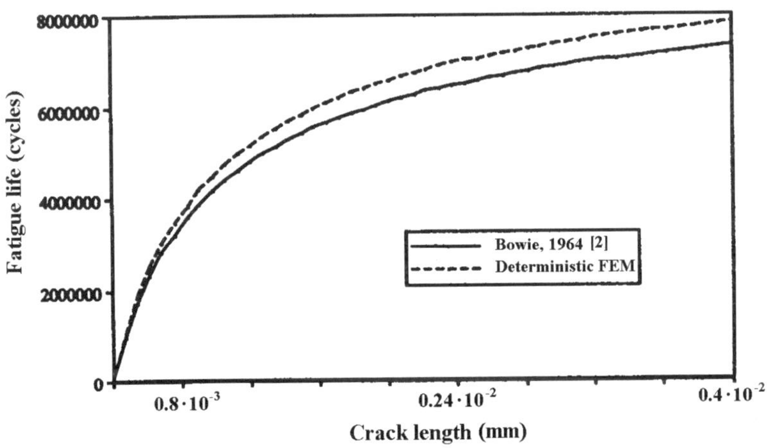

Figure 3: Comparison of the fatigue life between the reference and FEM solutions.

3.1 Fatigue crack growth optimization

Above all, the crack direction law has to be discretized into "n" points along the crack path. At each discretization point, the crack direction equals, as follows

$$z_k = \theta_k^T \kappa_k = 0 \qquad k = 1, \ldots, n \tag{4}$$

κ and θ mean vectors of both the stress intensity factors and the angle between the tangent to the crack-tip and the x-axis, where κ_k and θ_k stand for κ and θ interpreted at $\xi=\xi_k$, k = 1, …, n. Like this, at each crack path discretization point, the new crack direction is recalculated and the crack is next permitted to propagate to the next point.

3.1.1 Optimization problem statement

The determination of the reliability index by the first-order probability theory is represented being a constrained optimization problem. The resultant nonlinear programming problem consists of determining the value of the primitive correlated random variables, b, and the generalized displacements, δ_i, i = 1, …, n, which minimize the distance from the origin to the limit-state surface in the independent standard normal space. The minimum distance from the origin to the limit-state surface in the independent standard normal space is called the reliability index, β, i.e.

$$\beta = r^T r \tag{5}$$

where r is a vector of independent standard normal random variables.

The minimization is submitted to the equality constraints being based on the crack growth direction law and equilibrium, i.e., eqn. (4) and further

$$K_i \delta_i = f_i \qquad \text{no sum on i,} \quad i = 1, \ldots, n \tag{6}$$

respectively, where K_i and f_i are the enriched stiffness matrix and external force vector, and the generalized displacements have the form

$$\delta_i = \begin{bmatrix} d \\ \overline{\kappa} \end{bmatrix}_i \qquad i = 1, \ldots, n \tag{7}$$

where d denotes the usual nodal displacement vector. An enriched element approach, which has the near crack-tip located singular strain field, is employed to obtain the eqns (6) and (7).

The minimization is also subject to the following inequality constraint

$$T - T_s \leq 0 \tag{8}$$

(i.e., the performance function being on the limit state surface is a constraint in the optimization problem).

4 Uncertainty simulation perturbation technique

Generally, uncertainties can emerge in material properties, specified surface values, the geometry of the body, or any combination of these factors. When extending the formulation to embrace uncertainties, we differentiate between various classes' problems (see [4]). The philosophy of randomness or uncertainty may be distinguished like this:

(i) uncertainty in view of system parameters leads to a "spread" or density of possible parameter values, i.e., random variable simulation
(ii) uncertainty with regard to the behaviour of parameters as time-dependent processes, where the statistics of each process may be either-time-dependent or –independent, i.e., random or stochastic function model, and
(iii) uncertainty as to the characteristics of parameter properties as functions of spatial coordinates where, again, the statistics of the functional behaviour may be either space dependent or independent, i.e., random or stochastic field model.

An instance of considerable interest in engineering is one where the statistics of the material parameters μ and ν are available in the way that their mean values are known and deviations relative to the mean are not too large. We limit this problem to a body of one material with given surface values and geometry. Afterwards, the unknown surface displacements and tractions can be expanded in a Taylor series about the mean values.

4.1 Beam stiffness

The definition of two independent material properties, for the most part μ and ν (the shear modulus and Poisson's ratio), is needed for linear elastic analysis of beams. There exist several circumstances when an accurate definition of μ and ν is impossible. This ends in great difficulties in the analysis routine. Such events are e.g.: (i) reinforced (or plain) concrete beams, (ii) equivalent (built up) beams.

Table 1: Beam material properties.

$\bar{\mu}$	μ_1	μ_2	$\bar{\nu}$	ν_1	ν_2
1.0	0.75	1.25	0.3	0.2	0.4

5 Conclusion

For engineering problems both, theoretical and experimental, the transformation models serve as a tool to determine the response of the structure to loading. A selection of the model depends on the requirement of a solution, the accuracy required and on the nature of the reliability condition (from the standpoint of the carrying capacity or serviceability), and suchlike. The assignment of the scientific field – structural mechanics – is to work out and improve the simulations in the way that they faithfully replicate the actual effectiveness of the systems, components and elements of the structures and efficiently allow to monitor the relation "loading-response" for the construction.

The technique presented to derive fatigue crack growth reliability produces very useful outcomes, is computationally effective, and may be extended to constituents of many forms. The method is suitable for mixed-mode fatigue crack growth. The aptitude of this method to stipulate the probability of fatigue collapse owing to uncertainties in the element geometry, applied loads, material properties, and crack geometry is of primary importance to the design engineers of various specialist fields. An inherent shortcoming of the methodology is that a very accurate calculation of the stress intensity factors is necessary so as to obtain reasonable precision for the reliability index. The results demonstrated prove the initial crack length to be a critical parameter. Taking into account that crack lengths below the threshold of a visitation limit are assumed to show a considerable quantity scatter, this makes it imperative that the life expectance of a construction be interpreted from a stochastic standpoint. But still, except for some difficulties that may emerge when remeshing, the procedure is practicable to arbitrary geometries.

Acknowledgement

The author gratefully acknowledges the financial support of the presented research by the Grant Agency of the Czech Republic (project No. 103/03/0655).

References

[1] Marek, P., Guštar, M., Anagnos, T., *Simulation-Based Reliability Assessment for Structural Engineers*, CRC Press, Inc.: Boca Raton, Florida, 1995.

[2] Bowie, O.L., Rectangular tensile sheet with symmetric edge cracks. *Journal Applied Mechanics*, ASME 31, 1964.

[3] Besterfield, G.H., Liu, W.K., Lawrence, M. & Belytschko, T.B., Fatigue Crack Growth Reliability by Probabilistic Finite Elements (Chapter 16). *Computational Mechanics of Probabilistic and Reliability Analysis*, eds. W.K. Liu & T. Belytschko, Elmepress Internat: Lausanne, pp. 345-369, 1989.

[4] Ettouney, M., Benaroya, H. & Wright, J., *Probabilistic Boundary Element Method* (Chapter 5), ibid, pp. 142-165, 1989.

[5] Benaroya, H., Rehak, M., Finite Element Methods and Probabilistic Structural Analysis - A Selective Review, *ASME Applied Mechanics Review*, **41(5)**, 1988.

[6] Ahmad, S., Banerjee, P.K., Time-domain transient elastodynamic analysis of 3-D solids by BEM, *Int Journal for Num Meth in Eng*, **26**, pp. 1709-1728, 1988.

Rapid re-analysis in BEM elastostatic calculations

J. Trevelyan & D. J. Scales
School of Engineering, University of Durham, UK

Abstract

The real time solution and updating of stress contours for small elasticity problems allows the rapid and interactive evaluation of multiple design proposals and the convergence on an optimum solution in a significantly reduced time. The increased level of interactivity allows the user to see the potentially complex interactions between multiple geometric features. A number of strategies are proposed and compared with current techniques. Boundary element integrals are accelerated by the use of Look-Up Tables (LUT's) containing pre-computed integrals. These can be employed in the matrix assembly and the internal point calculations for both flat and arc elements. The effect of problem size on propagated error has been investigated and the effect of interpolation has been considered. An error analysis has been performed and a suitable scheme for LUT refinement proposed. The approach has been benchmarked against regular Gauss-Legendre quadrature. The large storage requirement of the LUT's has been considered and the use of a least squares surface fit to the LUT's has been investigated and the computational cost compared with Gauss-Legendre quadrature. Iterative re-solution of the problem is accelerated by the use of an approximate LU preconditioner that is updated as required by a secondary thread of execution. The proposed preconditioner has been benchmarked against other preconditioners and reduces solve times to typically less than 30% of the direct solve time. The proposed techniques achieve the aim of providing a real-time update of contours for small problems on a PC.
Keywords: rapid re-analysis, integrations, surface fitting.

1 Introduction

Accelerating the stress analysis of structures allows the rapid consideration of multiple designs for a variety of criteria, and provides for efficient solution of

WIT Transactions on Modelling and Simulation, Vol 42, © 2006 WIT Press
www.witpress.com, ISSN 1743-355X (on-line)
doi:10.2495/BE06027

problems requiring multiple runs, e.g. structural optimisation algorithms. The motivation for the current work is the rapid re-analysis of a modified problem to allow for real-time display of results contours as a design geometry is interactively perturbed. In such a case, the governing matrix system and its solution vector are both available for the unperturbed system, and it is required to modify appropriate terms and find the revised solution very rapidly to provide a real-time animation of contours.

For completeness a brief overview of the boundary element method will be presented; more detail can be found in, for instance, Becker [1]. Consider a domain Ω with boundary Γ upon which displacement and traction conditions are applied. We can solve this problem for displacements, u, and tractions, t, using the standard direct formulation of the boundary integral equation

$$c(p)u_i(p)+\int_\Gamma t_{ij}^*(p,Q)u_j(Q)\mathrm{d}\Gamma(Q)=\int_\Gamma u_{ij}^*(p,Q)t_j(Q)\mathrm{d}\Gamma(Q) \tag{1}$$

where p is a collocation point on Γ, $c(p)$ is a scalar multiplier depending on the boundary geometry, i and j are coordinate direction indices and u^* and t^* are the fundamental solutions. In 2D elasticity problems these are represented by

$$u_{ij}^* = C_1\left\{C_2 \ln\left[\frac{1}{r}\right]\delta_{ij} + \frac{\partial r}{\partial x_i}\frac{\partial r}{\partial x_j}\right\} \tag{2}$$

and

$$t_{ij}^* = C_3\left[\left(\frac{\partial r}{\partial n}\right)\left(C_4\delta_{ij} + 2\frac{\partial r}{\partial x_i}\frac{\partial r}{\partial x_j}\right) + C_4\left(\frac{\partial r}{\partial x_j}n_i - \frac{\partial r}{\partial x_i}n_j\right)\right] \tag{3}$$

where $r = |Q - p|$, n is the unit outward normal at Q, δ_{ij} is the Kronecker delta and C_k, k=1,…,4 are coefficients based on material properties. Accommodating the boundary element discretisation into (1) yields

$$c(p)u_i(p)+\sum_{elements}\int_{-1}^{1} t_{ij}^*(p,Q)\boldsymbol{N}^T J d\xi \boldsymbol{u} = \sum_{elements}\int_{-1}^{1} u_{ij}^*(p,Q)\boldsymbol{N}^T J d\xi \boldsymbol{t} \tag{4}$$

where the vector $\boldsymbol{N}$ contains the Lagrange polynomial shape functions and J is the Jacobian of the isoparametric mapping $Q=(x,y) \rightarrow \xi(-1 \le \xi \le 1)$. Vectors $\boldsymbol{u}$ and $\boldsymbol{t}$ contain the nodal displacements and tractions respectively. It is the efficient approximate evaluation of the integrals in (4) that is the subject of this paper. Integrals similar to those in (4) are also fundamental to the data recovery at internal points. Typically the integrals are evaluated by the use of a variable order Gauss-Legendre quadrature [2 ,3]. A number of authors [4, 5] have investigated the required order of Gauss quadrature dependent on the relative scale of the problem, i.e. r_m/L.

We refer to the general case of a collocation (or source) point – field element pair as depicted in Figure 1, and seek a value of the integral $h_{ijk}(r_m, L, \phi, \theta)$ and $g_{ijk}(r_m, L, \phi, \theta)$, being respectively the integrals on the left and right hand sides of (4) evaluated for the node k (k = 0, 1, 2) on the quadratic element. Both angular variables are measured anticlockwise from the global Cartesian x-axis. L is the length of the field element and r_m is the value of r evaluated at the mid-node of the quadratic element, which is centrally located and also defines $\xi = 0$.

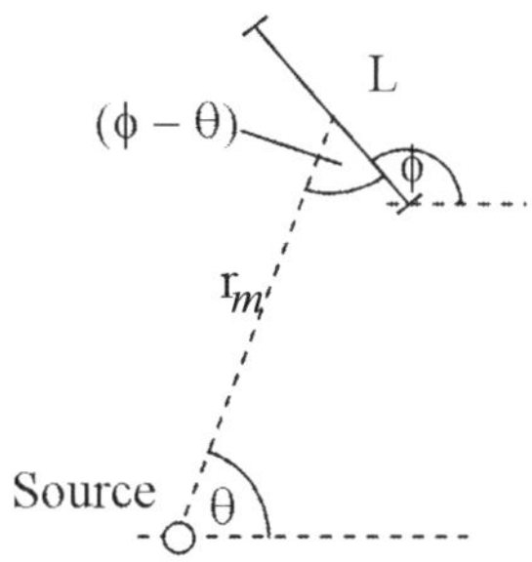

Figure 1: Geometric parameters for general collocation point – field element pair.

We are concerned in this work with small to medium sized problems in plane stress and plane strain. Scales and Trevelyan [6] found that the three major contributors to the computational cost of solving this type of problem are typically approximately equal contributors. They are

1. Evaluation of the boundary integrals to assemble the system matrix
2. Solution of this problem for the unknown displacements and tractions
3. Solution at a sufficient number of internal points to allow accurate contour plots

Literature on acceleration of BEM computations concentrates almost entirely on consideration of item 2, because generally the motivation for these works is the accelerated solution of very large systems, for which of course the solver phase dominates. Trevelyan *et al.* [7] considered the optimisation of an iterative solver for problems in re-analysis. This solver has been used in the current work to ensure a rapid solution of the matrix equations. The current paper focuses on the efficient computation of the integrals central to items 1 and 3.

Trevelyan and Wang [8] proposed the use of Look-up Tables (LUT's) to reduce the computational cost of integrating equation (4) for flat elements. The use of LUT's was later extended [6, 7] to allow the use of LUT's for arc based elements and further extended for integrating the derivative kernels in internal point stress calculations. LUT's can be created for any integral based on four parameters (r_m, L, ϕ, θ). However, this gives rise to excessive memory requirements, and it was found that a complete set of integrals could be stored in a LUT defined in terms of only two parameters (r_m/L, ϕ–θ). In this two-parameter form adjustments need to be made for element length, and the efficiency suffers from the requirement to transform the LUT values to take account of the values of θ and ϕ. Scales and Trevelyan [6] developed an optimal LUT refinement strategy, and showed that large gains can be made through the application of LUT's for both stress and displacement calculations. However, in the proposed method 1GB of RAM was required for storage of LUT's so the method might be considered to be limited to the higher end of today's PC market.

In this paper we present an approach that promises to outperform LUT's in both computational efficiency and memory requirements. It is based on the fitting of analytical expressions to approximate the surfaces describing the values of integrals h_{ijk} and g_{ijk} in the parametric space (ϕ, θ, r_m/L).

2 Least squares basis functions

The variation of integrals h_{ijk} and g_{ijk} in the three-dimensional parametric space (ϕ, θ, r_m/L) may be seen as the description of a four-dimensional surface. Initial analysis considered the variation in the integrals along individual lines through the 3D space. A least squares fit allowed a number of basis functions of importance to be determined and then extended to higher dimensional fits. For example, it is clear from the analysis that variation in g_{ijk} is dominated by logarithmic behaviour in r_m/L for $i = j$, and this is not unexpected because of the logarithmic term in (2). This is illustrated in Figure 2 for the integral g_{120} for the case $\phi = 90°$, $\theta = 0°$. Similarly it is found that variation in the angular direction θ is largely oscillatory with an underlying Sin($2\theta + \psi$) behaviour, where ψ defines the phase which is determined by ϕ. Unfortunately the sinusoidal behaviour is not simple, and Figure 3 illustrates the distorted sinusoid describing the variation in g_{120} with θ for the case $\phi = 90°$, $r_m/L = 3$. The level of distortion varies with the case being investigated.

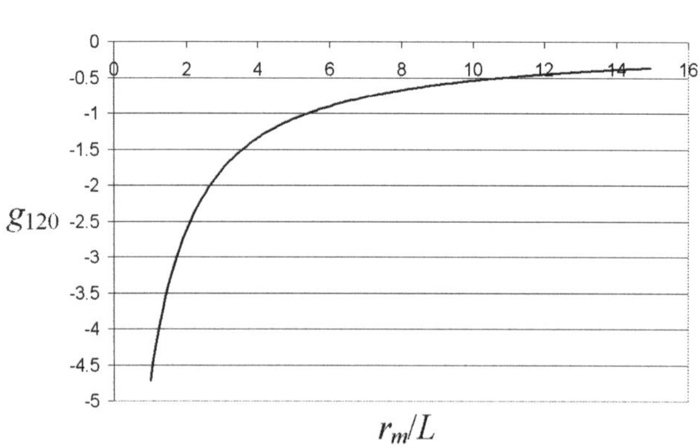

Figure 2: Line fit in r_m/L to integral g_{120} for the case $\phi = 90°$, $\theta = 0°$.

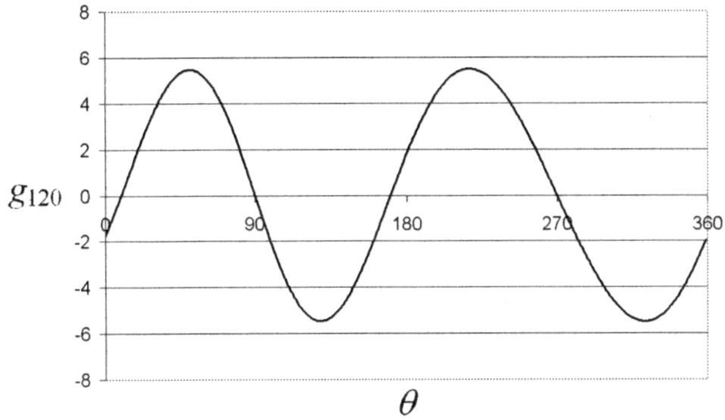

Figure 3: Line fit in θ to integral g_{120} for the case $\phi = 90°$, $r_m/L = 3$.

Our analysis suggests that a good fit on a surface level can be achieved by using a least squares fit based on basis functions as follows in the r_m/L and θ directions:

$$Integral_{r_m/L} = f\left(1, \left(\ln\left(\frac{r_m}{L}\right)\right)^q, \left(\frac{r_m}{L}\right)^{-q}\right), q = 1,\dots,4 \tag{5}$$

$$Integral_\theta = f(1, \sin(q\theta), \cos(q\theta)), q = 1,\dots,6 \tag{6}$$

Work on the full four-dimensional surface fits is in progress. However, we focus in this paper on the simpler case of a three-dimensional surface fit in the 2D parameter subspace (θ, r_m/L). This represents a plane of constant ϕ through the full space, and we focus on the particularly common element orientations of $\phi = 0°, 90°, 180°, 270°$.

3 Results

In this paper the surfaces describing two integrals have been considered,

1. g_{121} term for $\phi = 90°$
2. h_{111} term for $\phi = 0°$

The basis functions presented in (5) and (6) for line fitting have been progressed to the surface fitting stage, including all of the possible products of a basis function from (5) and a basis function from (6), giving a least squares fit using 117 functions. However, it is desirable from the viewpoint of computational efficiency to reduce this set to a considerably smaller number. Fortunately this is possible for the surfaces we are considering, and each surface may be approximated by a linear combination of just a few of the basis functions.

Two approaches have been used to reduce the number of basis functions to a minimum. Firstly, we subdivide the subspace in the r_m/L direction so that we fit over strips ($0 \le \theta \le 360°$, $\eta_l \le r_m/L \le \eta_{l+1}$, $1 \le l \le l_{max}$). The results presented in this paper are for $3 \leq r_m/L \leq 4$. Secondly we use a progressive reduction scheme that in its initial form is expressed as follows:

1. Initialise $n = 117$
2. Perform least squares fit using n basis functions
3. For each term determine its importance by weighting the L_2 norm of the basis function by its least squares coefficient
4. Reject the basis function having least importance and set $n := n - 1$
5. If $n > 1$ return to step 2

In general we aim for a target accuracy of 0.1% in each integral [6] and select the final number of basis functions to be the smallest that meets this criterion. However, we consider some fits to be desirable in which the error is as much as 0.25% if the final value of n is very low. The error will vary over the parameter space, and so we compute this in an L_2 sense, defining a relative L_2 error norm ε as

$$\varepsilon = \frac{|z_0 - z|_{L2}}{|z_0|_{L2}}$$

where z is the value of the integral found from the least squares fit and z_0 is the value of the same integral found through high order Gauss-Legendre integration, which we deem to be close to the exact. In order to relate ε to the expected maximum error in an individual integral, a series of numerical tests was performed for each integral C. This resulted in a simple relationship being established between these error indicators. The relationship for the integral g_{121} has linear characteristics and is presented in Figure 4.

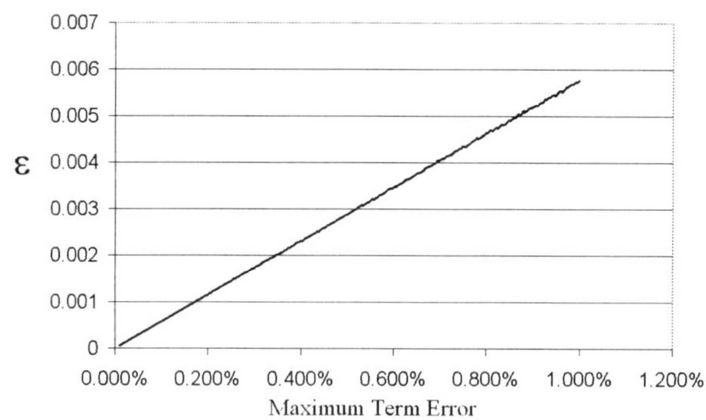

Figure 4: Relationship between ε and maximum error in an integral term.

Using this relationship we can plot the maximum percentage in each term against the number of functions used in the surface fit. From figure 5 we can see that the error increases at specific locations (marked), at which stages functions critical to achieving a good fit are removed from the least squares basis. We identify the basis functions of greatest importance in this way, and in the final stage, undertake a least squares fit using a basis comprising these functions.

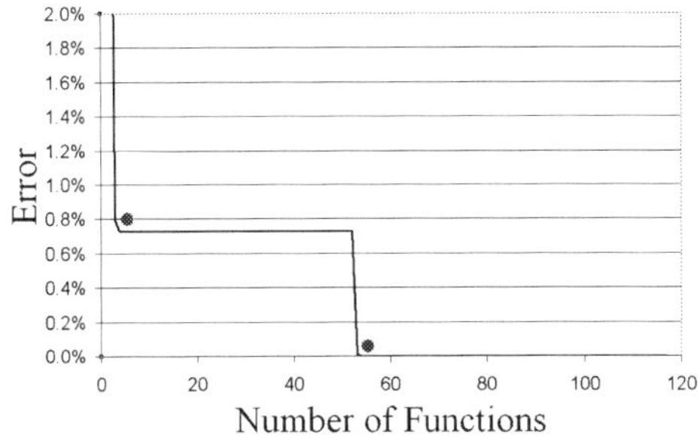

Figure 5: Error in g_{121} as number of fitting functions is reduced.

The reduction scheme is therefore augmented to its final form:

1. Initialise $n = 117$
2. Perform least squares fit using n basis functions
3. For each term determine its importance by weighting the L_2 norm of the basis function by its least squares coefficient
4. Reject the basis function having least importance and set $n := n - 1$
5. If $n > 1$ return to step 2
6. From a plot of the form of figure 5 determine the most important basis functions
7. Final least squares surface fit using only the most important functions

Results show that a fit with only 5 functions is accurate to within 0.009% per term for the integral g_{121}. The expression for this term is found to be

$$g_{121} = \begin{Bmatrix} \sin(2\theta)\left[0.30\left(\frac{L}{r_m}\right) - 1.55\left(\frac{L}{r_m}\right)^2 + 3.55\left(\frac{L}{r_m}\right)^3 - 3.04\left(\frac{L}{r_m}\right)^4\right] \\ -1.07\times10^{-3}\sin(4\theta)\left(\frac{L}{r_m}\right)^2 \end{Bmatrix} \times 10^{-5} \quad (7)$$

Figure 6 shows a graphical comparison of the integral surfaces generated by high order Gauss-Legendre and by the least squares fit.

A similar analysis for the term h_{111} for the case $\phi = 0°$ shows the most economical fit to be provided by $n = 3$ with a maximum term error of 0.0164%. Again the subspace ($0 \le \theta \le 360°$, $3 \le r_m/L \le 4$) is used. The expression for the surface is

$$h_{111} = 0.209\sin(\theta)\left(\frac{L}{r_m}\right) + 0.101\sin(3\theta)\left(\frac{L}{r_m}\right) + 0.015\sin(5\theta)\left(\frac{L}{r_m}\right)^3 \quad (8)$$

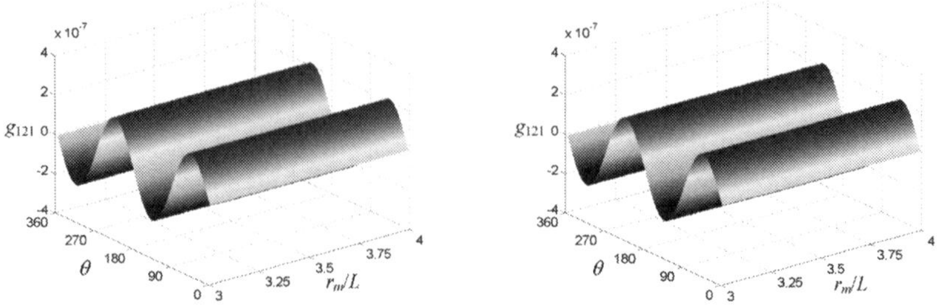

(a) (b)

Figure 6: g_{121} surface (a) Gauss-Legendre (b) least squares fit.

Figure 7 shows a graphical comparison of the integral surfaces generated by high order Gauss-Legendre and by the least squares fit.

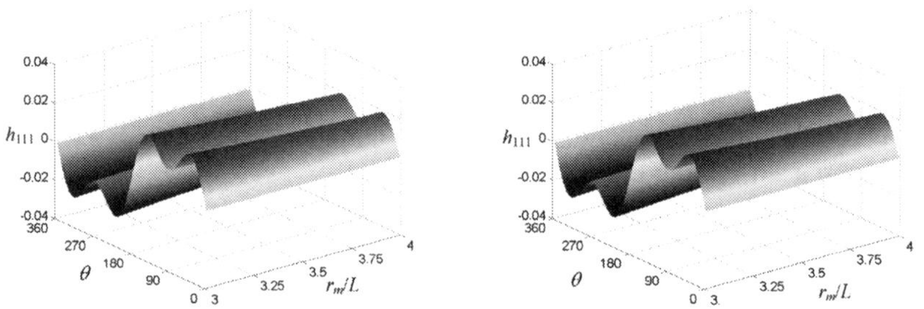

(a) (b)

Figure 7: h_{111} surface (a) Gauss-Legendre (b) least squares fit.

The computational efficiency of the method proposed has been compared with 2nd Order Gauss-Legendre quadrature for the datasets concerned by considering the required floating point operations and their relative efficiency [6]. This is represented in Table 1, which shows that for the two terms calculated we can reduce the runtime by one half and two thirds respectively. Readers will be able to scale the benefits easily to compare against higher order Gauss-Legendre integration. It is possible to extend this method for all of the h_{ijk} and g_{ijk} terms although certain terms, namely end-nodes ($k = 0, 2$), require larger number of fitting functions to meet the required target error.

Table 1: Comparison of computational efficiency.

Method	Normalised Runtime g_{121}	Normalised Runtime h_{111}
2nd order Gauss-Legendre	1.000	1.000
Surface Fits	0.457	0.384

The extension to the derivative kernels involved in computation of internal point stresses conditions is also possible though care will have to be taken over the least squares fits since these integrals exhibit only C^0 continuity.

4 Conclusions

A method has been proposed for rapid computation of boundary integrals in 2D elasticity that overcomes the RAM storage requirement from which Look-up Table approaches can suffer. At the time of writing, the proposed method has been investigated for the integrations involving the u^* and t^* kernels. Other kernels, notably the derivative kernels used in internal point stress evaluation, may also be handled in this way.

The method proposed reduces the integration time for the cases studied by 55% and 62% over a conventional 2nd order Gauss-Legendre technique. Of course, these savings will increase in an easily predictable way if compared against a higher order numerical scheme. It is noted, however, that this saving will reduce somewhat when the end-node shape functions are introduced as they require higher order fitting.

Acknowledgements

The authors would like to thank BAE Systems and EPSRC for sponsoring this work under an Industrial CASE award.

References

[1] A.A. Becker, The Boundary Element Method in Engineering: A complete course, McGraw Hill, London, 1992.

[2] J.H. Kane, A. Gupta and S. Saigal (1989), Reusable Intrinsic Sample Point (RISP) Algorithm for the Efficient Numerical Integration of Three Dimensional Curved Boundary Elements, *Int. J. for Num. Meth. in Engg.*, **28** (1989) 1661-1676.

[3] J.C.F. Telles (1987), A self-adaptive coordinate transformation for efficient numerical evaluation of general boundary element integrals, *Int. J. for Num. Meth. in Engg.* **24** (1987) 959-973.

[4] Bu S, Davies T. Effective evaluation of non-singular integrals in 3D BEM. *Advances in Engineering Software* 1995; **23**: 121–128.

[5] Eberwien U, Duenser C, Moser W. Efficient calculation of internal results in 2D elasticity BEM. *Engineering Analysis with Boundary Elements* 2005; **29**: 447–453.

[6] D.J. Scales and J. Trevelyan, Rapid re-analysis in 2D BEM elastostatic calculations, in K. Chen (Ed.) Advances in Boundary Integral Methods, UKBIM 5, Liverpool University Press, 2005, p.153-162.

[7] J. Trevelyan, D.J. Scales, R. Morris and G.E. Bird, Acceleration of boundary element computations in reanalysis of problems in elasticity, in

Z.H. Yao, M.W. Yuan, W.X. Zhong (Eds.) Computational Mechanics: Abstract (Volume 2), WCCM VI in conjunction with APCOM 04, Tsinghua University Press and Springer-Verlag, 2004, p.44.

[8] J. Trevelyan and P. Wang, Interactive re-analysis in mechanical design evolution. Part 2: Rapid evaluation of boundary integrals, *Computers and Structures*, **79** (2001) 939-951.

Section 7
Dynamics and vibrations

3D wave field scattered by thin elastic bodies buried in an elastic medium using the Traction Boundary Element Method

P. Amado Mendes & A. Tadeu
Department of Civil Engineering, University of Coimbra, Portugal

Abstract

A combined formulation based on the Traction Boundary Element Method (TBEM) and the Boundary Element Method (BEM) is proposed to model three-dimensional wave scattering in solid media, which contain thin elastic bodies whose geometry remains the same along one direction. Although the classical BEM models degenerate in the presence of thin heterogeneities, this combined formulation, in which one side of the body is discretized with the BEM and the other side with the TBEM formulation, is able to model the wave field in the vicinity of these heterogeneities. Arbitrary-shaped and oriented elastic thin bodies, even with no thickness, may be modelled using this formulation. Analytical integrations are performed to evaluate the hypersingular integrals. The proposed model is applied to two different thin heterogeneities buried in an elastic unbounded medium.
Keywords: wave propagation, elastic scattering, thin elastic inclusion, BEM, TBEM, two-and-a-half-dimensional problem.

1 Introduction

The problem of modeling the scattering of elastic waves by an elastic inclusion, a cavity or a crack has been addressed in different fields of research related to the remote detection, location and identification of heterogeneities, delaminations or anomalies. A complete understanding of how waves propagate from a generic source to a receiver in a homogeneous elastic medium and in the vicinity of irregularities, the so-called direct problem analysis, is required for the correct interpretation of many of those testing techniques [1–4].

WIT Transactions on Modelling and Simulation, Vol 42, © 2006 WIT Press
www.witpress.com, ISSN 1743-355X (on-line)
doi:10.2495/BE06028

Since analytical solutions have only been derived for objects with simple geometry [5], several numerical methods have been proposed over the years to study the detection of defects and heterogeneities by elastic wave scattering. Among these numerical methods, the Boundary Element Method (BEM) seems quite suitable for wave propagation modeling in homogeneous unbounded systems [6,7]. However, when the heterogeneity is very thin, such as a crack or almost imperceptible defect, the conventional direct BEM degenerates.

The Indirect Boundary Element formulation [8], the Traction Boundary Integral Equation Method [9] and the Traction Boundary Element Method (TBEM) [10] are among the numerical methods that solve the thin-body difficulty. These formulations require the solution of hypersingular integrals, and different techniques have been proposed to cope with this complexity [11,12]. Prosper & Kausel [10] proposed an indirect approach for the analytical evaluation of integrals with hypersingular kernels for plane-strain cases in a two-dimensional (2D) medium. Another boundary element formulation, mainly adopted for the analysis of crack problems, is referred to as the Dual Boundary Element Method. In this technique, a single region formulation is achieved by combining the displacement boundary integral equation discretizing one of the crack surfaces and the traction boundary integral equation used to discretize the opposite crack surface [13].

This paper describes a combined formulation based on the TBEM and the BEM that is used to model the three-dimensional (3D) wave scattering in solid media, which contain thin elastic bodies whose geometry remains the same along one direction. Firstly, the 3D problem formulation and the boundary elements formulations are presented. Afterwards, the implementation of this model is successfully verified for the case of an elastic circular inclusion by using analytical solutions, already known for simple geometry cases.

An application of the proposed model is presented for a softer and a harder thin heterogeneity buried in an elastic unbounded medium. As the model is initially formulated in the frequency domain, time results are obtained by applying inverse fast Fourier transformations to the frequency responses. Time results are presented for the displacement fields and both the wave pattern evolution and the perturbation due to the presence of the thin bodies in the elastic medium can be clearly observed.

2 Problem formulation

An infinite 2D elastic inclusion is buried in an unbounded homogeneous isotropic elastic medium, with no intrinsic attenuation. The exterior medium (with density ρ_1) allows shear wave and compressional wave velocities of β_1 and α_1, respectively. The inclusion is aligned along the z axis and perfectly filled with a different solid material with density ρ_2, shear wave and compressional wave velocities of β_2 and α_2, respectively. The host medium is excited by a harmonic dilatational 3D point source, placed at O $(x_s, y_s, 0)$, which oscillates with frequency ω and emits an incident field that can be given by a dilatational potential ϕ,

$$\phi_{\text{inc}} = \frac{A\,\mathrm{e}^{\mathrm{i}\frac{\omega}{\alpha_1}\left(\alpha_1 t - \sqrt{(x-x_s)^2+(y-y_s)^2+z^2}\right)}}{\sqrt{(x-x_s)^2+(y-y_s)^2+z^2}}, \tag{1}$$

with 'inc' referring to the incident field, A the wave amplitude and $\mathrm{i}=\sqrt{-1}$.

As the geometry of this problem does not change along the z direction, applying a Fourier transformation along that direction allows the 3D solution to be expressed as a summation of 2D problems, each one solved for a specific spatial wavenumber, k_α,

$$\phi_{\text{inc}}(\omega,x,y,z) = \frac{2\pi}{L_{vs}} \sum_{m=-\infty}^{\infty} \hat{\phi}_{\text{inc}}(\omega,x,y,k_{zm}), \tag{2}$$

where $\hat{\phi}_{\text{inc}}(\omega,x,y,k_{zm}) = {}^{-\mathrm{i}A}\!/\!_2\, \mathrm{H}_0\left(k_{\alpha 1}\sqrt{(x-x_s)^2+(y-y_s)^2}\right)\mathrm{e}^{-\mathrm{i}k_{zm}z}$, k_{zm} is the axial (longitudinal) wavenumber defined by $k_{zm} = {}^{2\pi}\!/\!_{L_{vs}}\, m$, the effective wavenumbers are given by $k_{\alpha 1} = \sqrt{{}^{\omega^2}\!/\!_{\alpha_1^2} - k_{zm}^2}$, with $\mathrm{Im}(k_{\alpha 1})<0$, $\mathrm{H}_n(\ldots)$ are second Hankel functions of order n, and L_{vs} is the distance between virtual point sources periodically placed along z. If the space between sources is kept large enough to prevent spatial contamination from the virtual sources, that summation converges and an approximation can be given by a finite sum of terms [8]. The pure 2D problem corresponds to $k_z = 0$.

Solutions in the time domain are obtained by applying an inverse Fourier transform to the frequency domain responses, where the dynamic excitation source is modelled as a Ricker pulse. To avoid the aliasing phenomena and to minimize the contribution of the periodic virtual sources, complex frequencies with an imaginary part of the form $\omega_c = \omega - \mathrm{i}\eta$ (with $\eta = 0.7\Delta\omega$) are used [8].

3 Boundary integral formulations

A combined formulation based on the TBEM and the BEM, will be presented to model the 3D wave scattering in solid media which contain thin elastic bodies whose geometry remains the same along one direction. Prior to the combined formulation, both boundary elements methods are presented.

3.1 Boundary element formulation

An elastic body bounded by a surface S is buried in a homogeneous infinite elastic medium, which is illuminated by a spatially sinusoidal harmonic line load located in the exterior solid medium at x_s, with axial wavenumber k_z. The wave propagation in this system is governed by boundary integral equations, which can be derived by applying the reciprocity theorem [6], leading firstly to:

along boundary surface S *, in the host elastic domain*

$$c_{ij}u_i\left(\boldsymbol{x}_0,\omega\right)=\int_S t_j\left(\boldsymbol{x},n_n,\omega\right)G_{ij}\left(\boldsymbol{x},\boldsymbol{x}_0,\omega\right)\mathrm{d}s - \int_S u_j\left(\boldsymbol{x},\omega\right)H_{ij}\left(\boldsymbol{x},n_n,\boldsymbol{x}_0,\omega\right)\mathrm{d}s+u_i^{\mathrm{inc}}\left(\boldsymbol{x}_s,\boldsymbol{x}_0,\omega\right), \tag{3}$$

where $i,j=1,2$ correspond respectively to the normal and tangential directions relative to the inclusion surface, and $i,j=3$ to the z direction. $H_{ij}\left(\boldsymbol{x},n_n,x_0,\omega\right)$ and $G_{ij}\left(\boldsymbol{x},\boldsymbol{x}_0,\omega\right)$ represent the fundamental traction and displacement solutions (Green's functions) in direction j on boundary S, at $\boldsymbol{x}$, due to a unit point force applied in direction i at collocation point, $\boldsymbol{x}_0$. $u_j(\boldsymbol{x},\omega)$ and $t_j(\boldsymbol{x},n_n,\omega)$ refer to the displacement and traction unknowns in direction j at $\boldsymbol{x}$. $u_i^{\mathrm{inc}}(x_s,x_0,\omega)$ corresponds to the displacement incident field at the collocation point along direction i. The coefficient c_{ij} is equal to $\delta_{ij}/2$, with δ_{ij} the Kronecker delta, for a smooth boundary. The unit outward normal at the boundary at x is defined by the vector $n_n=\left(\cos\theta_n,\sin\theta_n\right)$.

The Green's functions for loads and displacements in the x, y and z directions in an elastic medium have been derived and presented by Tadeu and Kausel [14]. The derivatives of these Green's functions make it possible to find the following tractions along the x, y and z directions, in the solid medium,

$$H_{rx}=2\mu\left[\frac{\alpha^2}{2\beta^2}\frac{\partial G_{rx}}{\partial x}+\left(\frac{\alpha^2}{2\beta^2}-1\right)\left(\frac{\partial G_{ry}}{\partial y}+\frac{\partial G_{rz}}{\partial z}\right)\right]\cos\theta_n+\mu\left[\frac{\partial G_{ry}}{\partial x}+\frac{\partial G_{rx}}{\partial y}\right]\sin\theta_n$$

$$H_{ry}=2\mu\left[\left(\frac{\alpha^2}{2\beta^2}-1\right)\left(\frac{\partial G_{rx}}{\partial x}+\frac{\partial G_{rz}}{\partial z}\right)+\frac{\alpha^2}{2\beta^2}\frac{\partial G_{ry}}{\partial y}\right]\sin\theta_n+\mu\left[\frac{\partial G_{ry}}{\partial x}+\frac{\partial G_{rx}}{\partial y}\right]\cos\theta_n$$

$$H_{rz}=\mu\left[\frac{\partial G_{rx}}{\partial z}+\frac{\partial G_{rz}}{\partial x}\right]\cos\theta_n+\mu\left[\frac{\partial G_{ry}}{\partial z}+\frac{\partial G_{rz}}{\partial y}\right]\sin\theta_n$$

where $H_{rt}=H_{rt}\left(\boldsymbol{x},n_n,x_0,\omega\right)$, $G_{rt}=G_{rt}\left(\boldsymbol{x},x_0,\omega\right)$ and $r,t=x,y,z$. These expressions can be combined so as to give $H_{ij}\left(\boldsymbol{x},n_n,x_0,\omega\right)$ in the normal and tangential directions. In these equations, $\mu=\rho_1\,\beta_1^2$.

A similar boundary integral equation to that in eqn. (3) is defined along the boundary in the interior domain, where the elastic properties are those of the internal inclusion medium. The computation of these two boundary integral equations requires the discretization of both the boundary and boundary values. A system of linear equations is obtained by applying a virtual load to each node on the boundary in turn, and these can be solved to determine the nodal tractions and displacements. When the element to be integrated is the loaded one, the necessary integrations are performed analytically [15, 16], but when the element to be integrated is not the loaded element a numerical Gaussian quadrature scheme is applied.

3.2 Traction boundary element formulation

In the presence of a thin elastic inclusion, the previous BEM formulation degenerates. The TBEM can be derived [10] to cope with this limitation, leading to the following boundary integral equation:

along boundary surface S, *in the host elastic domain*

$$\begin{aligned} c_{ij} t_j(\mathbf{x}_0, n_n, \omega) + a_{ij}\, u_i(\mathbf{x}_0, \omega) = & \int_S t_j(\mathbf{x}, n_n, \omega)\, \overline{G}_{ij}(\mathbf{x}, n_n, \mathbf{x}_0, \omega)\, \mathrm{d}s \\ & - \int_S u_j(\mathbf{x}, \omega)\, \overline{H}_{ij}(\mathbf{x}, n_n, \mathbf{x}_0, \omega)\, \mathrm{d}s + \overline{u}_i^{\text{inc}}(\mathbf{x}_s, \mathbf{x}_0, n_n, \omega) \end{aligned}, \tag{4}$$

where $i, j = 1,2$ refer respectively to the normal and tangential directions relative to the inclusion surface, and $i, j = 3$ to the z direction. This equation can be understood as resulting from the application of dipoles (dynamic doublets) to the previous displacement boundary integral. Coefficients c_{ij} are defined as described above while a_{ij} are zero for piecewise straight boundary elements. Green's functions $\overline{G}_{ij}(\mathbf{x}, n_n, \mathbf{x}_0, \omega)$ and $\overline{H}_{ij}(\mathbf{x}, n_n, \mathbf{x}_0, \omega)$ are given by the application of the traction operator to $G_{ij}(\mathbf{x}, \mathbf{x}_0, \omega)$ and $H_{ij}(\mathbf{x}, n_n, \mathbf{x}_0, \omega)$, which can be seen as the combination of the derivatives of the displacement boundary integral, in order to x, y and z, so as to obtain stresses $\overline{G}_{ij}(\mathbf{x}, n_n, \mathbf{x}_0, \omega)$ and $\overline{H}_{ij}(\mathbf{x}, n_n, \mathbf{x}_0, \omega)$. Along the boundary element, at x, after performing the equilibrium of stresses, the following equations are expressed for x, y and z generated by loads also applied along the x, y and z directions:

$$\overline{G}_{rx} = 2\mu \left[\frac{\alpha^2}{2\beta^2} \frac{\partial G_{rx}}{\partial x} + \left(\frac{\alpha^2}{2\beta^2} - 1 \right) \left(\frac{\partial G_{ry}}{\partial y} + \frac{\partial G_{rz}}{\partial z} \right) \right] \cos\theta_0 + \mu \left[\frac{\partial G_{ry}}{\partial x} + \frac{\partial G_{rx}}{\partial y} \right] \sin\theta_0$$

$$\overline{G}_{ry} = 2\mu \left[\left(\frac{\alpha^2}{2\beta^2} - 1 \right) \left(\frac{\partial G_{rx}}{\partial x} + \frac{\partial G_{rz}}{\partial z} \right) + \frac{\alpha^2}{2\beta^2} \frac{\partial G_{ry}}{\partial y} \right] \sin\theta_0 + \mu \left[\frac{\partial G_{ry}}{\partial x} + \frac{\partial G_{rx}}{\partial y} \right] \cos\theta_0$$

$$\overline{G}_{rz} = \mu \left[\frac{\partial G_{rx}}{\partial z} + \frac{\partial G_{rz}}{\partial x} \right] \cos\theta_0 + \mu \left[\frac{\partial G_{ry}}{\partial z} + \frac{\partial G_{rz}}{\partial y} \right] \sin\theta_0$$

$$\overline{H}_{rx} = 2\mu \left[\frac{\alpha^2}{2\beta^2} \frac{\partial H_{rx}}{\partial x} + \left(\frac{\alpha^2}{2\beta^2} - 1 \right) \left(\frac{\partial H_{ry}}{\partial y} + \frac{\partial H_{rz}}{\partial z} \right) \right] \cos\theta_0 + \mu \left[\frac{\partial H_{ry}}{\partial x} + \frac{\partial H_{rx}}{\partial y} \right] \sin\theta_0$$

$$\overline{H}_{ry} = 2\mu \left[\left(\frac{\alpha^2}{2\beta^2} - 1 \right) \left(\frac{\partial H_{rx}}{\partial x} + \frac{\partial H_{rz}}{\partial z} \right) + \frac{\alpha^2}{2\beta^2} \frac{\partial H_{ry}}{\partial y} \right] \sin\theta_0 + \mu \left[\frac{\partial H_{ry}}{\partial x} + \frac{\partial H_{rx}}{\partial y} \right] \cos\theta_0$$

$$\overline{H}_{rz} = \mu \left[\frac{\partial H_{rx}}{\partial z} + \frac{\partial H_{rz}}{\partial x} \right] \cos\theta_0 + \mu \left[\frac{\partial H_{ry}}{\partial z} + \frac{\partial H_{rz}}{\partial y} \right] \sin\theta_0$$

with $n_0 = (\cos\theta_0, \sin\theta_0)$ defining the unit outward normal at the collocation point, $\mathbf{x}_0$, $\overline{G}_{rt} = \overline{G}_{rt}(\mathbf{x}, n_n, \mathbf{x}_0, \omega)$, $G_{rt} = G_{rt}(\mathbf{x}, \mathbf{x}_0, \omega)$, $\overline{H}_{rt} = \overline{H}_{rt}(\mathbf{x}, n_n, \mathbf{x}_0, \omega)$, $H_{rt} = H_{rt}(\mathbf{x}, n_n, \mathbf{x}_0, \omega)$ and $r, t = x, y, z$.

As with $\overline{G}_{rt}$ and $\overline{H}_{rt}$, the incident field component (referring to stresses) is evaluated by comparable expressions:

$$\bar{u}_x^{\text{inc}} = 2\mu\left[\frac{\alpha^2}{2\beta^2}\frac{\partial u_x^{\text{inc}}}{\partial x} + \left(\frac{\alpha^2}{2\beta^2} - 1\right)\left(\frac{\partial u_y^{\text{inc}}}{\partial y} + \frac{\partial u_z^{\text{inc}}}{\partial z}\right)\right]\cos\theta_0 + \mu\left[\frac{\partial u_y^{\text{inc}}}{\partial x} + \frac{\partial u_x^{\text{inc}}}{\partial y}\right]\sin\theta_0$$

$$\bar{u}_y^{\text{inc}} = 2\mu\left[\left(\frac{\alpha^2}{2\beta^2} - 1\right)\left(\frac{\partial u_x^{\text{inc}}}{\partial x} + \frac{\partial u_z^{\text{inc}}}{\partial z}\right) + \frac{\alpha^2}{2\beta^2}\frac{\partial u_y^{\text{inc}}}{\partial y}\right]\sin\theta_0 + \mu\left[\frac{\partial u_y^{\text{inc}}}{\partial x} + \frac{\partial u_x^{\text{inc}}}{\partial y}\right]\cos\theta_0$$

$$\bar{u}_z^{\text{inc}} = \mu\left[\frac{\partial u_x^{\text{inc}}}{\partial z} + \frac{\partial u_z^{\text{inc}}}{\partial x}\right]\cos\theta_0 + \mu\left[\frac{\partial u_y^{\text{inc}}}{\partial z} + \frac{\partial u_z^{\text{inc}}}{\partial y}\right]\sin\theta_0$$

where $\bar{u}_r^{\text{inc}} = \bar{u}_r^{\text{inc}}(\boldsymbol{x}_s, \boldsymbol{x}_0, n_n, \omega)$, $u_r^{\text{inc}} = u_r^{\text{inc}}(\boldsymbol{x}_s, \boldsymbol{x}_0, \omega)$ and $r = x, y, z$.

Fundamental solutions $\bar{G}_{ij}(\boldsymbol{x}, n_n, \boldsymbol{x}_0, \omega)$, $\bar{H}_{ij}(\boldsymbol{x}, n_n, \boldsymbol{x}_0, \omega)$ and $\bar{u}_i^{\text{inc}}(\boldsymbol{x}_s, \boldsymbol{x}_0, n_n, \omega)$ in the normal and tangential directions are obtained by combining the previous expressions.

As before, in the boundary element formulation, a similar boundary integral equation can be obtained along the boundary in the interior elastic domain. The solutions of the two traction boundary integral equations are evaluated, as before, by discretizing the boundary into N straight boundary elements, with one nodal point in the middle of each element. This procedure leads to a set of integrations, which are performed using a numerical Gaussian quadrature scheme, if the element to be integrated is not the loaded element. If the element to be integrated is the loaded one, hypersingular integrals are obtained which are evaluated through an indirect approach, which consists of defining the dynamic equilibrium of an isolated semi-cylinder defined above the boundary of each boundary element. Their derivation was presented by Tadeu et al. [17].

3.3 Combined dual BEM formulation

The two boundary elements formulations just described can be combined to address the same problems and the case of a thin elastic inclusion. This is achieved by loading part of the boundary surface with monopole loads, and the remaining part with dipole loads. The thin solid bodies will then be solved by means of a closed surface.

4 Verification of the BEM formulations

The BEM results are verified by comparison with the analytical results for a scattered wavefield generated by an elastic cylindrical inclusion [5]. The point harmonic load exciting the host medium is applied at O $(0.0\,\text{m}, -0.125\,\text{m}, 0.0\,\text{m})$ and two lines of receivers are placed in the host medium, at R$_1$ $(0.075\,\text{m}, -0.025\,\text{m})$, and inside the inclusion, at R$_2$ $(0.025\,\text{m}, 0.025\,\text{m})$, as illustrated in Figure 1a. The responses given by the BEM models are successfully compared with the analytical results in Figures 1b-c, when $k_z = 25\,\text{rad/m}$. Both the real and imaginary parts of the responses can be observed, with the analytical solution represented by solid and broken lines.

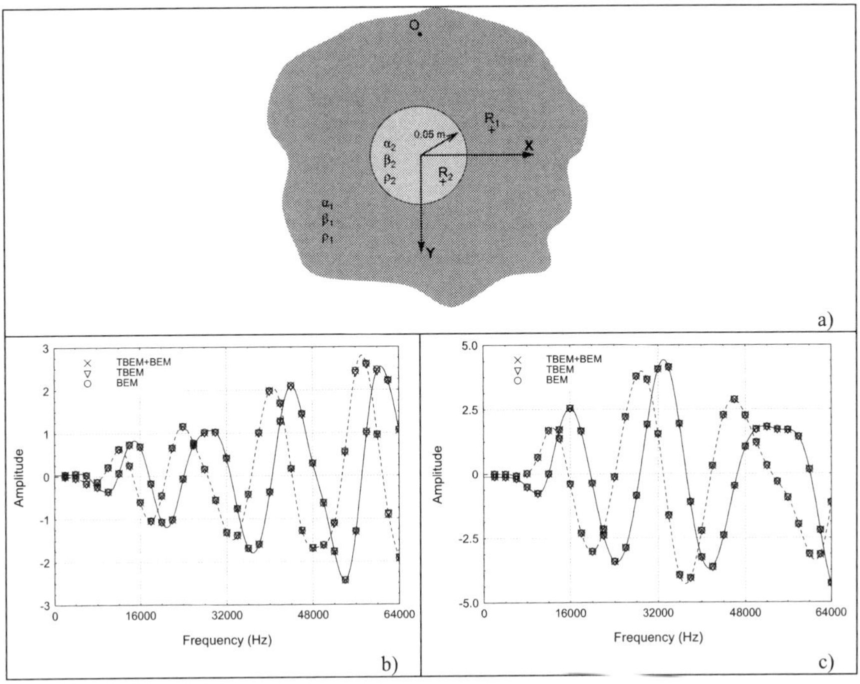

Figure 1: Verification of the BEM algorithms: a) Geometry of the circular elastic inclusion; b-c) uy displacement at receivers R_1 (b) and R_2 (c).

5 Numerical examples

The combined Dual BEM formulation is used to assess the scattering of elastic waves generated by two thin 2D inclusions filled with different solid materials, placed in an unbounded host medium. In the first case, a weaker 5 mm thick solid inclusion is made of cork ($180\,kg/m^3$), allowing the propagation of dilatational and shear waves with velocities of 288.7 m/s and 204.1 m/s, respectively. In the second example, the same thin inclusion is now made of steel ($7850\,kg/m^3$), which permits the corresponding wave propagation velocities of 5970.0 m/s and 3191.1 m/s, respectively. The host elastic medium is homogeneous ($2140\,kg/m^3$) and allows a dilatational wave velocity of 2696.5 m/s and a shear wave velocity of 1451.7 m/s. A harmonic line load is placed at point O and excites the elastic medium near the 2D inclusions (Figure 2).

The TBEM formulation is used to discretize the upper part of the inclusion's surface and the BEM formulation is applied to discretize the lower part. A relation between the wavelength (λ) and the length of the boundary elements (L), is set equal to 10 to determine a suitable number of boundary elements. Computations are performed in the frequency domain from 2000 Hz to 256000 Hz with a frequency increment of 2000 Hz. The time responses are modelled by a Ricker pulse with a characteristic frequency of 75000 Hz. The horizontal thin inclusions are 5 mm thick and their extremities are represented by 5 mm diameter semi-circumferences (see Figure 2 for global geometry and a close-up of the round extremities).

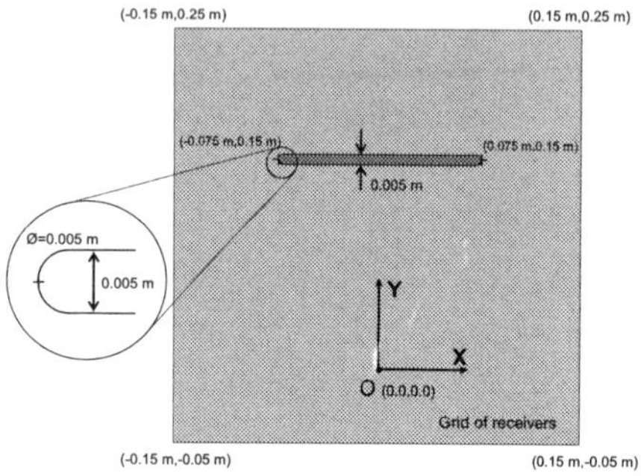

Figure 2: Elastic 5 mm thick horizontal inclusion: problem geometry.

The computations are performed in grids of evenly spaced receivers along x and y directions at 0.003 m. The displacement components in the y direction scattered by the softer and harder thin inclusions are shown at different time instants in Figure 3.

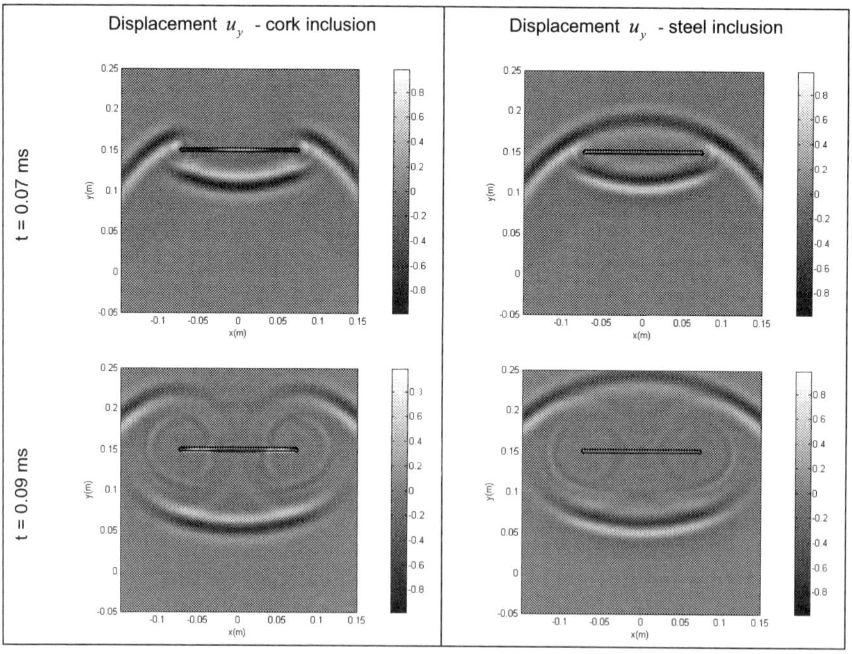

Figure 3: Elastic scattering by different 5 mm thick horizontal inclusions: y- component displacements (u_y) at two time instants.

The waves exciting the elastic medium propagate in all directions. When the incident pulses hit the elastic inclusions they are partly reflected back as P- and S-waves. Some energy is transmitted to the opposite side of the inclusions after passing through the cork and steel media as P- and S-waves. These waves generate multiple reflections on the inclusions' upper and lower surfaces. In this process P- and S-waves are generated simultaneously, and these propagate away from the inclusion in the elastic medium.

The main differences in the observed wave fields are related to the dilatational and shear wave propagation velocities in the solid media inside both horizontal inclusions. For the case of the softer inclusion, a significant number of the incident pulses are diffracted around the heterogeneity, and at the later time instants waves can still be seen trapped inside the slower elastic medium. For the case of the harder inclusion, the pulses that pass through the heterogeneity are slightly attenuated and the pulses visible inside the harder medium are quickly transmitted to the exterior elastic medium. It is interesting to notice that the phase shift (in relation to the incident pulses) of the pulses reflected back from the buried inclusions only occurs for the harder inclusion example. This is due to the relative differences in elastic wave propagation speeds between the host elastic medium and the solid materials that completely fill the thin inclusions.

6 Conclusions

A combined Dual BEM algorithm has been implemented in order to model the scattering of 3D elastic waves generated by thin elastic inclusions buried in an unbounded 2D host medium. This model is formulated in the frequency domain and expressed in terms of waves with different wavenumbers in the z direction, considering the 2-1/2D geometry of the problem. The numerical model was successfully verified by comparing its displacement field, scattered by a circular cylindrical elastic inclusion, with results obtained analytically and by a BEM and a TBEM model. Two numerical examples are presented, where the wave scattering in the vicinity of two different elastic inclusions is computed by the combined Dual BEM proposed model.

References

[1] Achenbach, J.D., Modeling for quantitative non-destructive evaluation. *Ultrasonics*, **40**, pp. 1-10, 2002.

[2] Sánchez-Sesma, F.J., Diffraction of elastic waves by three dimensional surface irregularities. *Bull. Seism. Soc. Am.*, **73**, pp. 1621-1636, 1983.

[3] Tadeu, A.J.B., Kausel, E. & Vrettos, C., Scattering of Waves by Subterranean Structures Via the Boundary Element Method. *Journal of Soil Dynamics and Earthquake Engineering*, **15(6)**, pp. 387-397, 1996.

[4] Liu, G.R. & Chen, S.C., Flaw detection in sandwich plates based on time-harmonic response using genetic algorithm. *Comput. Methods Appl. Mech. Engrg.* **190**, pp. 5505-5514, 2001.

[5] Pao, Y.H. & Mow, C.C., *Diffraction of Elastic Waves and Dynamic Stress Concentrations*, Crane and Russak, 1973.

[6] Manolis, G.D. & Beskos, D.E., *Boundary Element Methods in Elastodynamics*, Unwin Hyman (sold to Chapman and Hall), London, 1988.

[7] Tadeu, A., António, J. & Kausel, E., 3D Scattering of waves by a cylindrical irregular cavity of infinite length in a homogeneous elastic medium. *Computer Methods in Applied Mechanics and Engineering*, **191(27-28)**, pp. 3015-3033, 2002.

[8] Pointer, T., Liu, E. & Hudson, J.A., Numerical modeling of seismic waves scattered by hydrofractures: application of the indirect boundary element method. *Geophys. J. Int.*, **135**, pp. 289-303, 1998.

[9] Sládek, V. & Sládek, J., A boundary integral equation method for dynamic crack problems, *Engineering Fracture Mechanics*, **27(3)**, pp. 269-277, 1987.

[10] Prosper, D. & Kausel, E., Wave scattering by cracks in laminated media, *Proc. of the Int. Conf. on Computational Engineering and Science ICES'01*, Puerto Vallarta, Mexico, 19-25/08/2001. eds. S. N. Atluri, T. Nishioka & M. Kikuchi, Tech Science Press, 2001.

[11] Rudolphi, T.J., The use of simple solutions in the regularisation of hypersingular boundary integral equations. *Math. Comp. Modelling*, **15**, pp. 269-278, 1991.

[12] Watson, J.O., Singular boundary elements for the analysis of cracks in plane strain. *I. J. Num. Meths. Engg.* **38**, pp. 2389-2411, 1995.

[13] Aliabadi, M.H., A new generation of boundary element methods in fracture mechanics. *International Journal of Fracture*, **86**, pp. 91-125, 1997.

[14] Tadeu, A. & Kausel, E., Green's functions for two-and-a-half dimensional elastodynamic problems. *J. Eng. Mech.*, ASCE, **126(10)**, pp. 1093-1097, 2000.

[15] Tadeu, A., Santos, P. & Kausel, E., Closed-form integration of singular terms for constant, linear and quadratic boundary elements - Part I: SH wave propagation. *Eng. Anal. Bound. Elem.*, **23(8)**, pp. 671-681, 1999.

[16] Tadeu, A., Santos, P. & Kausel, E., Closed-form integration of singular terms for constant, linear and quadratic boundary elements - Part II: SV P wave propagation. *Eng. Anal. Bound. Elem.*, **23(9)**, pp. 757-768, 1999.

[17] Tadeu, A., Amado Mendes, P. & António, J., 3D elastic wave propagation modelling in the presence of 2D fluid filled thin inclusions. *Eng. Anal. Bound. Elem.*, **30(3)**, pp. 176-193, 2006.

Wave propagation in elastic and poroelastic media in the frequency domain by the boundary element method

M. A. C. Ferro[1] & W. J. Mansur[2]
[1]*Military Institute of Engineering, Rio de Janeiro, RJ, Brazil*
[2]*Department of Civil Engineering,*
COPPE/Federal University of Rio de Janeiro, CP 68506,
CEP 21945-970, Rio de Janeiro, RJ, Brazil

Abstract

This work presents a methodology to solve dynamic elastic and poroelastic problems in the frequency-domain approach, i.e., problems of the propagation of waves in an elastic and a saturated poroelastic media and the coupling between them. The methodology presented is concerned mainly with seismic wave propagation; in one of the examples the stability of a concrete dam is studied. The formulation of the governing equations for poroelastic isotropic and saturated media was proposed by Biot in time-domain. Dominguez presented a formulation in frequency-domain, using the analogy between dynamic poroelasticity and thermoelasticity. In the present paper the subregion technique, which is frequently used in the Boundary Element Method, was used to obtain the coupling between the elastic and poroelastic media. In order to validate the proposed formulation four examples were analyzed. The obtained numerical results are compared with the studies of other authors and they were quite good.
Keywords: frequency-domain, elasticity, poroelasticity.

1 Introduction

The analysis of the seismic response of dams and other structures is a very important problem of interest in geophysics and engineering fields. This subject is of special interest in geophysics and engineering areas, e.g., soil and rock mechanics, soil-structure interaction, etc. Wave propagation in isotropic elastic

WIT Transactions on Modelling and Simulation, Vol 42, © 2006 WIT Press
www.witpress.com, ISSN 1743-355X (on-line)
doi:10.2495/BE06029

and isotropic saturated poroelastic media are governed by Euler and Biot's equations in the time-domain approach, respectively. Applying the Fourier Transform into these equations, they become transformed equations in the frequency-domain and the wave propagation phenomena can be analyzed in this domain. The Boundary Element Method was used to solve the problem numerically, in terms of solid displacement and fluid stress for poroelastic problems and displacement and traction for elastic problems. In the present paper the subregion technique, which is frequently used in the Boundary Element Method, was used to obtain the coupling between the elastic and poroelastic media. The system of equations related to this problem is ill-conditioned and the Singular Value Decomposition method (SVD) was implemented to solve it.

2 Governing equations

2.1 Boundary element formulation in elastic medium

Using the Boundary Element Method for plane elastic problems, [1,6,7,10], the unknown displacements and tractions vector are **u** and **t** and the fundamental solution tensors are denoted by $\mathbf{u}^*$ and $\mathbf{t}^*$ (for the time-domain approach, see Mansur [12]). Considering now that the boundary under study is discretized using constant elements and it is divided into N straight- line elements the basic equations become

$$\sum_{j=1}^{N} H^{ij} u^j = \sum_{j=1}^{N} G^{ij} t^j \tag{1}$$

The fundamental solutions are

$$u_{lk}^* = \frac{1}{2\pi\rho c_2^2}\left[\psi\delta_{lk} - \chi r_{,l} r_{,k}\right] \tag{2}$$

and

$$t_{lk}^* = \frac{1}{2\pi}\left[\left(\frac{d\psi}{dr} - \frac{1}{r}\chi\right)\left(\delta_{lk}\frac{\partial r}{\partial n} + r_{,k} n_l\right) - \frac{2}{r}\chi\left(n_k r_{,l} - 2 r_{,l} r_{,k}\frac{\partial r}{\partial n}\right),\right.$$
$$\left. - 2\frac{d\chi}{dr} r_{,l} r_{,k}\frac{\partial r}{\partial n} + \left(\frac{c_1^2}{c_2^2} - 2\right)\left(\frac{d\psi}{dr} - \frac{d\chi}{dr} - \frac{1}{r}\chi\right) r_{,k} n_{,k}\right] \tag{3}$$

where

$$\psi = K_0\left(\frac{i\omega r}{c_2}\right) + \frac{c_2}{i\omega r}\left[K_1\left(\frac{i\omega r}{c_2}\right) - \frac{c_2}{c_1} K_1\left(\frac{i\omega r}{c_1}\right)\right] \tag{4}$$

and

$$\chi = K_2\left(\frac{i\omega r}{c_2}\right) + \frac{c_2^2}{c_1^2} K_2\left(\frac{i\omega r}{c_1}\right) \tag{5}$$

c_1 and c_2 are the longitudinal and rotational velocities of plane waves; ρ is the mass density; K_0, K_1 and K_2 are modified Bessel functions of second kind and zero, first and second order, respectively.

2.2 Boundary element formulation in poroelastic medium

The equilibrium equations for linear dynamic poroelasticity were proposed by Biot [2,3,4,5] in the time-domain approach. Dominguez [8,9] presented a formulation in the frequency-domain, starting from Biot's original equations, which are

$$\mu\Delta u_i + (\lambda+\mu)e_{,i} + \tau_{,i}\left(\frac{Q}{R} + \frac{i\omega b + \rho_{12}\omega^2}{i\omega b - \omega^2\rho_{22}}\right),$$
$$+ u_i\omega^2\left(\frac{\omega^2\left(-\rho_{11}\rho_{22}+\rho_{12}^2\right) + i\omega b(\rho_{11}+\rho_{22}+2\rho_{12})}{i\omega b - \omega^2\rho_{22}}\right), \tag{6}$$
$$+ X_i + \frac{i\omega b + \rho_{12}\omega^2}{i\omega b - \omega^2\rho_{22}}X_i^{'} = 0$$

and

$$\Delta\tau + \frac{\tau}{R}\left(-i\omega b + \omega^2\rho_{22}\right) + e\left[i\omega b\left(1+\frac{Q}{R}\right) + \omega^2\left(\rho_{12} - \rho_{22}\frac{Q}{R}\right)\right] + X_{i,i}^{'} = 0 \tag{7}$$

where $\mathbf{u_i}$ are solid displacements and $\boldsymbol{\tau}$ is the fluid stress. μ and λ are elastic constants; Q and R are poroelastic constants; $e = u_{i,i}$; b is a dissipation constant; ρ_{11}, ρ_{12} and ρ_{22} are mass densities; X_i and $X^{'}_i$ are body forces acting in the solid and the fluid, respectively.

The same basic equations (1) can be used in poroelastic medium, being **u** and **t** the unknown variables for displacements and stresses

$$u = \begin{Bmatrix} u_1 \\ u_2 \\ \tau \end{Bmatrix} \quad \text{and} \quad t = \begin{Bmatrix} t_1 \\ t_2 \\ U_n \end{Bmatrix} \tag{8}$$

where U_n are the fluid displacements.

The fundamental solution tensors $\mathbf{u}^*$ and $\mathbf{t}^*$ are

$$u^* = \begin{bmatrix} u_{11}^* & u_{21}^* & -\tau_1^* \\ u_{12}^* & u_{22}^* & -\tau_2^* \\ u_{13}^* & u_{23}^* & -\tau_3^* \end{bmatrix} \quad \text{and} \quad t^* = \begin{bmatrix} t_{11}^* & t_{21}^* & -U_{n1}^* \\ t_{12}^* & t_{22}^* & -U_{n2}^* \\ t_{13}^* & t_{23}^* & -U_{n3}^* \end{bmatrix} \tag{9}$$

The computations of the eighteen terms of the tensors $\mathbf{u}^*$ and $\mathbf{t}^*$ are given in Dominguez [10].

3 The coupling between elastic and poroelastic media

In order to couple elastic and poroelastic domains, the subregion technique was used. In the figure 1, Ω_1 and Ω_2 are elastic and poroelastic domains, Γ_1 and Γ_2 are only elastic or poroelastic boundaries, Γ_{12} is the boundary between both media.

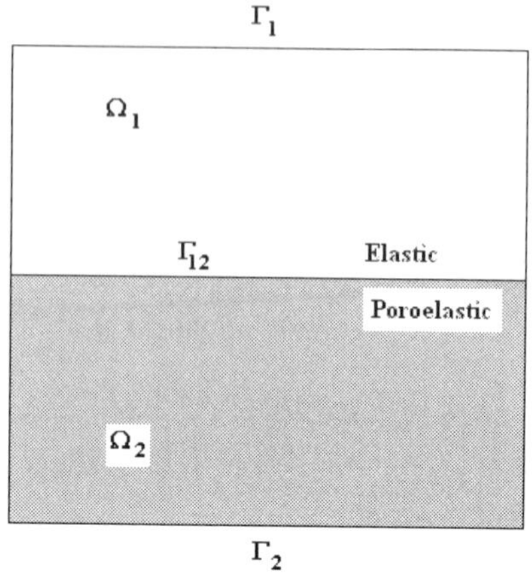

Figure 1: Elastic and poroelastic media.

This coupling is illustrated by equations (10), and was proposed by Ferro [11]

$$\begin{bmatrix} H^1 & 0 & H^{12} & 0 & G^{12} \\ 0 & H^2 & H_s^{21} & H_f^{21} & G^{21} \end{bmatrix} \begin{Bmatrix} u^1 \\ u^2 \\ u^{12} \\ U_n^{21} \\ t^{21} \end{Bmatrix} = \begin{bmatrix} G^1 & 0 \\ 0 & G^2 \end{bmatrix} \begin{Bmatrix} t^1 \\ t^2 \end{Bmatrix} \tag{10}$$

where the superscript indices 1 and 2 are relatives to the elastic or poroelastic media; 12 is relative to the interface between them coming from the elastic medium; 21 is relative to the interface coming from the poroelastic medium; s and f are related to solid and fluid phases in the poroelastic medium.

The system of equations (10) is ill-conditioned and the SVD Method was used to solve it.

4 Numerical examples

Four numerical examples are analyzed with constant boundary elements. The first and the second were analyzed by Dominguez [9,10]. They correspond to a square domain of a saturated porous media, see figure 2. The surface is the only drained side. The boundary conditions to the first example are

$$\begin{aligned} &t_1 = u_2 = U_n = 0 \quad \text{at} \quad x_2 = 0; \\ &t_1 = 0;\; t_2 = -T_0;\; \tau = 0 \;\text{ at }\; x_2 = L \\ &u_1 = t_2 = U_n = 0 \;\text{ at }\; x_1 = 0 \;\text{ and }\; x_1 = L \end{aligned} \tag{11}$$

The material properties are μ = 6x10^9 N/m^2; λ = 4x10^9 N/m^2 ; R = 0.444x10^9 N/m^2; Q = 1.399x10^9 N/m^2; b = 0.19x10^9 Ns/m^4, ρ_{11} = 2418 kg/m^3; ρ_{22} = 340 kg/m^3 and ρ_{12} = -150 kg/m^3.

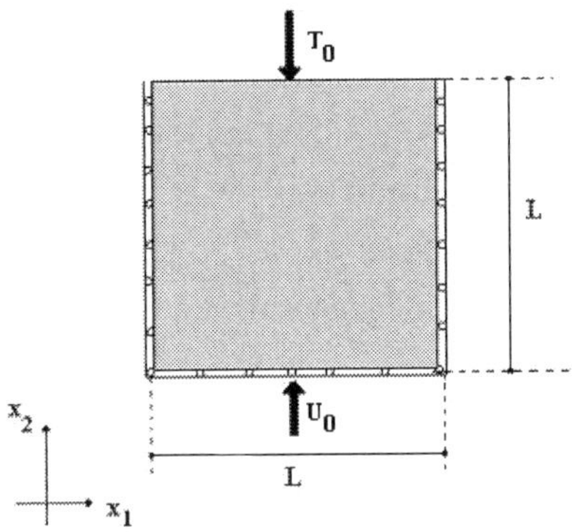

Figure 2: Square domain for the first and second examples.

Figure 3 shows the displacement at the top of the domain versus dimensionless frequency. The displacement is normalized by the static displacement of an elastic region $\left(u_2 E_u / T_0 L\right)$ and the frequency by the first natural frequency

$$\omega_1 = \frac{\pi}{2L}\sqrt{E_u / \rho} \quad \text{where} \quad E_u = \frac{2\mu(1-\nu_u)}{(1-2\nu_u)}; \; \rho = \rho_{11} + \rho_{22} + \rho_{12} \; \text{and} \; \nu_u = 0.33$$

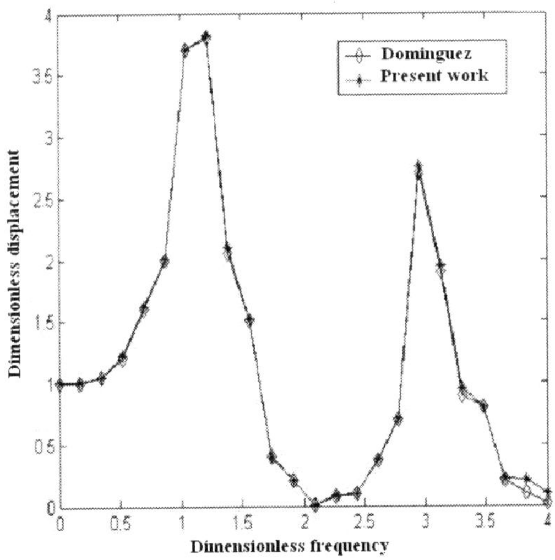

Figure 3: Displacement at the top for the first example.

In the second example, the geometry and material properties are the same as in the first one and the boundary conditions are

$$\begin{aligned} &u_1 = t_2 = U_n = 0 \quad \text{at} \quad x_1 = L; \\ &t_1 = 0; \; u_2 = U_0; \; U_n = -U_0 \quad \text{at} \quad x_2 = 0 \\ &u_1 = t_2 = U_n = 0 \quad \text{at} \quad x_1 = 0 \\ &t_1 = t_2 = \tau = 0 \quad \text{at} \quad x_2 = L \end{aligned} \tag{12}$$

The displacement is normalized by u_2 / U_0 and the frequency in the same way as the in first example. The displacement at the top of the square domain is shown at the figure 4.

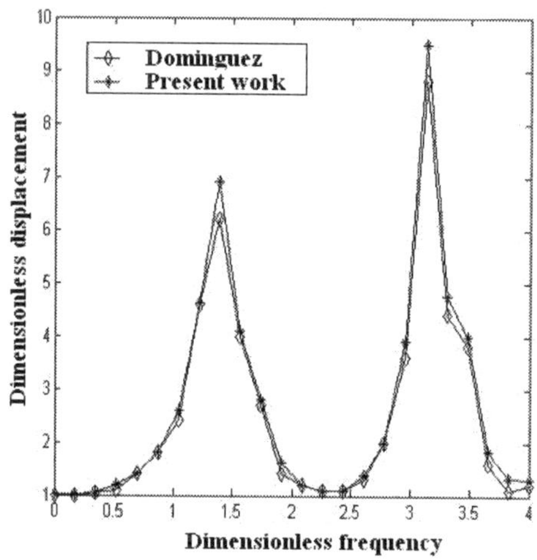

Figure 4: Displacement at the top versus frequency for the second example.

The third and the fourth examples represent a dam- foundation system subjected to a harmonic displacement at the dam crest, see Medina [13] and Dominguez [10]. The difference between these examples is the depth of the soil layer.

The model is shown at the figure 5, and it presents a concrete-dam (upper triangle) over a rock foundation (rectangle) with 18.8 horizontal dimension and H_f as depth.

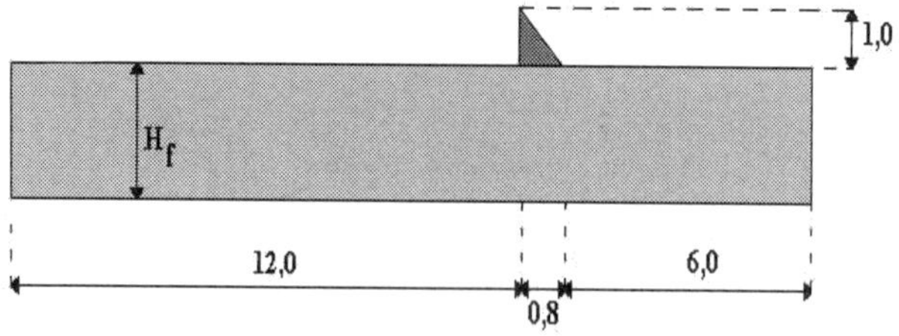

Figure 5: Dam-foundation system for third and fourth examples.

The concrete properties are ρ = 2480 kg/m^3; E = 27.5 GPa; ν = 0.20; ζ (hysteretic damping) = 0.05 = 5%. The soil properties are ρ_{11} = 2548 kg/m^3; ρ_{12} = -59.4 kg/m^3; ρ_{22} = 311.4 kg/m^3; λ = 20.625 GPa; μ = 10.3125 GPa; Q = 36.67 GPa; R = 15.71 GPa and ς (hysteretic damping) = 0.05 = 5%. The boundary conditions are

$u_2 = U_n = \cos\left(\frac{\omega H_f}{c_p}\right)$ and $u_1 = 0$ at the soil base; $t_1 = t_2 = 0$ at the boundaries of the concrete structure; $t_1 = t_2 = \tau = 0$ at the soil surface (except the base). The vertical displacement at top of the concrete structure is plotted in figure 6, with H_f = 1.0. The displacement is normalized as $|u_2| = \mathrm{mod}(u_2(\omega) - 1)$ and the frequency as ω/ω_1, where ω_1 = 9550 rad/s.

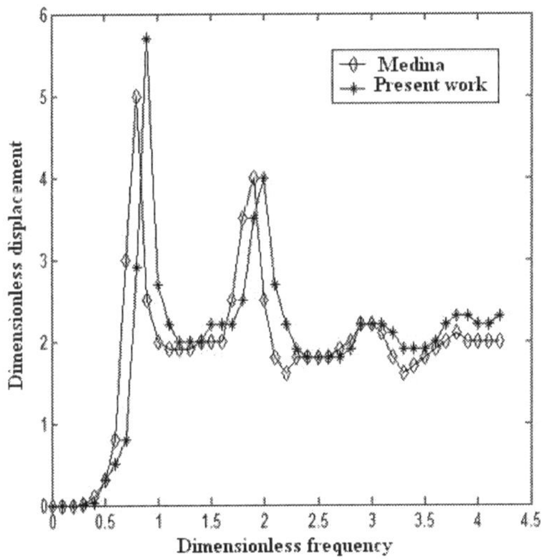

Figure 6: Displacement at the crest of the concrete structure for the third example.

The fourth example has the same geometric and material properties of the third one, except for the soil depth, whose value now is 2.0. The figure 7 shows the displacement at top of the concrete structure with H_f = 2.0.

5 Conclusions

In the present work, an original formulation was developed with the aim of solving wave propagation problems in elastic and poroelastic media in the frequency-domain. The coupling between these media was done using the subregion technique, witch is a usually procedure adopted in the BEM and the

Singular Value Decomposition method was implemented to solve the systems of equations. Four numerical examples were analyzed and compared with other authors' studies and the obtained responses were quite good.

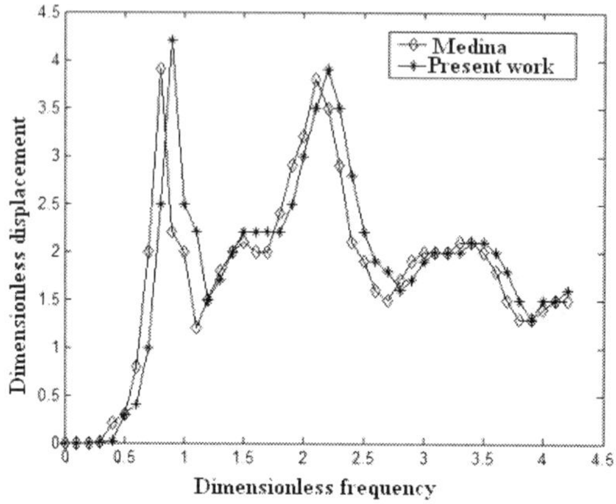

Figure 7: Displacement at the crest of the concrete structure for the fourth example.

References

[1] Banerjee, P. K., The boundary element method in engineering, Berkshire, England. McGraw-Hill Book Company, 1994.

[2] Biot, M. A., 1956a, Theory of propagation of elastic waves in a fluid-saturated porous solid. I. Low- frequency range, Journal of Applied Physics, Nr 28. pp. 168-178.

[3] Biot, M. A., 1956b, Theory of propagation of elastic waves in a fluid-saturated porous solid. I. High- frequency range, Journal of Applied Physics. Nr 28, pp. 179-191.

[4] Biot, M. A., 1956c, General solution of the equations of elasticity and consolidation for a porous material, Journal of Applied Mechanics. Nr 78, pp. 91-96.

[5] Biot, M. A., Mechanics of deformation and acoustic propagation in porous media, Journal of Applied Physics, Nr. 33, pp. 1482-1498, 1962.

[6] Brebbia, C. A., Telles, J. C. F., Wrobel, L. C., Boundary element techniques – Theory and applications in engineering. Berlin, German. Springer-Verlag Berlin, 1984.

[7] Brebbia, C. A., Dominguez, J., Boundary Elements – An Introductory Course. Southampton, England. McGraw-Hill Book Company, 1989.

[8] Dominguez, J., 1991, An integral formulation for dynamic poroelasticity, Transactions of the ASME, Vol. 58, pp. 588-591, 1991.

[9] Dominguez, J., Boundary element approach for dynamic poroelastic problems, International Journal for Numerical Methods in Engineering, Vol. 35, pp. 307-324, 1992.

[10] Dominguez, J., Boundary elements in dynamics. London. Elsevier Science Publishers Ltd, 1993.
[11] Ferro, M. A. C., Poroelasticidade dinâmica acoplada usando o método dos elementos de contorno, COPPE/UFRJ, D. Sc. Thesis, 1992.
[12] Mansur, W. J., A time-stepping technique to solve wave propagation problems using the boundary element method, Ph. D. Thesis. Southampton, England. University of Southampton, 1983.
[13] Medina, F., Analisis de la repuesta sismica de presas incluyendo efetos de interaccion suelo-agua-estructura. Cátedra de Estructuras. Escuela Técnica Superior de Ingenieros Industriales, Universidad de Sevilla, 1987.

A space-time boundary element method for 3D elastodynamic analysis

J. X. Zhou & T. G. Davies
Department of Civil Engineering, Glasgow University, Glasgow, U.K.

Abstract

The classical BEM approach for elastodynamics, which employs a finite difference methodology in time (with piece-wise analytic integration in the time domain) and boundary element discretisation in space, can produce poor results for elastodynamic problems where high gradients occur, such as wave fronts. High gradient areas evolve over time and their locations are unknown a priori. Such high gradients can neither be captured by mesh refinement in advance, nor can they be properly approximated by ordinary lower order polynomial shape functions. In this paper,we propose a novel method which interpolates both spatial and temporal domains. A posteriori error estimation formula in space-time is developed to locate the moving wave front. An h- hierarchical adaptive scheme is used to capture the wave fronts accurately and to forestall generation of spurious oscillations there. An numerical examples is given to demonstrate the power and scope of the method.
Keywords: time domain BEM, Elastodynamics, error estimation, adaptive mesh refinement.

1 Introduction

Since the time domain Boundary Element formulation for elastodynamics and scalar wave propagation was firstly developed (Mansur and Brebbia [1]. Banerjee and Kobayashi [2]), algorithmic has been the major numerical difficulty. To address this problem, various spatial and temporal interpolation schemes have been implemented to improve accuracy and stability, such as combinations of constant, linear and quadratic functions (Dominguez [3], Mansur and Carrer [4]), B-splines interpolation schemes (Rizos and Karabalis [5]), quadratic time interpolation schemes (Wang and Wang [6]). Other strategies to improve stability include

WIT Transactions on Modelling and Simulation, Vol 42, © 2006 WIT Press
www.witpress.com, ISSN 1743-355X (on-line)
doi:10.2495/BE06030

the linear θ method (Yu and Mansur [7]), the time discontinuous traction method (Mansur and Carrer [4]), the half-step method (Birgisson *et al* in [8]), and an integral formulation combining several time step (Marrero and Dominguez [9]). However, none of these methods address the problem of approximating wave fronts. As Peirce and Siebrits [10]) and Frangi and Novati [11]) pointed out, the reason why the time-domain BEM often produce poor results is that wave fronts are unknown *a priori*. They propagate and can not be properly approximated by ordinary shape functions. Spurious oscillations or even instabilities are often observed in these regions. Alternatively, dispersion occurs if lower-order approximations or bigger time steps are used to damp these oscillations. These high gradient areas must be tracked and accurately modelled by adaptive schemes. Many adaptive methods in FEM and BEM have been proposed to address general problems of this nature. Two classes of adaptive methods for dynamic problems can be distinguished; namely, (i) moving mesh methods and (ii) static mesh with h-, p- or hp-refinement.

In moving mesh methods, nodes evolve in the space-time domain and discretization of the governing equations is coupled with the moving mesh. The advantage of this method is that it uses fewer nodes and larger time steps if the nodes travel smoothly. But consequent mesh distortions can introduce considerable difficulties. Above all, controlling the movement of mesh itself is a computationally intensive task.

In static mesh methods the location of original mesh is fixed. The h adaptive method involves adding more nodes when they are necessary and removing them when they are no longer needed. The p adaptive method captures rapidly changing functions by refining the shape functions. For example, in order to accommodate discontinuities, the Discontinuous Galerkin method was developed [12]. However, it does not solve the problem of rapidly moving high gradients in general. Farhat *et al.* [13] suggest using the discontinuous enrichment method to address multi-scale problem. Chessa and Belytschko [14] proposed a space-time element of X-FEM to capture arbitrary discontinuities in the time domain. Yue and Robbins [15] observed that adaptive schemes should be applied to both space and time domains.

In this paper we propose a novel approach, in which the space-time domain is discretized in a true sense, combined with posteriori error estimation and h adaptive technology to locate the moving wave fronts and enrich the mesh nearby. The rest of the paper is organized as follows: in Section 2, the space-time approach for 3D elastodynamics BEM is introduced. The error estimation for BEM analysis and the adaptive scheme are presented in Section 3. An examples of the application of the space-time BEM and h- adaptive scheme is given in Section 4.

2 Space-time approach for 3D elastodynamics in the BEM

Conventionally, the BEM employs a finite difference methodology in time and boundary element discretisation in space. However, regions of high stress or high strain evolve over time and exhibit a high degree of localization at any instant in time. Using smaller time steps for all regions is computationally wasteful.

Thus, building a BEM model in which the space-time continuum is discreted in a true sense, combined with an adaptive scheme in both space and time domain should be more efficient.

2.1 Boundary integral formulation for space-time BEM elastodynamics

The governing equations of elastodynamics are:

$$(c_1^2 - c_2^2)\boldsymbol{u}_{i,ij} + c_2^2\boldsymbol{u}_{j,ii} + \boldsymbol{f}_j = \ddot{\boldsymbol{u}}_j \tag{1}$$

with initial and boundary conditions

$$\begin{aligned} u_i\left(\boldsymbol{x},0^+\right)|_\Gamma &= u_{0i}\left(\boldsymbol{x}\right) & \dot{u}_i\left(\boldsymbol{x},0^+\right)|_\Gamma &= v_{0i}\left(\boldsymbol{x}\right) \\ u_i\left(\boldsymbol{x},t\right)|_{\Gamma_1} &= \overline{u}_i\left(\boldsymbol{x},t\right) & p_i\left(\boldsymbol{x},t\right) &= |_{\Gamma_2}\,\overline{p}_i\left(\boldsymbol{x},t\right) \end{aligned} \tag{2}$$

where boundary $\Gamma = \Gamma_1 + \Gamma_2$, $u(\boldsymbol{x},t)$ is the displacement in 4D space-time,$p(\boldsymbol{x},t)$ are tractions on the boundary, $\boldsymbol{x}$ is a 3D vector (x,y,z) , $f_j = b_j/\rho, b_j$ is the body force, ρ is the density, c_1 is the dilatational wave speed and c_2 is the shear wave speed. Assuming zero body forces and initial conditions, the 3D elastodynamic BEM integral equation can be written as:

$$\begin{aligned} c_{lk}^i u_k^i\left(\boldsymbol{x}^i,t\right) = &\int_\Gamma u_{lk}^*\left(\boldsymbol{x},t-\tau;\boldsymbol{x}^i\right) * p_k\left(\boldsymbol{x},\tau\right) d\Gamma\left(\boldsymbol{x},\tau\right) \\ &- \int_\Gamma p_{lk}^*\left(\boldsymbol{x},t-\tau;\boldsymbol{x}^i\right) * u_k(\boldsymbol{x},\tau) d\Gamma(\boldsymbol{x},\tau) \end{aligned} \tag{3}$$

where, $\boldsymbol{x}$ is a vector in 3D space. The Green's function u_{lk}^* is given by

$$\begin{aligned} u_{lk}^*\left(\boldsymbol{x},t;\boldsymbol{x}^i\right) = &\frac{1}{4\pi\rho}\{\frac{t}{r^2}\left(\frac{3r_{,l}r_{,k}}{r} - \frac{\delta_{lk}}{r}\right)[H\left(t-\frac{r}{c_1}\right) - H\left(t-\frac{r}{c_2}\right)]\} \\ &+ \frac{1}{4\pi\rho}\{\frac{r_{,l}r_{,k}}{r}[\frac{1}{c_1^2}\delta(t-\frac{r}{c_1}) - \frac{1}{c_2^2}\delta(t-\frac{r}{c_2})] \\ &+ \frac{\delta_{lk}}{rc_2^2}\delta(t-\frac{r}{c_2})\} \end{aligned} \tag{4}$$

where $r = \left|\boldsymbol{x}-\boldsymbol{x}^i\right|$, $H(x)$ is the Heaviside function and $\delta(x)$ is the Dirac-delta function

$p_{lk}^*\left(\boldsymbol{x},t;\boldsymbol{x}^i\right)$ can be obtained from the relationship $p_{lk}^* = \sigma_{ijk}^* n_j$.

$$p_{lk}^*\left(\boldsymbol{x},t;\boldsymbol{x}^i\right) = \frac{1}{4\pi}\{\left(\frac{\partial r}{\partial n}\delta_{lk} + r_{,k}n_l\right)A + r_{,l}r_{,k}\frac{\partial r}{\partial n}B + r_{,l}n_kC\} \tag{5}$$

where the details of coefficients A, B, C can be found in Mansur and Brebbia [1].

In equation (3), the integrals of the products of $p_k(\boldsymbol{x},\tau)$ and the Heaviside function $H\left(t-\frac{r}{c}\right)$ or the Dirac-delta function $\delta\left(t-\frac{r}{c}\right)$ in space-time become spatial integrals of $p_k(\boldsymbol{x},\tau)$ over the boundary. Most integrals of $u_k(\boldsymbol{x},\tau)$ and fundamental traction solution can be similarly reduced to spatial integrals. The integral of the product of displacement $u_k(\boldsymbol{x},\tau)$ and the time derivative of the Dirac-delta function reduces to:

$$\int \frac{\partial\delta(t-\frac{r}{c_1})}{\partial\tau} u_k(\boldsymbol{x},\tau)d\Gamma(\boldsymbol{x},\tau) = \int \frac{\partial u_k(\boldsymbol{x},t-\frac{r}{c_1})}{\partial\tau} d\Gamma(\boldsymbol{x}) \tag{6}$$

Here, we introduced a new interpolation scheme. Not only the spatial domain is interpolated via shape functions, but now both spatial and temporal domains are interpolated using $N^k(\xi,\eta,\tau)$ as a space-time interpolation function. Thus:

$$u_k = \sum_j N^k(\xi,\eta,\tau)\, u_k^j; \qquad p_k = \sum_j N^k(\xi,\eta,\tau)\, p_k^j;$$
$$t_k = \sum_j N^k(\xi,\eta,\tau)\, t_k^j; \tag{7}$$

Then, distance in space-time is defined as:

$$|ct-r| = \left|c(t_i - \sum N^k(\xi,\eta,\tau)\cdot t_j^k) - r\right| = |c(t_i - \Phi(\xi,\eta,\tau)) - r| \tag{8}$$

where, $\sum N^k(\xi,\eta,\tau)\cdot t_j^k = \Phi(\xi,\eta,\tau)$.

After this numerical approximation, equation (3) takes the form

$$c_{lk}^i u_k^i = \sum_{j=1}^{N} \int_{\Gamma_j} u_{lk}^* \cdot N^j(\xi,\eta,\tau)\, p_k^j d\Gamma(\boldsymbol{x},\tau)$$
$$- \int_{\Gamma_j} p_{lk}^* \cdot N^j(\xi,\eta,\tau)\, u_k^j d\Gamma(\boldsymbol{x},\tau) \tag{9}$$

Letting

$$\begin{aligned} G_{lk}^{ij} &= \textstyle\int_{\Gamma_j} u_{lk}^* \cdot N^j(\xi,\eta,\tau)\, d\Gamma(\boldsymbol{x},\tau) \\ \hat{H}_{lk}^{ij} &= \textstyle\int_{\Gamma_j} p_{lk}^* \cdot N^j(\xi,\eta,\tau)\, d\Gamma(\boldsymbol{x},\tau) \end{aligned} \tag{10}$$

and adopting the notation:

$$H_{lk}^{ij} = \begin{cases} \hat{H}^{ij} & i \neq j \\ \hat{H}^{ij} + c_{lk}^i & i = j \end{cases} \tag{11}$$

then the system of equations for all boundary nodes can be expressed in matrix form as

$$\sum_{j=1}^{N} H_{lk}^{ij} u_k^j = \sum_{j=1}^{N} G_{lk}^{ij} p_k^j \tag{12}$$

where N is total number of nodes in space-time domain.

2.2 Time and space integration

Piecewise linear interpolation functions or piecewise quadratic functions are used for tractions and displacements. In 3D space-time, eight nodes must be defined for linear interpolation. Denoting the node numbers by the subscript α, we can express all eight shape functions ($\alpha = 1 - 8$) in the form

$$N_\alpha(\xi_1, \xi_2, \xi_3) = \frac{1}{8}(1 + \xi_1^\alpha \xi_1)(1 + \xi_2^\alpha \xi_2)(1 + \xi_3^\alpha \xi_3) \tag{13}$$

where (ξ_1, ξ_2, ξ_3) denotes the intrinsic coordinates of the αth node. For quadratic interpolation in space-time, the element is defined by twenty nodes. For the eight corner nodes ($\alpha = 1 - 8$) , the shape functions are

$$N_\alpha(\xi_1, \xi_2, \xi_3) = \frac{1}{8}(1+\xi_1^\alpha \xi_1)(1+\xi_2^\alpha \xi_2)(1+\xi_3^\alpha \xi_3)(\xi_1^\alpha \xi_1+\xi_2^\alpha \xi_2+\xi_3^\alpha \xi_3-2) \tag{14}$$

and for the twelve remaining mid-side nodes ($\alpha = 9 - 20$), the shape functions are

$$\begin{aligned} N_\alpha(\xi_1, \xi_2, \xi_3) =& \frac{1}{4}(1 + \xi_1^\alpha \xi_1)(1 + \xi_2^\alpha \xi_2)(1 + \xi_3^\alpha \xi_3) \\ & \times (1 + [\xi_1^{\alpha 2} - 1]\xi_1^2 + [\xi_2^{\alpha 2} - 1]\xi_2^2 + [\xi_3^{\alpha 2} - 1]\xi_3^2) \end{aligned} \tag{15}$$

If we divide space-time $\Gamma(\boldsymbol{x}, t)$ into n parts along the time axis, the integral terms in Eq. 9 for the mth element (t_{m-1}, t_m) will be the integral over the parts of elements that lie within two concentric spherical surfaces of radius $r_m = c(t_n - t_m)$ and $r_{m+1} = c(t_n - t_{m+1})$, with $c = c_1$ or $c = c_2$ depending on which term is being integrated.

If the source node is not contained in the space-tim elements to be integrated, a standard $3 \times 3 \times 3$ Gauss quadrature is used for integration. Otherwise, if the source node is in the space-time elements, a singular term in the integral will appear and should be treated carefully. For displacement singularity, the singularity is $O(1/r)$ and can be treated by employing an element subdivision technique to divide original elements into several tetrahedra. Then each tetrahedron subelement can be mapped into cubic intrinsic element space where the weak singularity is nullified and the integral can be performed using normal Gauss quadrature (Gao and Davies [16]). For traction singularity of $O(1/r^2)$, an indirect method is used by employing rigid body motion condition to calculate the c_{lk}^j coefficient and H_{lk}^{ii} is computed by using Eq. 11.

By looping over each node and element, all these terms are calculated and assembled into matrices $[G]$ and $[H]$. Once the matrices $[G]$ and $[H]$ have been obtained, we can then solve the matrix equation using standard techniques.

3 Error indicators and adaptive scheme for space-time BEM

3.1 Error estimation in space-time boundary

The high-gradient areas are usually strongly localized, it is important to have an efficient method to find these localizations . There are two widely-used methods of error indicators (see the survey paper by Kita and Kamiya [17]). One is based on error estimates which are developed from theory of numerical error analysis. For example, the results obtained with coarse and refined mesh are used to calculate errors for each node. Another strategy is to deduce the boundary integral equation of some defined residues and then solve the equations again. However, this latter method is very time-consuming. Here, we use straight-forward error estimation based on comparison between coarse and refined meshes. The advantage of this strategy is that we combine two processes of error estimation and adaptivity by continuing the refinement of elements whose errors are large while stopping the refinement of elements whose errors are small.

The nodal errors are defined as follows:

$$
\begin{aligned}
e_u &= u_i - u_i^h = \delta u_i \\
e_p &= p_i - p_i^h = \delta p_i
\end{aligned}
\tag{16}
$$

where u_i, p_i are values obtained from the refined mesh and u_i^h, p_i^h are values obtained from the coarse mesh. The L_2 norm error of the elements can be expressed as

$$
\|e_u\| = \left[\int \left(\sum_{j=1}^{K} N^j \delta u^j \right)^2 d\Gamma \right]^{\frac{1}{2}} \qquad \|e_p\| = \left[\int \left(\sum_{j=1}^{K} N^j \delta p^j \right)^2 d\Gamma \right]^{\frac{1}{2}}
\tag{17}
$$

where N^j are the shape functions used in the space-time domain, and the errors in nodal displacements and tractions are obtained from above Eq. (14).

In order to avoid wasting time on areas with near-zero fluctuation, we define the relative element error by using the local element error weighted by the sum of displacement and traction.

$$
\eta_i^u = \frac{\|e_u\|_i}{\|u\|} \qquad \eta_i^p = \frac{\|e_p\|_i}{\|p\|}
\tag{18}
$$

where

$$
\|u\| = \left[\int \left(\sum_{j=1}^{K} N^j i u^j \right)^2 d\Gamma \right]^{\frac{1}{2}} \qquad \|p\| = \left[\int \left(\sum_{j=1}^{K} N^j p^j \right)^2 d\Gamma \right]^{\frac{1}{2}}
\tag{19}
$$

3.2 Error indicator and adaptive process

Since space-time is not symmetric in time axis, it would be improper to use global error indicators to decide where to refine the space-time elements. Numerical errors which happen at early times accumulate and can not be recovered by refining the elements at later times. Therefore, we use an error estimation and an adaptive scheme for each segment in time. It is well known that a proper space-time ratio β is essential for stability (Bathe [18], Frangi and Novati [11]) and hence one constraint on the adaptive scheme is to keep this ratio to unity.

The adaptive process is:

Step 1: Apply the coarse mesh on the space-time boundary.

Step 2: From the first time step,refine all elements at the first time step by halving both spatial size and time step.

Step 3: Carry out the BEM analysis for both configurations; Compute local node errors e_u, e_p; compute the relative element error η_i^u, η_i^p. Then, locate the maximum local error:

$$\eta_{\max}^u = Max(\eta_i^u) \qquad \eta_{\max}^p = Max(\eta_i^p) \qquad i = 1, \ldots, M \tag{20}$$

where *M* is the total number of elements within this time step.

Step 4: If $\eta_{\max}^u + \eta_{\max}^p < \bar{\eta}_{tol}$ ($\bar{\eta}_{tol}$ is the prescribed global error tolerance), stop the adaptive process and move to the next time step. If the final time is reached, stop the algorithm. Otherwise mark those elements in space-time if the following conditions are satisfied:

$$\eta^u > \beta_s \cdot \eta_{\max}^u, \qquad \eta^p > \beta_s \cdot \eta_{\max}^p \tag{21}$$

whereβ_s is the threshold for the refinement criteria, usually set to 0.5 – 0.7

Step 5. Use an h- hierarchical adaptive scheme to enrich the approximation for those elements as Eq. 22 , then back to step 2.

$$u = u^G + u^L = \sum N^{Gj} u^{Gj} + \sum N^{Lj} u^{Lj} \tag{22}$$

where $N^{Gj}u^{Gj}$ are the displacements and shape functions in the coarse mesh while $N^{Lj}u^{Lj}$ are for the refined one respectively. The obvious advantage of this adaptive scheme is that we keep the original coefficients matrix intact and just add more new elements to form the new one. We could write Eq. 12 in the matrix form for the refined mesh as follows:

$$\begin{bmatrix} H^{GG} & H^{GL} \\ H^{GL} & H^{LL} \end{bmatrix} \begin{pmatrix} u^G \\ u^L \end{pmatrix} = \begin{bmatrix} G^{GG} & G^{GL} \\ G^{GL} & G^{LL} \end{bmatrix} \begin{pmatrix} p^G \\ p^L \end{pmatrix} \tag{23}$$

4 Numerical examples

The numerical example here is the simple plane wave problem where a strip of height L = 8 m, extending indefinitely in the y direction, with a uniform pressure *P(t)* applied on the upper surface, where *P(t)* is a Heaviside function. Wave speed is 200 m/s. This is shown in Fig. 1.

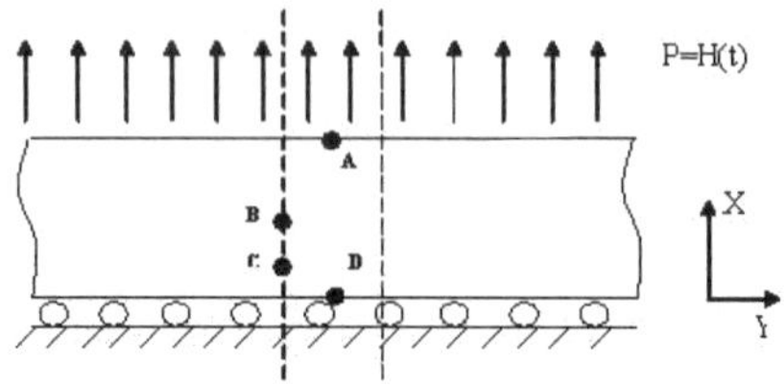

Figure 1: Plane wave problem.

This problem is essentially 1D in space since it is symmetric in the y direction and we have an analytical solution as the benchmark. Here we solve it in 2D space-time by taking a strip with 8×4 m and meshing it in the space-time domain. The boundary was initially discretized into 600 constant and linear mixed elements in space-time. The space-time ratio $\beta = 1$, mesh size is 4 m and the time step is 0.02 s. Then we initiate the adaptive scheme by refining first time step in the mesh. The mesh is illustrated in Fig. 2.

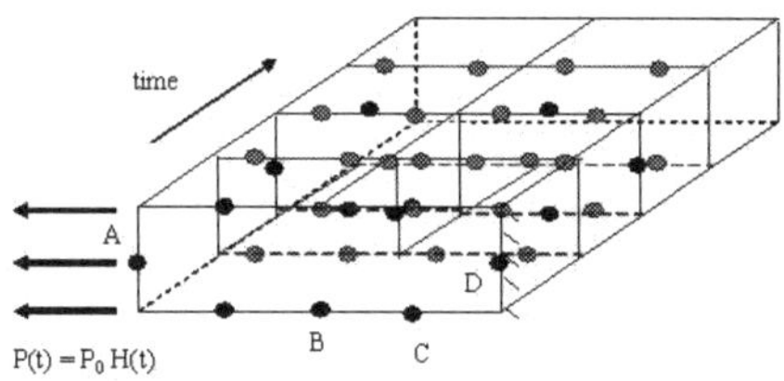

Figure 2: h-adaptive scheme in space-time.

Refining the mesh yields the graph of relative element error η_i^u, η_i^p in Fig. 3. We see that the error estimation algorithm identifies the moving wave front. Based on this, we use smaller space-time elements near the wave fronts and re-compute. Figure 4 illustrates the improvement near to the wave fronts obtained by this process.

5 Conclusion

For transient loading, BEM has some advantages over other numerical methods because boundary-only discretisation means that mesh dispersion is less significant. The biggest problem is to locate high gradient areas and to approximate them with high accuracy and stability. We have proposed an h- adaptive scheme on a mesh in the space-time domain to address these problems. Given that high stresses / strains are mainly localized, their positions can be detected in the space-time using error estimation technology. An h- adaptive scheme is employed to

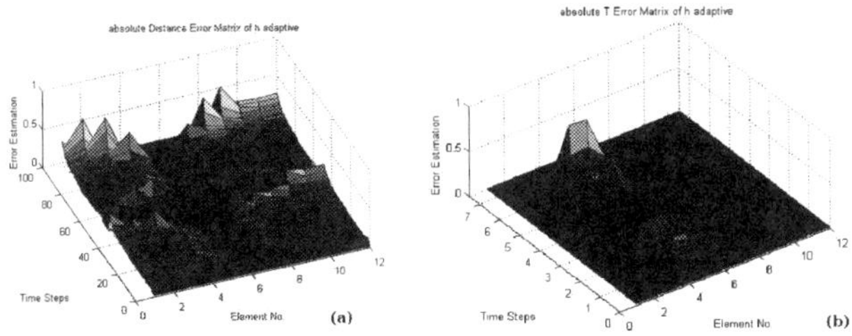

Figure 3: Relative element error of u (a), relative element error of p (b).

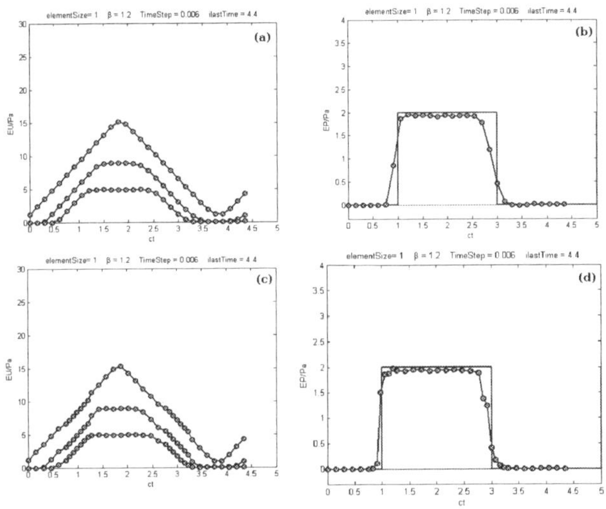

Figure 4: Improved results from adaptive scheme. (a) Displacements at points A, B, C; (b) Traction at point D without adaptive scheme; (c) Displacements at points A, B, C; (d) Traction at point D with adaptive scheme.

undertake a multilevel analysis near to those areas. The numerical results of 2D wave propagation are given to demonstrate its potential. Further research in process focuses on implementing more flexible triangular elements in 4D space-time elastodynamic problems.

References

[1] Mansur, W. & Brebbia, C.A., *Topics in Boundary Element Research*, volume Vol. 2: Time-dependent and Vibration Problems. Springer-Verlag, 1985.

[2] Banerjee, P.K. & Kobayashi, S., (eds.) *Advanced dynamic analysis by boundary element methods*, volume Vol.7, Developments in Boundary element Methods. Elsevier Applied Science, 1992.

[3] Dominguez, J., *Boundary elements in dynamics*. Southampton/Lodon: Computational mechanics publications, 1993.

[4] Mansur, W.J. & Carrer, J., Time discontinuous linear traction approximation in time-domain bem scalar wave propagation analysis. *International Journal for Numerical Methods in Engineering*, **42(4)**, pp. 667–683, 1998.

[5] Rizos, D. & Karabalis, D., Advanced direct time domain bem formulation for general 3-d elastodynamic problems. *Computational Mechanics*, **15(3)**, pp. 249–269, 1994.

[6] Wang, C.C. & Wang, H.C., Two-dimensional elastodynamic transient analysis by ql time-domain bem formulation. *International Journal for Numerical Methods in Engineering*, **39(6)**, pp. 951–985, 1996.

[7] Yu, G.Y. & Mansur, W., Linear θ method applied to 2d time-domain bem. *Communications in Numerical Methods in Engineering*, **14(12)**, pp. 1171–1179, 1998.

[8] Birgisson, B., Siebrits, E. & Peirce, A.P., Elastodynamic direct boundary element methods with enhanced numerical stability properties. *International Journal for Numerical Methods in Engineering*, **46(6)**, pp. : 871–888, 1999.

[9] Marrero, M. & Dominguez, J., Numerical behavior of time domain bem for three-dimensional transient elastodynamic problems. *Engineering Analysis with Boundary Elements*, **27(1)**, pp. 39–48, 2003.

[10] Peirce, A. & Siebrits, E., Stability analysis and design of time-stepping schemes for general elastodynamic boundary element models. *International Journal for Numerical Methods in Engineering*, **40(2)**, pp. 319 – 342, 1997.

[11] Frangi, A. & Novati, G., On the numerical stability of time-domain elastodynamic analyses by bem. *Computer Methods in Applied Mechanics and Engineering*, **173(3-4)**, pp. 403–417, 1999.

[12] Krivodonova, L. & Xin, J., Shock detection and limiting with discontinuous galerkin methods for hyperbolic conservation laws. *Applied Numerical Mathematics*, **48(3-4)**, pp. 323–338, 2004.

[13] Farhat, C., Hararib, I. & Francac, L.P., The discontinuous enrichment method. *Computer Methods in Applied Mechanics and Engineering*, **190(48)**, pp. 6455–6479, 2001.

[14] Chessa, J. & Belytschko, T., Arbitrary discontinuities in space-time finite elements by level sets and x-fem. *International Journal for Numerical Methods in Engineering*, **61(15)**, pp. 2595–2614, 2004.

[15] Yue, Z. & Robbins, D. Jr, Adaptive superposition of finite element meshes in elastodynamic problems. *Journal for Numerical Methods in Engineering*, **63(11)**, pp. 1604–1635, 2005.

[16] Gao, X. & Davies, T.G., *Boundary Element Programming in Mechanics*. Cambridge University Press, 2002.

[17] Kita, E. & Kamiya, N., An overview, error estimation and adaptive mesh refinement in bem. *Engineering Analysis with Boundary Elements*, **25(7)**, pp. 479–495, 2001.

[18] Bathe, K.J., *Finite Element Procedures*. Prentice Hall: Englewood Cliffs, New Jersey, 1996.

Section 8
Damage fracture and mechanics

Wave motion through cracked, functionally graded materials by BEM

G. D. Manolis[1], T. V. Rangelov[2] & P. S. Dineva[3]
[1]*Department of Civil Engineering, Aristotle University, Thessaloniki, Greece*
[2]*Institute of Mathematics and Informatics, Bulgarian Academy of Sciences, Sofia, Bulgaria*
[3]*Institute of Mechanics, Bulgarian Academy of Sciences, Sofia, Bulgaria*

Abstract

Elastic waves in cracked, functionally graded materials (FGM) with elastic parameters that are continuous functions of a single spatial co-ordinate are studied herein under conditions of plane strain and for time-harmonic incident pressure (P) and vertically polarized shear (SV) waves. The FGM has a fixed Poisson's ratio, while both shear modulus and density profiles vary proportionally. The method of solution is the boundary element method (BEM). The necessary Green's functions for the infinite plane are derived in closed-form using functional transformation methods. Subsequently, a non-hypersingular, traction-type BEM is developed using parabolic boundary elements, supplemented with special crack-tip elements for handling crack edges. The methodology is validated against benchmark problems and then used to study wave scattering phenomena around a crack in an infinitely extending FGM.

1 Introduction

Abrupt change in material properties across interfaces between layers in composites and other materials may result in large inter-laminar stresses leading to delamination phenomena. One way to overcome these effects is to use FGM, which are inhomogeneous materials with continuously varying material properties. However, defects and cracks are commonly present in FGM, both during the manufacturing process and under service conditions. This calls for advanced numerical methods to assist in the development of ultrasonic and other

WIT Transactions on Modelling and Simulation, Vol 42, © 2006 WIT Press
www.witpress.com, ISSN 1743-355X (on-line)
doi:10.2495/BE06031

techniques for analyzing FGM [1,2]. In general, wave motion phenomena in inhomogeneous media have been attracting attention in recent years [3], given important applications in diverse fields such as material science, seismology, etc. Within the context of geological media, discontinuities such as cracks, inclusions, fractures and faults that have different length scales must be taken into account [3]. Thus, the present work is a continuation of earlier author efforts [4,5] to look at wave scattering phenomena by cracks in inhomogeneous/ anisotropic continua from a dynamic fracture mechanics perspective [6,7].

2 Problem statement

Consider an infinite elastic plane that contains a traction-free crack $S_{cr} = S_{cr}^{+} \cup S_{cr}^{-}$, swept by time-harmonic, P- and SV- waves travelling at frequency ω (see Fig. 1). The material parameters of the plane are functions of a single spatial coordinates, namely $\mu(x) = \lambda(x) = h(x)\mu_0$, and Poisson's ratio has a fixed value of $\nu = 0.25$. The particular 'material' function $h(x)$ considered here is of the exponential type, i.e., $h(x_2) = e^{2ax_2}$. Furthermore, the material density profile remains proportional to the shear modulus profile as $\rho(x)/\mu(x) = \rho_0/\mu_0$. More precisely, $\rho(x) = \rho_0 h(x)$, where $\mu_0 > 0$, $\rho_0 > 0$ are values at the reference horizontal surface, while a is a constant.

The governing equations of motion for this problem, in the absence of body forces, are

$$\sigma_{ij,j}(x,\omega) + \rho(x)\omega^2 u_i(x,\omega) = 0 \tag{1}$$

where $\sigma_{ij,j}(x,\omega) = \{\lambda(x) u_{k,k}(x,\omega)\}_{,i} + \{\mu(x)(u_{i,j}(x,\omega) + u_{j,i}(x,\omega))\}_{,j}$ is the static stress equilibrium operator and u_i is the displacement vector. As a first step in recovering a Green's function for eqn (1), the following functional transformation is introduced

$$u_i(x,\omega) = h^{-1/2}(x) U_i(x,\omega) \tag{2}$$

where $U_i(x,\omega)$ is a displacement solution for the dynamic equilibrium equations expressed in terms of the equivalent homogenous medium case as $\Sigma_{ij,j} + \gamma_i U_i = 0$. In the above, $\gamma_1 = \rho_0\omega^2 - \mu_0 a^2$, $\gamma_2 = \rho_0\omega^2 - 3\mu_0 a^2$, the corresponding stress equilibrium operator is $\Sigma_{ij,j} = \mu_0 U_{i,jj} + 2\mu_0 U_{j,ij}$, and all counters range as $i, j = 1,2$.

The total wave fields in the cracked medium can be expressed as

$$u_i^t = u_i^{in} + u_i^{sc} \qquad t_i^t = t_i^{in} + t_i^{sc} \tag{3}$$

where u_i^{in}, t_i^{in} are displacement and tractions generated by the incident wave field, while u_i^{sc}, t_i^{sc} are scattered by the crack. The interior crack itself is traction free, i.e.,

$$t_i^{in} + t_i^{sc} = 0 \text{ or } t_i^{sc} = -t_i^{in} \text{ at } x = (x_1, x_2) \in S_{cr} \tag{4}$$

where tractions are defined as $t_i = \sigma_{ij} n_j$, with n_j the unit normal vector on S_{cr}^{+}. Also, stresses are $\sigma_{ij} = C_{ijkm} u_{k,m}$, and then $C_{ijkm} = h(x_2) C_{ijkm}^{0}$, $C_{ijkm}^{0} = \mu_0 \left(\delta_{ij}\delta_{km} + \delta_{ik}\delta_{jm} + \delta_{im}\delta_{jk} \right)$ is the elastic stress tensor for the homogeneous continuum.

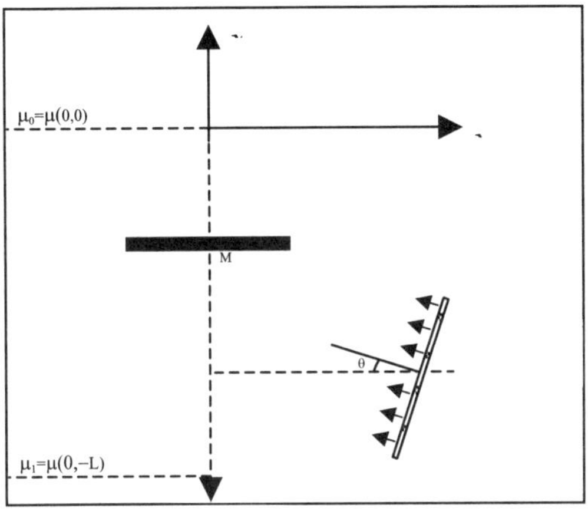

Figure 1: Elastic wave scattering by a crack in an elastic, isotropic FGM.

The boundary-value problem (BVP) consists of eqn (1) and the boundary conditions (BC) of eqn (4). In addition, the following conditions must be satisfied: (a) At the crack-tips, the asymptotic behavior of both displacement and traction vectors as the reference radial distance $r_e \to 0$ is of the order $O(\sqrt{r_e})$ and $O(1/\sqrt{r_e})$), respectively. (b) The scattered wave field must also satisfy a Sommerfeld-type radiation BC.

3 Displacement and traction Green's functions

A transformed fundamental solution $U_{ij}^{*}(x,\xi)$ solves the equation of motion in the form

$$\Sigma_{ijk,j}^{*} + \gamma_i U_{ik}^{*} = -h^{-1/2}(\xi)\delta(x-\xi) \tag{5}$$

Similarly, the fundamental solution $u_{ij}^{*}(x,\xi) = h^{-1/2}(x) U_{ij}^{*}$ solves eqs (1) as

$$\sigma_{ijk,j}^{*x} + \gamma_i u_{ik}^{*} = -\delta(x-\xi) \tag{6}$$

In terms of notation, superscript x in the stress equilibrium operator $\sigma_{ijk,j}^{*x}$ indicates differentiation with respect to coordinate x and $\delta(x,\xi) = \delta(x_1 - \xi_1)\delta(x_2 - \xi_2)$ is Dirac's delta function in 2D.

Equation (5) is solved by the Radon transform, which is defined as follows:

$$\hat{f}(s,m) = R(f) = \int_{<x,m>=s} f(x)dx = \int f(x)\delta(s - <x,m>)dx \tag{7}$$

The inverse Radon transform is $R^{-1}(\hat{f}) = R^*(K\hat{f})$, where

$$\tilde{f}(s,m) = K\hat{f} = \int_{-\infty}^{\infty} \frac{\partial_\sigma f(\sigma,m)}{s-\sigma} d\sigma ,$$

and

$$R^*(\tilde{f}) = \frac{1}{4\pi^2} \int_{|m|=1} \tilde{f}(s,m)\Big|_{s=<x,m>} dm \tag{8}$$

requires integration over the circle with unit radius. After re-casting the problem in matrix notation, the Radon transformed displacement vector is defined as $\hat{U} = R(U)$, $L(m) = \mu_0 \begin{pmatrix} 3m_1^2 + m_2^2 & 2m_1 m_2 \\ 2m_1 m_2 & m_1^2 + 3m_2^2 \end{pmatrix}$, with $|m| = 1$. Since we have that the transform of the forcing function is $R(\delta(x,\xi)) = \int \delta(x,\xi)\delta(s- < x,m >)dx = \delta(s- < m,\xi >)$, application of the Radon transform to both sides of the equation of motion yields

$$(L(m)\partial_s^2 + \Gamma)\hat{U} = -h^{-1/2}(\xi)\delta(s- < m,\xi >)I_2 \tag{9}$$

where matrix $\Gamma = \{\gamma_j^i\}, \gamma_1^1 = \gamma_1, \gamma_2^2 = \gamma_2, \gamma_2^1 = \gamma_1^2 = 0$, and I_2 is the unit matrix in 2D. In order to solve eqn (9), it is necessary to transform matrices $L(m)$ and Γ into canonical form using transform matrix $T = \{t_j^i\}$. Once this has been accomplished by using the solutions for a series of eigenvalue problems, the original system uncouples and yields two equations in the form

$$\left[\partial_s^2 + \eta\right] \hat{Y} = \delta(s-\tau) f \tag{10}$$

with $\hat{y} = \alpha f e^{ik|s-\tau|}$, $\alpha = \frac{i}{2k}$, $k = \sqrt{\eta}$. Completing now the first part of the inverse Radon transform gives

$$\begin{aligned} \tilde{y}(s) = K(\hat{y}) &= \alpha f\left[i\pi e^{ikz} - 2\left(ci(kz)\cos(kz) + si(kz)\sin(kz)\right)\right]\Big|_{z=|s-\tau|} \\ \partial_s \tilde{y} &= \alpha f\left[-\pi k e^{ikz} - \frac{2}{z} + 2k\left(ci(kz)\sin(kz) - si(kz)\cos(kz)\right)\right]\Big|_{z=|s-\tau|} \operatorname{sgn}(s-\tau) \end{aligned} \tag{11}$$

where $\mathrm{ci}(z) = -\int_z^{\infty} \frac{\cos t}{t} dt$, $\mathrm{si}(z) = -\int_z^{\infty} \frac{\sin t}{t} dt$ are the cosine and sine functions. Keeping in mind the previously applied eigenvalue transformations, the inverse Radon transform of the displacement vector is

$$U^* = R^{-1}(\hat{U}^*) = \frac{1}{4\pi^2} \int_{|m|=1} \tilde{U}^*(z)\Big|_{z=|\langle m,x-\xi\rangle|} dm \tag{12}$$

Finally, by defining $K(\hat{Y}) = \begin{pmatrix} f_1^1 \tilde{u}_1 & f_1^2 \tilde{u}_1 \\ f_2^1 \tilde{u}_2 & f_2^2 \tilde{u}_2 \end{pmatrix}$, where $\tilde{u}_i$ is the vector form of $\tilde{y}$ with the two wave numbers k_i, *i=1,2* (note that in terms of more standard notation $k_1 = k_p, k_2 = k_s$) being the eigenvalues of $L^{-1/2}\Gamma L^{1/2}$, the fundamental solution of eqn (12) and its spatial derivative assume the following form:

$$U^*(x,\xi) = \frac{1}{4\pi^2}\int_{|m|=1}\begin{pmatrix} t_1^1 & t_1^2 \\ t_2^1 & t_2^2 \end{pmatrix}\begin{pmatrix} f_1^1\tilde{u}_1 & f_1^2\tilde{u}_1 \\ f_2^1\tilde{u}_2 & f_2^2\tilde{u}_2 \end{pmatrix}\Bigg|_{z=|\langle m,x-\xi\rangle|} dm$$

$$U^*_{,k}(x,\xi) = \frac{1}{4\pi^2}\int_{|m|=1}\begin{pmatrix} t_1^1 & t_1^2 \\ t_2^1 & t_2^2 \end{pmatrix}\begin{pmatrix} f_1^1\partial_z\tilde{u}_1 & f_1^2\partial_z\tilde{u}_1 \\ f_2^1\partial_z\tilde{u}_2 & f_2^2\partial_z\tilde{u}_2 \end{pmatrix}\Bigg|_{z=|\langle m,x-\xi\rangle|} m_k \,\mathrm{sgn}(\langle m,x-\xi\rangle)\, dm \tag{13}$$

What is now left is a simple inverse algebraic transform that will return the displacement vector to the original, physical domain. Thus, by invoking eqn (2), we obtain the final form of the fundamental solution (Green's function) for eqn (1) as

$$u^*_{ij}(x,\xi) = h^{-1/2}(x)h^{-1/2}(\xi)U^*_{ij}(x,\xi) \tag{14}$$

and similar expressions are derived in Ref. [5] for the corresponding strain and stress tensors, where *(x,ξ)* is the source-receiver pair. Finally, an asymptotic expansions for the Green's function is also derived in Ref. [5].

4 Incident P- and SV- wave fields

In the inhomogeneous continuum, an incident planar P-wave can be recovered as a solution of the system of eqs (1) in the form $u_j^{in}(x,\omega) = h^{-1/2}(x)U_j^{in}(x,\omega)$. The corresponding incident tractions are given as

$$\begin{aligned} t_j^{in}(x,\omega) &= C_{ijkl}(x)n_i(x)(h^{-1/2}(x)U_k^{in}(x,\omega))_{,l} = \\ &= h(x)C^0_{ijkl}n_i(x)\left[-\frac{1}{2}h^{-3/2}(x)h_{,l}(x)U_k^{in}(x,\omega) + h^{-1/2}(x)U_{k,l}^{in}(x,\omega)\right] \end{aligned} \tag{15}$$

The transformed incident displacement U^{in} satisfies the equation of motion in the form

$$\left[\mu_0\begin{pmatrix} 3\partial_1^2+\partial_2^2 & 2\partial_{12}^2 \\ 2\partial_{12}^2 & \partial_1^2+3\partial_2^2 \end{pmatrix} + \begin{pmatrix} \gamma_1 & 0 \\ 0 & \gamma_2 \end{pmatrix}\right]U^{in} = 0 \tag{16}$$

For a P-wave with normal incidence, we have $U^{in}(x,\omega) = D^{in}\begin{Bmatrix} 0 \\ 1 \end{Bmatrix} e^{ik_p x_2}$, where D^{in} is incident wave amplitude and $k_p = k_1 = \sqrt{(\rho_0/3\mu_0)\omega^2 - a^2}$ is the wave number. Assuming k_p is real places a restriction on the frequency of propagation in the form $\omega^2 > 3\mu_0 a^2/\rho_0$. Finally,

$$u^{in} = D^{in}\begin{pmatrix} 0 \\ 1 \end{pmatrix} e^{(-a+ik_p)x_2}, \quad t^{in} = D^{in}\begin{pmatrix} 0 \\ 1 \end{pmatrix} 3\mu_0\left[-a+ik_p\right] e^{(a+ik_p)x_2} \tag{17}$$

Similar expressions can be derived for the normally incident SV- wave [5].

5 BIEM formulation and numerical implementation

Following Ref. [6], we formulate the following system of non-hypersingular, traction-type BEM at field point (the receiver) $x \in S_{cr}$:

$$t_i^{in}(x)\Big|_{x\in Sa} = C_{ijkl}(x)n_j(x)\int_{S_{cr}}([\sigma^{*y}_{p\eta k}(y,x)\Delta u_{p,\eta}(y) \\ -\rho(y)\omega^2 u^*_{dk}(y,x)\Delta u_d(y)]\delta_{\lambda l}n_\lambda(y) \\ -\sigma^{*y}_{m\lambda k}(y,x)\Delta u_{m,l}(y)n_\lambda(y))dS_y \tag{18}$$

In the above, $\sigma^{*z}_{kji}(y,x) = C_{kjml}(z)u^{*z}_{mi,l}(y,x)$ is the fundamental stress tensor derived from the fundamental strain tensor $u^{*z}_{mi,l} = \partial/\partial z_l \ u^*_{mi}$, where either $z = y$ or $z = x$. The field and source points are labeled x and y , respectively, and the unknown quantities here are the COD Δu_i . The above integro-differential equation is numerically solved by discretizing the interior crack surfaces using parabolic-type, three-noded boundary elements (BE). In addition, special edge-type BE are introduced to satisfy the crack-tip boundary conditions. This discretization scheme satisfies Hölder continuity for the displacements and tractions at internal nodes only. At odd-numbered BE nodes, the collocation points are shifted inwards, but the element nodes themselves remain at the same positions, so as to satisfy Hölder continuity for tractions and tangential derivatives of displacements [7]. Overall method accuracy is dependent on the precision by which surface integrals are evaluated over a given BE. Regular integrals are computed by standard 32-point Gaussian quadrature. Singular integrals are solved analytically by using asymptotic expansions of the Green's function and its derivatives for small arguments as $r \to 0$, where r is the distance between the source and field point.

The BC for the wave scattering problem in question are satisfied by the discrete form of the BEM, a process that yields an algebraic system of equations in complex form. This system is solved by Gauss elimination, and numerical values for the unknown COD on the crack surfaces are thus obtained. From these values, dynamic SIF are calculated using the well known traction formula [7], which is being adapted here for the inhomogeneous case and then normalized by its static value.

6 Numerical results

We consider a FGM with 'background' material properties of $\mu_0 = \lambda_0 = 180.10^6 N/m^2$ and density $\rho_0 = 2000 kg/m^3$. Two basic shear modulus profiles are examined, namely a stiffening / softening with increasing depth from reference surface Ox_1. Specifically, we have $\mu(0,-L) = 1,5\mu_0 = 270.10^6 Pa$ and $\mu(0,-L) = 0.5\mu_0 = 90.10^6 Pa$ in the former and latter cases, where $L = 520\ m$ is a depth scale over which material properties vary.

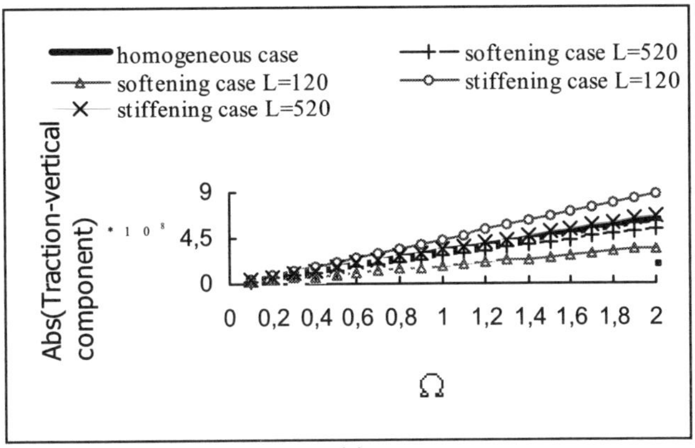

(a)

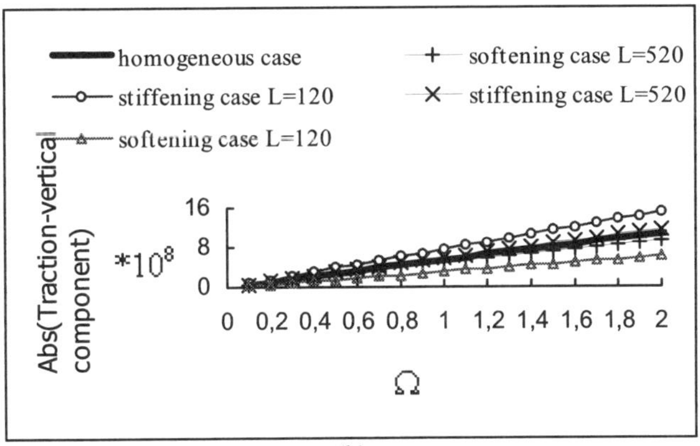

(b)

Figure 2: Vertical component of incident traction on the crack surface for normally incident (a) P-wave and (b) SV-wave in infinite FGM plane.

The test example chosen to validate the BEM is that appearing in Chen and Sih [8], who analytically solved for the SIF around an interior crack in an infinite, homogeneous plane subjected to harmonic waves under normal incidence. Figure 2 plots the vertical component of the incident P- and SV- wave traction field versus normalized frequency $\Omega = \omega\ \tilde{a}/C_s$, where ω is the frequency of propagation (in rad/sec), $\tilde{a}$ is the crack half-length (in m) and C_s is the shear wave speed (in m/sec). Also, Fig. 3 shows the SIF for normally incident P- and SV- waves. The results generated by the BIEM for $a = 0$ correspond to the homogenous material, in which case we observe good agreement with Ref. [8]. The following observations are now made:

(a) The effect of inhomogeneity is more pronounced at higher frequencies of vibration.
(b) Changes in the depth scale for material variability influence the dynamic load that develops along the crack surface in different ways. The largest differences with respect to the homogeneous 'background' medium are obtained for small depth scales, i.e., the rapidly softening or stiffening profiles. Quantitatively, the differences observed in the tractions for the exponentially softening / stiffening materials at an incident P-wave frequency of $\Omega = 2$ and for depth scales of $L = 520\ m$ and $L = 120\ m$ are 23% and 150%, respectively. These numbers become 24% and 155%, respectively, for the case of an incident SV-wave.
(c) The type of incident wave becomes important when the material through which it propagates is inhomogeneous. For instance, at $\Omega = 2$, the difference observed between the homogeneous background case and the softening material is 13% and 77% for the P- and SV- wave types, respectively.

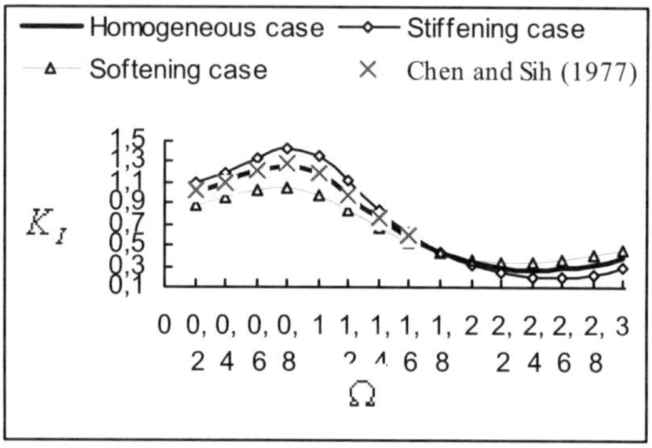

(a)

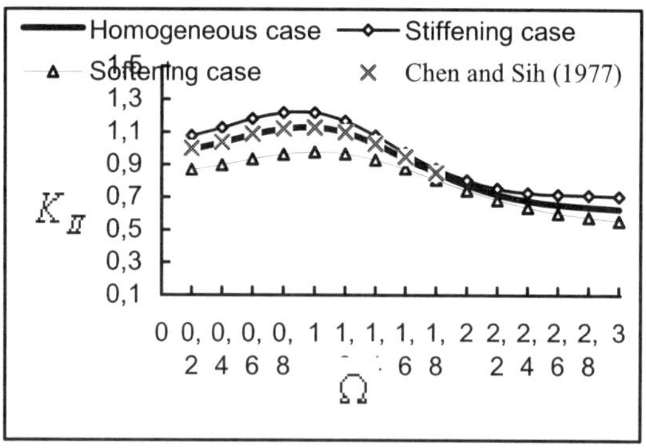

(b)

Figure 3: Normalized (a) SIF-I for normal P-wave incidence and (b) SIF-II for normal SV-wave incidence vs. frequency in a cracked, infinite FGM plane.

As previously mentioned, Fig. 3(a) shows the SIF-I (mode 1) normalized with respect to its static value versus Ω for an incident P-wave and the three types of materials (homogeneous, stiffening and softening) at depth scale $L = 520\ m$. Although inhomogeneity does not change the basic shape of the SIF curve, there are otherwise large numerical differences in the results obtained for the above three cases. As example, the difference observed between the SIF-I curve peaks for the soft and stiff materials at $\Omega = 0.8$ is 40%. Furthermore, Fig. 3(b) shows the normalized SIF-II (mode 2) versus Ω for the incident SV-wave. The same trend as before is again observed, with the difference between peaks in the SIF-II curves for the soft and stiff materials at $\Omega = 0.8$ registering as 28%.

7 Conclusions

A plane strain, time-harmonic elastodynamic analysis for FGM with an exponential spatial variation of the elastic parameters was presented here. The analysis employed a non-hypersingular, traction-type BEM with Green's functions that were separately obtained by functional transform methods. The basic problem that was solved addressed a crack buried in infinite sheet of this FGM under incident, time harmonic P- and SV- waves. The results showed that the SCF at the crack tips and the scattered displacement far-field are strongly influenced by the presence of inhomogeneity.

Acknowledgement

The authors acknowledge support through the 'Joint Research & Technology Program 2004-2006', Project No. BgGr-11/2005Y

References

[1] Yue, Z.Q., Xiao, H.T., Tham, L.G., Boundary element analysis of crack problems in functionally graded materials. *Int. J. Solids Str.*, 40, 3273-3291, 2003.

[2] Zhang, C., Sladek, J., Sladek, V., Effect of material gradients on transient dynamic mode-III. Stress intensity factor in FGM. *Int. J. Solids Str.*, 40, 5252-5270, 2003.

[3] DeHoop, A.T., *Handbook of Radiation and Scattering of Waves.* Academic Press, London, 1995.

[4] Rangelov, T.V., Manolis, G.D., Dineva, P.S., Elastodynamic fundamental solutions for certain families of 2D inhomogeneous anisotropic domains: basic derivations, *Eur. J. Mech. / A Solids*, 24, 820-836, 2005.

[5] Dineva P.S., Rangelov, T.V. and Manolis, G.D., Elastic wave propagation in a class of cracked, functionally graded materials by BIEM, *Comp. Mech.*, under review, 2005.

[6] Zhang, C., Gross, D., *On Wave Propagation in Elastic Solids with Cracks.* CMP, Southampton, 1998.

[7] Aliabadi, M., Rooke, D., *Numerical Fracture Mechanics*. CMP, Southampton, 1991.

[8] Chen, E.P., Sih, G.C., Scattering waves about stationary and moving cracks. *Mechanics of Fracture: Elastodynamic Crack Problems*, G.C. Sih, Noordhoff, Leyden, 1977.

WIT Transactions on Modelling and Simulation, Vol 42, © 2006 WIT Press
www.witpress.com, ISSN 1743-355X (on-line)

Boundary element formulation applied to multi-fractured bodies

E. D. Leonel, O. B. R. Lovón & W. S. Venturini
São Carlos School of Engineering, University of São Paulo, Brazil

Abstract

In this work, the performance of the boundary element method applied to multi-fractured bodies is analysed. The algebraic equations are written either using only displacement equations for collocations defined along the crack surfaces keeping a very small distance between them to avoid singularities or using displacement and traction representations. Adaptive schemes are employed to adjust the crack advance direction and to refine the elements near the tip. A remeshing procedure is also adopted to appropriately reduce the boundary approximation influence for elements distant enough from the crack tip. Examples of multi-fractured bodies that are loaded to the rupture are shown to illustrate the applicability of the proposed scheme.
Keywords: boundary elements, linear fracture mechanics.

1 Introduction

Analysis of fractured solids is a very common problem in engineering. The BEM has demonstrated to be the most accurate numerical technique for the analysis of this kind of problem. In fracture mechanics analysis the dimensionality reduction of BEM is clear, as only boundary discretization will be required. Moreover, internal points are needed only to approximate the crack line, but without requiring remeshing.

Cruse [1] was the first to use boundary integral methods to study cracks. After 35 years of use that attempt the method has been improved and has become the most efficient numerical technique to model linear and non-linear cracks. During these four decades many formulations have been tested with accurate results, as can be seen in a state of art published by Aliabadi [2]. Among these works we have to point out some interesting works that are often used to extract stress

WIT Transactions on Modelling and Simulation, Vol 42, © 2006 WIT Press
www.witpress.com, ISSN 1743-355X (on-line)
doi:10.2495/BE06032

intensity factors or to model crack propagation: Blandford et al. [3], Cruse [4]; Portela et al. [5], in which classical singular formulations using sub-regions, the green function for cracks and dual BEM (DBEM) formulations have been proposed, respectively. We would like to mention some interesting alternatives proposed by the senior author that have demonstrated to be accurate enough for practical applications in engineering: Venturini [6] using sub-regions governed by elastoplástica criterion; Venturini [7] using dipoles to enforce the crack displacements governed by cohesive cracks; Manzoli and Venturini [8] using field displacement approximation with strong discontinuity; and Leite and Venturini [9], using narrow regions with rigidity going to zero.

In this article we are going to use DBEM for elastic fracture mechanics to analyse crack growth in multi-fractured bodies. The size and the direction of the crack advance is evaluated by an adaptive scheme for which the error can be specified. To avoid large number of unknowns along the crack surfaces, a remeshing technique is used to reduce the total number of elements. The examples analysed deal with bodies containing random micro-cracks distributions that will localise to a major rupture surface.

2 Integral equations

Let us consider a 2D elastic domain Ω (Figure 1) limited externally by the boundary Γ and containing several internal crack surfaces Γ^f, that may reach the external boundary, representing the micro-cracks or micro-voids which may appear during the fabrication process (concrete casting for example). For this solid the following integral representation of displacements can be written [10]:

$$c_{ij}\,u_j = -\int_{\Gamma} p_{ij}^{*} u_j \,d\Gamma - \sum_{k=1}^{N_f} \int_{\Gamma_k^f} p_{ij}^{*} u_j^f \,d\Gamma + \int_{\Gamma} u_{ij}^{*} p_j \,d\Gamma + \sum_{k=1}^{N_f} \int_{\Gamma_k^f} u_{ij}^{*} p_j^f \,d\Gamma \tag{1}$$

where u_j^f and p_j^f represent displacements and tractions along the fracture surfaces, u_j and p_j are boundary values of displacements and tractions, u_{ij}^* and p_{ij}^* are Kelvin's fundamental solutions, c_{ij} is the well-known free term that is given in terms of the boundary geometry, and N_f is the number of pre-existing cracks.

Similarly, we can derive the traction integral representation for smooth collocation points:

$$1/2\,p_j = -\int_{\Gamma} S_{ij}^{*} u_j \,d\Gamma - \sum_{k=1}^{N_f} \int_{\Gamma_k^f} S_{ij}^{*} u_j^f \,d\Gamma + \int_{\Gamma} D_{ij}^{*} p_j \,d\Gamma + \sum_{k=1}^{N_f} \int_{\Gamma_k^f} D_{ij}^{*} p_j^f \,d\Gamma \tag{2}$$

where S_{ij}^* and D_{ij}^* are fundamental values for the stress equation [10].

To equations (1) and (2) one has to add new integral terms to take into account the advance of the cracks that may increase.

Crack analysis can be carried out by using the two equations given above. If a small but finite gap is preserved between both sides of all fracture one may use independently eqn (1) or (2). In a recent work, it has been demonstrated that for small gaps, 10^{-5} the element size the system can be solved and the results in terms of stress intensity factor are still reliable [9]. The dual boundary element is defined when one employs the displacement and traction representations, eqns (1) and (2), at each pair of collocation points defined at opposite crack surfaces, for which the gap may goes to zero.

3 Algebraic equations

As usual for boundary elements one has to divide the external boundary and the crack surfaces into elements and approximate displacements and tractions by using convenient shape functions. Although quadratic elements are usually recommended to approximate the displacement and traction fields along the crack surfaces, we preferred simpler linear approximation with adaptive scheme that guarantees the required accuracy for facture analysis. For collocation along the boundary, we have used only displacement representations, allowing the use of continuous and discontinuous elements. Algebraic equations written from eqn (2) can also be used but the results are clear worse for narrow domains.

As the traction equation requires the continuity of the displacement derivatives, discontinuous elements have to be adopted. When the crack intersects the boundary the intersected element is divided in two new elements that must allow the boundary displacement and traction discontinuities due to the crack.

In this work we are carrying out all boundary and crack element integrals analytically. This is always important in crack analysis to keep the stability of the system even when thousands of iterations or load increments are required. Moreover the scheme based on the use of displacement or traction equations only with small gaps between the crack surfaces cannot be accurate without the analytical integration schemes.

The algebraic equations will be displayed in three different blocks: the boundary equations; pre-existing crack equations (displacement and traction relations); and new crack equations that will be written during the analysis. Thus, the following system of algebraic equations represented by using line and column blocks can be written as follows,

$$\begin{bmatrix} H_{bb} & H_{bf} & H_{ba} \\ H_{fb} & H_{ff} & H_{fa} \\ H_{ab} & H_{af} & H_{aa} \end{bmatrix} \begin{Bmatrix} U_b \\ u_f \\ u_a \end{Bmatrix} = \begin{bmatrix} G_{bb} & G_{bf} & G_{ba} \\ G_{fb} & G_{ff} & G_{fa} \\ G_{ab} & G_{af} & G_{aa} \end{bmatrix} \begin{Bmatrix} P_b \\ p_f \\ p_a \end{Bmatrix} \tag{3}$$

where the subscripts b, f and a indicate collocation points and values defined on the boundary elements, along the pre-existing crack elements and the new crack elements, respectively; U_b and P_b represent nodal displacements and tractions along the boundary, u_f, u_a, p_f and p_a contain the displacement and traction crack values, and matrices H and G are obtained by integrating properly the corresponding kernels along boundary and crack elements.

For a solid with pre-existing cracks or micro-cracks, the size of the two first blocks are very large. First, because having a fine boundary discretization is convenient to reduce the boundary errors according to a convenient tolerance. Then because the number of pre-existing cracks can be very large, with minimum of four crack element each. We can solve the two first block lines and then the iterative process to model the crack growth can deal only with the new equations. Thus, after applying the boundary conditions from the two first block lines we have:

$$\begin{bmatrix} A_{bb} & H_{bf} \\ A_{fb} & H_{ff} \end{bmatrix} \begin{Bmatrix} X_b \\ u_f \end{Bmatrix} = -\begin{bmatrix} H_{ba} \\ H_{fa} \end{bmatrix} \{u_a\} + \begin{Bmatrix} B_b \\ B_f \end{Bmatrix} + \begin{bmatrix} G_{bf} & G_{ba} \\ G_{ff} & G_{fa} \end{bmatrix} \begin{Bmatrix} p_f \\ p_a \end{Bmatrix} \tag{4}$$

where A_{jk} contains coefficients of H_{jk} or G_{jk}, B_k gives the prescribed boundary value contributions.

Equation (4) may be written in its reduced form as follows:

$$AX = -H_b u_a + B + Gp \tag{5}$$

where X contains boundary and crack surface displacement unknowns and p gives possible traction applied along the crack edge.

Solving eqn (5) in terms of to obtain boundary unknowns and pre-existing crack displacements and neglecting the crack traction gives:

$$X = X_b + R_b u_a \tag{6}$$

where X_b contains the X values before the crack growth, while R_b=-$A^{-1}H_b$ gives the crack growth effects.

Taking into account that we are dealing with a non-linear problem, eqn (6) can also be written in terms of increments. Then, for a given load increment we can obtain ΔX_b directly and the correction due to the new crack elements are given by $R_b\Delta u_a$, being Δu_a represented by the third block line of eqn (3), which after applying the boundary conditions becomes:

$$H_{aa}\Delta u_a = -A_f \Delta X + \Delta B_a \tag{7}$$

where A_f contains the third block line H_{ak} and G_{ak} terms according to the prescribed value and B_a takes into account the prescribed boundary values.

Then, the new crack displacement increments are obtained by replacing eqn (6) into eqn (7), as follows:

$$\Delta u_a = +S_a^{-1}\left[\Delta B_a - A_f \Delta X_b\right] \tag{8}$$

with

$$S_a = A_f R_b + H_{aa} \tag{9}$$

4 Fracture mechanic model

There are many techniques for extracting stress intensity factors (SIF). So far, some of them have already been extended to BEM formulations as quarter point elements and J-integrals. In this work we have used the displacement correlation technique that has been proven to be very efficient and simple. In fact, as we are dealing with very small elements near the tip, displacement difference of any point along the tip element can be used and the results will be very accurate. In this case, as we are using collocations defined at quarter points (discontinuous elements), the SIF values are obtained directly from the computed degrees of freedom by using the expressions:

$$K_I = COD\sqrt{8\pi / \ell_T}\, G/(\kappa + 1) \tag{10}$$

$$K_{II} = CSD\sqrt{8\pi / \ell_T}\, G/(\kappa + 1) \tag{11}$$

where G is the shear elastic modulus of the material, $\kappa = 3 - 4\upsilon$ for the plane strain case and $\kappa = (3 - 4\upsilon)/(1 + \upsilon)$ for plane stress case and *COD* and *CSD* are the Crack Open Displacements perpendicular and parallel to the crack direction, respectively.

For the case of multi-fractured bodies, one has to take into account both stress intensity factors. Moreover, the propagation direction is another parameter that must be known to allow the analysis to be performed. Although several criteria have already been proposed to predict the crack growth, as those presented in references [11–13], we have implemented the simple technique known as the maximum principal stress direction propose by Erdogan and Sih [14]. Following this criterion, the racks are assumed to grow in a direction θ perpendicular to the maximum principal stress at the crack tip, being the angle θ measured with respect to the previous crack direction and given by the following expression:

$$tan(\theta / 2) = \left[K_I / K_{II} \pm \sqrt{(K_I / K_{II})^2 + 8} \right] / 4 \tag{12}$$

After computing the propagation angle Erdogan and Sih [14] have proposed an equivalent stress intensity factor K_{Eq} to verify the stability of each individual crack, as follows:

$$K_{Eq} = K_I \cos^3(\theta / 2) - 3K_{II} \cos^2(\theta / 2)\sin(\theta / 2) \tag{13}$$

5 Adaptive schemes

Crack growth analysis deserves the introduction of appropriate boundary and crack surface adaptive schemes for two reasons: to assure accuracy and to reduce the number of degrees of freedom.

In this formulation we first use an adaptive scheme to obtain a reasonable boundary discretization that can assure that the boundary value errors will be within a range defined by a chosen tolerance. This procedure is applied only at the first step level without considering the crack growth effects. The generations of new crack usually does not introduce significant errors at the boundary values, but will dramatically reduce the efficiency of the whole procedure.

A simple error evaluation technique based on the strain energy variation has shown to be very efficient. We define the exact displacements and tractions of all boundary elements by averaging the element end values. Thus, for a given element exact displacement and traction at the element middle point are:

$$u_k^e = (u_k^1 + u_k^2) / 2 \tag{14}$$

where the superscripts 1 and 2 are the element end nodes.

The approximate value will be computed by using eqns (1) and (2) applied for the element middle point to give u_k^a and p_k^a. Then the error will be given by:

$$E = \left\{ \sum_{j=1}^{N_e} \ell_j \sum_{k=1}^{2} (u_k^{je} - u_k^{ja})^2 \right\} / \sum_{j=1}^{N_e} \ell_j \sum_{k=1}^{2} (u_k^{je})^2 \tag{15}$$

where ℓ_j is the element length.

The adaptive scheme that has demonstrated to be efficient is based on subdividing a quarter of the total elements where the computed values show larger variation.

The most important adaptive scheme is applied to the crack growth. An initial crack element length is chosen to control the crack growth process. Then, the number of elements to approximate the pre-existing cracks or micro-cracks is evaluated. No less than eight elements (four along each surface) will be defined independently on the crack size (micro-cracks can lead to very small elements). At this stage the adaptive process may be applied to compute an accurate stress intensity factor according to a chosen tolerance, although this initial refinement has very limited effects on the whole analysis.

After selecting the elements that will grow the directions (θ_f) of the new cracks are computed according to the chosen criterion. The convergence test is made by defining test growth lengths equal to half to the corresponding crack element size. Using this test crack element the crack parameters are evaluated including the new direction (θ_0). If the difference $|\theta_f - \theta_0|$ were less the chosen tolerance the crack element size is maintained. Otherwise, the process continues redefining the crack element sizes where the test fails.

Defined all crack advances with an increment, the accuracy of the stress intensity factor has to be checked and the crack tip element divided accordingly. This sub-division is made progressively by multiplying the number of crack advance elements by two. The sub-division stops when the difference between two successive values of the corresponding stress intensity factor is less than the tolerance.

The procedure above can lead to very accurate results for computing the stress intensity factors. However the number of degrees of freedom will be very high and the whole procedure will be inefficient. To reduce the computer time we propose a remeshing procedure to eliminate unnecessary degrees of freedom. Bearing in mind that the refinement is required only to compute the stress intensity factors, we assigned to each load increment a block of adaptive provisory algebraic relations. After the convergence, all intermediate nodes will be neglected and the original linear element is recovered to continue the analysis. Thus, after the convergence a block of eight algebraic relations per each new crack element is added to the permanent system of eqns (7). Although the timing to recover the linear element may be also chosen, the degree of freedom reduction enforced just behind the crack tip elements has demonstrated to be more efficient.

6 Examples

Three examples are presented to show the efficiency of the proposed schemes. The first example illustrates the simple adaptive model to define the boundary and crack discretizations to reach the errors specified in the analysis. A cantilever beam, shown in Figure 1, was selected for this illustration. The geometry including a small crack inside the domain is defined in the same figure. The adopted Poisson ration for this example is assumed $\upsilon = 0.5$. For the illustration purpose, the example is solved first to refine the boundary mesh without analysis the crack parameters. Table 1 shows the convergence in terms of displacement norms. After obtaining the boundary mesh (also shown in Figure 1) the crack refinement is adopted to reach the stress intensity factor K_I according to the specified tolerance.

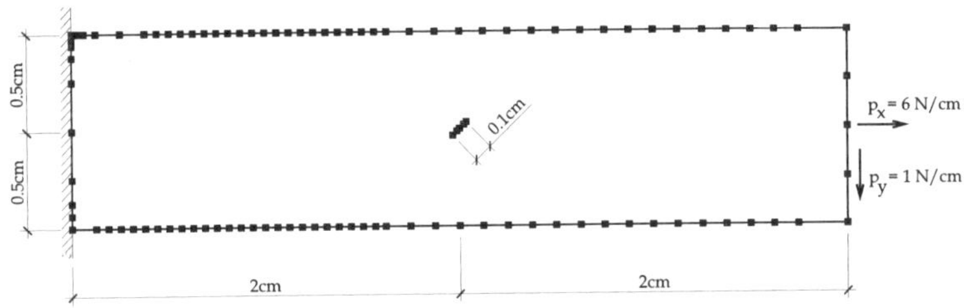

Figure 1: Discretized cantilever beam containing a crack at the centre.

Table 1: Adaptive process: a) norm of errors in displacements Δu; b) maximum vertical displacement u_{max}.

Nodes	Norm of Δu	$E.u_{max}$
20	4.17E-003	158.5
50	7.75E-004	233.0
100	2.26E-004	256.7
200	5.53E-005	266.0
400	1.33E-005	268.4

This analysis was carried out using displacement and traction equations as usual and also adopting only displacement equation for the collocations defined along the crack surfaces. In this case a gap $\delta=0.001\ cm$ was used to avoid the matrix singularity. Table 2 shows the stress intensity factors computed during the process to reach the specified tolerance (10-4) for the 100 boundary element mesh. The results given by using the singular formulation also converges to slightly different values of K_I and K_{II} (0.6% for K_I). These results improve when δ is reduced.

Table 2: Stress intensity factors convergence test.

Crack elements	K_I	K_{II}
8	2.9002E+00	1.3585E+00
28	2.9603E+00	1.3893E+00
42	2.9655E+00	1.3920E+00
52	2.9672E+00	1.3930E+00

The second example concerns the analysis of the rectangular cantilever beam (400cm × 200cm) shown in Figure 2. Fifty cracks were randomly distributed inside the domain (Figure 3). They may cut the boundary or remain entirely inside the domain. The young's modulus assumed for this problem was $E=2.1\ 10^4\ kN/cm^2$, while the Poisson ratio was $\upsilon = 0.3$. The material toughness was assumed equal to $K_c=1.11\ 10^4 kN/m^{-1.5}$. The load applied in 50 increments is given by prescribing both vertical and horizontal displacements along the beam right end: $\bar{u}_x = 5cm$ and $\bar{u}_y = 3cm$

For this kind of analysis, coalescence can be modeled if some dissipation area is assumed. When the distance between two cracks is small enough in such a way that the corresponding dissipation areas have common parts coalescence can be assumed and the crack lines joined together. This can be verified by assuming the dissipation area is given by the material internal length defined in the context of strain localization theories [15].

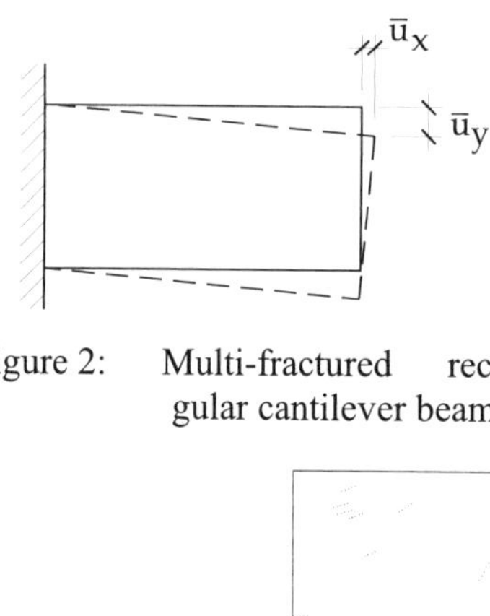

Figure 2: Multi-fractured rectangular cantilever beam.

Figure 3: Randomly distributed pre-existing cracks.

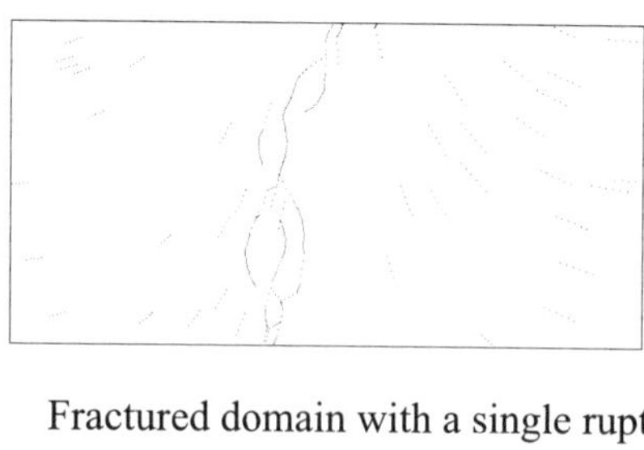

Figure 4: Fractured domain with a single rupture surface.

The crack growth analysis of this beam shows the complete rupture of the domain (Figure 4). As we have many cracks distributes inside the domain there were several coalescence until the definition of the rupture surface. Figure 5 shows the final displacement field illustrating the rigid body motion pattern of the two parts.

The third example analyzes a panel containing several rivets. The problem of crack propagation in butt-joint panel is one of the main challenges in aircraft engineering [16, 17]. Surprisingly, there is no BEM work dedicated to this important practical problem so far. The example shown here is still within the academic context, but it demonstrates that BEM can be applied to this complex problem.

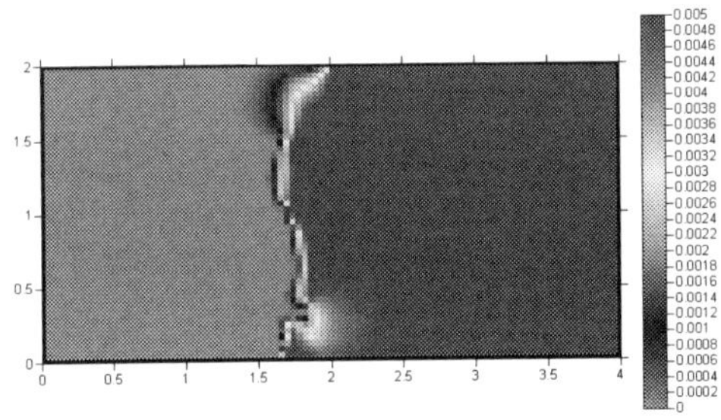

Figure 5: Final displacement field.

For this analysis we defined a square plate with a line of three rivets as depicted in Figure 6. The square side length is L=100cm while the diameter of the rivets is L/20. We have adopted the Young modulus $E = 2.1\ 10^9 kN/m^2$, Poisson ratio $\upsilon = 0.3$ and toughness $K_c = 1.04\ \ 10^5 kN/m^{-1.5}$.

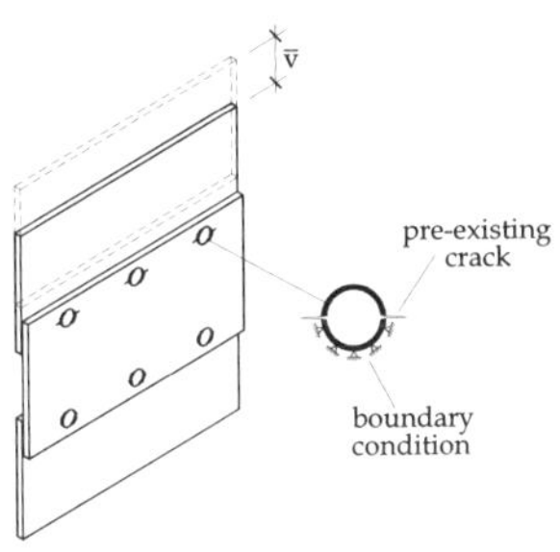

Figure 6: Plate with three rivets.

The analysis was carried out by applying constant displacement equal to $\bar{v} = 0.1cm$ in the direction y along the upper side. Displacements equal to zero have been also prescribed around the lower half part of the whole boundaries. The specimen was loaded by applying the prescribed displacements into forty uniform increments. First, the analysis was carried out by assuming pre-existing cracks with length equal to $L/100$ placed at both side of each rivet hole. Then, the analysis was repeated by assuming pre-existing micro-cracks with length equal to $L/500$.

The performance of the algorithm to model this crack growth problem has shown to be excellent. Although the exact position of the final fractured surface is very instable the result obtained by the two analyses are much closed as depicted in Figure 7. The force × displacement curve shown in Figure 8 also confirms the good agreement between the numerical solutions.

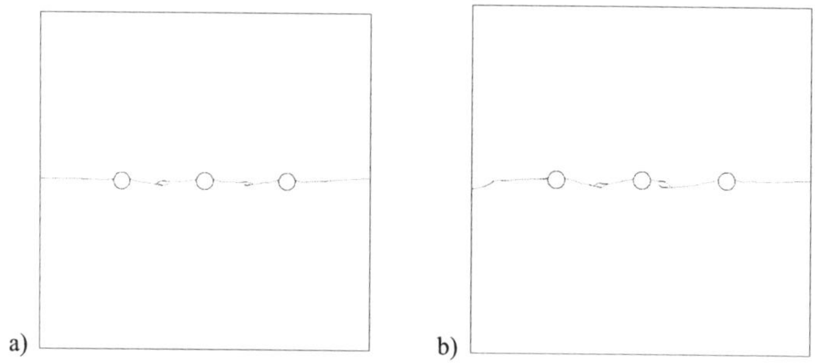

Figure 7: Crack growth pattern at the rupture with; a) with a pre-existing crack; b) with a pre-existing micro-crack.

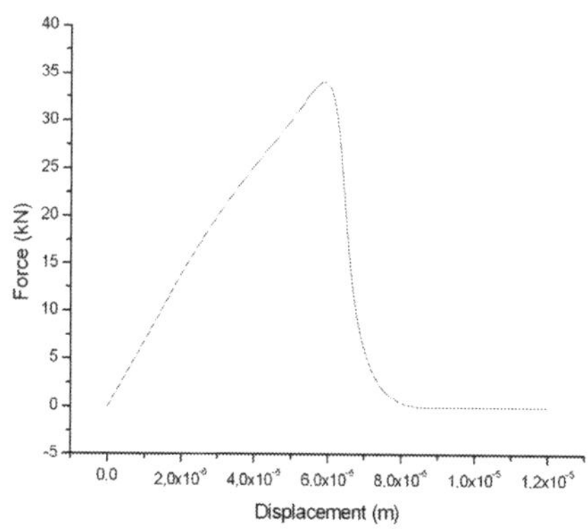

Figure 8: Load × displacement curve.

7 Conclusions

The work has shown a simple BEM scheme to model crack growth using low order elements. Several adaptive schemes together with accurate crack and boundary element integrals are enough to assure obtaining reliable results. The stability of the developed numerical algorithm was tested to analyze complex problems as multi fractured bodies and multiple site damage problems.

References

[1] Cruse, T.A., Lateral constraint in a cracked, three-dimensional elastic body. *International Journal of Fracture Mechanics*, **6**, pp. 326-328, 1970.

[2] Aliabadi, M.H., Boundary Element formulations in fracture mechanics. *Applied Mechanics Reviews*, **50**, pp. 83-96, 1997.

[3] Blandford, G.E., Ingraffea, A.R. & Ligget, J.A., Two-dimensional stress intensity factor computations using the boundary element method, *International Journal for Numerical Methods in Engineering*, **17 (3)**, pp. 387-404, 1981.

[4] Cruse, T.A., *Boundary Element Analysis in Computational Fracture Mechanics*. Kluwer Academic Publishers, Dordrecht, 1988.

[5] Portela, A., Aliabadi, M.H. & Rooke, D.P., Dual boundary element method: Efficient implementation for cracked problems. *International Journal for Numerical Methods in Engineering*, **33**, 1269-1287, 1992.

[6] Venturini, W. S., *Boundary element method in geomechanics*, Springer Verlag, 1983.

[7] Venturini, W.S., A new boundary element formulation for crack analysis. In: Brebbia, C.A., (ed.) *Boundary element method XVI*, Computational Mechanics Publications: Southampton and Boston, pp. 405-412, 1994.

[8] Manzoli, O.L. & Venturini, W.S., Uma formulação do MEC para a simulação numérica de descontinuidadcs fortes. *Revista Internacional de Métodos Numéricos para Cálculo y Diseño en Ingeniería*, **20 (3)**, pp. 215-234, 2004.

[9] Leite, L.G.S. & Venturini, W.S., Stiff and soft thin inclusions in two-dimensional solids by the boundary element method. *Engineering analysis with boundary elements*, **29(3)**, pp. 257-267, 2005.

[10] Brebbia, C.A. & Dominguez, J., *Boundary elements: an introductory course*, Computational Mechanics Publications: Southampton and Boston, 1992.

[11] Shi, G.C., Strain energy density factor applied to mixed mode crack problems, *Int. Journal Fract. Mechanics*, **10,** pp. 305-321, 1974.

[12] Hussain, M.A., Pu, S.U. & Underwood, J., Strain Energy release rate for a crack under combined mode I and II, *ASTM STP*, **560**, pp. 2-28, 1974.

[13] Theocaris, P.S. & Adrianopoulos, N.P., The T-Criterion applied to ductile fracture, *Int. Journal Fracture*, **20**, pp. R125-130, 1982.

[14] Erdogan, F. & Sih, G.C., On the crack extension in plates under plane loading and transverse shear. *Journal of Basic Engineering, Transaction of ASME*, **85**, pp. 519-527, 1963.

[15] Pijaudier-Cabot, G & Bazant, Z.P., Nonlocal damage theory, *J. Engng. Mech. ASCE*, **113**, pp.1512-1533, 1987

[16] Babuska, I. & Andersson, B. The splitting method as a tool for multiple damage analysis, *Siam Journal on Scientific Computing* **26 (4)**: pp. 1114-1145, 2005.

[17] Galatolo, R. & Nilsson, K-F, An experimental and numerical analysis of residual strength of butt-joints panels with multiple site damage, *Engineering Fracture Mechanics* **68**, pp. 1437-1461, 2001.

Penalty formulation of damage in classical composites

P. Procházka & J. Matyáš
Czech Technical University in Prague, Czech Republic

Abstract

The problem of damage in composites has long been studied by many authors. This paper is going to bring a new formulation, which is based on a combination of penalty and boundary element formulation. It is proved that this approach offers a combination of two very powerful tools. The penalty formulation can be very easily realized from the engineering point of view. It starts with artificial spring realization of contact. Such visualization provides a large range of different technical applications, involving simple and multiple coatings, artificial lay-out of contacts between fibers and matrix, and such like. The boundary element method, on the other hand, is very suitable for solving contact problems. The only problem when using the BEM is the nonlinear behavior of the matrix. Nevertheless such hurdles can be overcome by a well known trick: the introduction of a polarization tensor. By virtue of this tensor the influence of the fiber is eliminated (assuming the elastic behavior of the fibers) and the matrix is the only domain, which can be described as plastic. Moreover, attempts have shown that the plastic zones are concentrated only in small areas and a fine discretization of the domain of matrix is concentrated into these zones. A couple of examples accompany the theory.
Keynotes: classical composites, damage at interfacial zone, penalty formulation.

1 Introduction

In this paper a problem of debonding of fibers from matrix, the influence of loading choice, prevailingly in normal direction, and a range of damage along the interface of phases are studied. The generalized Mohr-Coulomb interfacial conditions are taken into account including the tensile stresses at the fiber-matrix interface to be excluded. The loading is introduced in the classical way in such a

WIT Transactions on Modelling and Simulation, Vol 42, © 2006 WIT Press
www.witpress.com, ISSN 1743-355X (on-line)
doi:10.2495/BE06033

way that unit overall load is supposed. The results are mutually compared, considering partial loadings, and evaluated in such a way that the properties of homogenized composite structure should be determined in the final stage of computations; it means the overall behavior of the composite aggregate should be described in the results of this paper.

In the paper, the complete solution of responses to unit normal loading is presented, including a discussion on particular cases. A unit cell model is used to study an effect of imperfectly bonded fiber-matrix interfaces in composites. A periodic structure of the fibers is assumed to simplify the discussion. The radial (normal) and the tangential tractions across the interface are continuous, but the displacements may be discontinuous at the interface fiber-matrix.

In Cartié, et al., mechanisms of crack bridging by composite and metallic rods is described. In Achenbach & Zhu, [2], the tractions are linearly dependent on jumps of the displacements (elastic law is assumed with coefficients of elasticity k_n - in normal - and k_t - in tangential directions, which are the stiffnesses of the interfacial zone). In our case similar denotation is used for spring model of interfaces, where k_n and k_t is used for spring stifnesses in normal and tangential direction with respect to the contact boundary. If the interface exhibits a partial debonding, the zero traction boundary conditions are invoked along debonding region. Moreover, Mohr-Coulomb conditions are applied along the fiber-matrix contact. Explicit solution of the problem of debonding composite is considered in Karihaloo & Vishvanathan, [3]. Tverggard in [4] is concern with nonlinear material behavior of matrix in disconnecting phases. Procházka, [5], solves homogenization of linear and of debonding composites using the BEM and Uzawa's algorithm. Pullout problem, which is very close to the problem solved in this paper, is described in Procházka & Sejnoha, [6].

The composite aggregate exhibits periodicity, which is very popular property to simplify considerations. 2D unit cell supplied with proper periodic boundary conditions can be used to represent composite structure. Since the linear behavior is assumed in both fiber and matrix, the analysis of the unit cell models is carried out using the BEM, and the penalty algorithm is used together with the advantage of pre-eliminated matrices (substructuring of the boundaries of fiber and matrix separately). Our objective is to examine the effects of imperfectly bonded interfaces on local stresses and on the overall response of the composite system.

A possible type of interface conditions is proposed in Prochazka & Sejnoha, [6]. It simulates the interfacial zone with linear behavior and debonding of the fiber-matrix system by cracking in the interfacial zone. They are feasible for direct contact between fiber and matrix simulation.

The unit cell technique is described in Suquet, [7].

2 Computational model: contact law

Consider a periodic composite structure under being loaded by normal overall stresses. Under this assumption, a problem of two elastic bodies in the unit cell, the geometry of its first quarter is depicted in Fig. 1, may be formulated.

The first body (fiber) in undeformed state occupies a circular domain Ω^f and the second body (matrix) occupies domain Ω^m. The common part of the boundaries of both domains is denoted by Γ_C. Because of symmetry (in the case of normal influences) and of antisymmetry (shear loading), only the first quarter can be considered in the analysis.

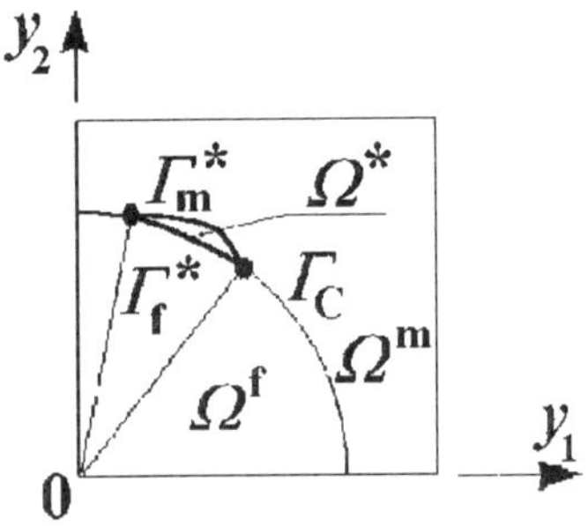

Figure 1: Geometry of the model.

Let **n** be the unit outward normal to Ω^f and **t** be the tangent to Γ_C with respect to Ω^f. Quantities u_n^i and u_t^i, $i = m, f$, defined on the contact Γ_C, are magnitudes of projections of the displacement vector to **n** and **t**, respectively. Superscript m denotes the quantities defined in the matrix, and f denotes the quantities defined in the fiber. Similarly, p_n and p_t are magnitudes of projections of tractions **p** to **n** and **t**, respectively; it holds $\mathbf{p} = \mathbf{p}^f = -\mathbf{p}^m$, $\mathbf{p} = (p_n, p_t)$ The transfer of elastic stresses from the matrix to the fiber has been assumed by the following cases of debonding rules:

a) An initial flaw Γ_{θ_C} on Γ_C is given. It means that $\mathbf{p}^f = -\mathbf{p}^m = \mathbf{0}$ on Γ_{θ_C}. This initial flaw is supposed because of very poor results from the boundary element method in the neighborhood of vertices, i.e. at the beginning and at the end of the contact Γ_C. On the rest of Γ_C the following conditions are fulfilled:

b) $u_n^m - u_n^f \equiv [u]_n \geq 0$;

c) $p_n \leq \sigma_+$, where σ_+ is the tensile strength; $|p_t| \leq -\tan\varphi p_n + \tau_b \kappa(-p_n)$, where $\tan\varphi$ is tangent of the internal friction of both materials (Coulomb friction), κ is the Heaviside function, τ_b is the shear strength or cohesion. $\tan\varphi$ and τ_b are given constants, being different for different couple of material on contact. These conditions describe the generalized Mohr-Coulomb law involving the exclusion of tension.

We concatenate the above conditions and generalize them to obtain a realistic model of the interfacial behavior of the zone between the matrix and the fibers. Then the problem can be formulated:

Problem A: It is to solve an elastic problem on both domains (fiber and matrix) on a unit cell subject to unit loading and the following interface conditions have to be fulfilled on the contact boundary Γ_C:

a) $[u]_n \geq 0$, $p_n \leq \sigma_+$, $[u]_n(\sigma_+ - p_n) = 0$

b) $|p_t| \leq -\tan\varphi p_n + \tau_b\kappa(-p_n)$; $\{|p_t| + \tan\varphi p_n - \tau_b\kappa(-p_n)\}|[u]_t| = 0$

c) if $|p_t| \geq -\tan\varphi p_n + \tau_b\kappa(-p_n)$ then $|p_t| = [-\tan\varphi p_n + \tau_b\kappa(-p_n)]sign(p_t)$.

The above mentioned problem is regarded as two-dimensional in this paper.

3 Variational formulation

Denote

$$H \equiv \{\boldsymbol{u} \in V;\ [u]_n < 0 \text{ a.e. on } \partial\Omega\}$$

The set H is a cone of admissible nodal displacements with respect to the essential boundary and contact conditions, V are displacements from the space of continuous functions.

Suppose that we disconnect both bodies under consideration, but keep the stress and deformation state in them "frozen". Using the condition a) to c) in *Problem A*, variational principle then leads to the following definition: Find minimum $\mathbf{u} = \{u_n, u_t\}$ and maximum $\mathbf{p} = \{p_n, p_t\}$ for the functional of entire energy E_{ent}:

$$\mathrm{E}_{\text{ent}} = \frac{1}{2}\Big(\int_{\Omega^f}\boldsymbol{\sigma}^{\mathrm{T}}\boldsymbol{\varepsilon}\,\mathrm{d}\Omega^f + \int_{\Omega^m}\boldsymbol{\sigma}^{\mathrm{T}}\boldsymbol{\varepsilon}\,\mathrm{d}\Omega^m\Big) - \int_{\partial\Omega}\boldsymbol{p}^{\mathrm{T}}\boldsymbol{u}\,\mathrm{d}\partial\Omega +$$
$$+ \int_{\Gamma_C}(\{|p_t| + \tan\varphi p_n - \tau_b\kappa(-p_n)\}|[\mathrm{u}]_{\mathrm{t}}|\,\mathrm{d}\Gamma + \int_{\Gamma_C}(p_n - \sigma_+)[u]_n\,\mathrm{d}\Gamma \quad (1)$$

where superscript T means transposition.

Let us write the total energy J of both bodies assuming them separately:

$$J(\boldsymbol{u},\boldsymbol{p}) = \Pi(\boldsymbol{u}) - I(\boldsymbol{u},\boldsymbol{p}), \quad (2)$$

where

$$\Pi(\boldsymbol{u}) = \frac{1}{2}a(\boldsymbol{u},\boldsymbol{u}) - \int_{\partial\Omega}(\boldsymbol{p})^{\mathrm{T}}\boldsymbol{u}\,\mathrm{d}\partial\Omega,$$
$$I(\boldsymbol{p},\boldsymbol{u}) = \int_{\Gamma_C}\{|p_t| + \tan\varphi p_n - \tau_b\kappa(-p_n)\}|[\mathrm{u}]_t|\,\mathrm{d}\Gamma + \int_{\Gamma_C}(p_n - \sigma_+)[u]_n\,\mathrm{d}\Gamma,$$
$$a(\boldsymbol{u},\boldsymbol{u}) = \int_{\Omega^f}\boldsymbol{\sigma}^{\mathrm{T}}\boldsymbol{\varepsilon}\,\mathrm{d}\Omega^f + \int_{\Omega^m}\boldsymbol{\sigma}^{\mathrm{T}}\boldsymbol{\varepsilon}\,\mathrm{d}\Omega^m \quad (3)$$

Latter variational formulation leads to Uzawa's algorithm, which is described, as above said, in [6], for example. It appears that Uzawa's algorithm is a powerful tool, but in the case of closed crack somehow unstable. This is why we introduce a new constraint.

From the above considerations it follows that the contact representation by spring stiffnesses k_n (normal direction) and k_t (tangential direction) can be characterized (normal and tangential directions are taken with respect to the surface of the fiber) and penalty like formulation can be obtained. If the spring stiffness is high, the bond of fibers to matrix is firm; if some of contact conditions are violated, the stiffness lowers its value. The impact of this formulation ensures always the solution, if the fiber is not disconnected from the matrix in all nodal points. In this sense the interfacial relations can be written as:

$$p_n = k_n[u]_n, \quad p_t = k_t[u]_t \tag{4}$$

and the interfacial energy I in (3) is written as:

$$I(\mathbf{p},\mathbf{u}) = \int_{\Gamma_C} \left\{ \begin{array}{l} (k_n[u]_n^2 + k_t[u]_t^2) + \\ + \frac{1}{2}\left([u]_n(k_n \tan\phi \,|[u]_t| - 2\sigma_+) + |[u]_t|\,(k_n \tan\phi[u]_n - 2\tau_b\kappa(-p_n)\right) \end{array} \right\} \mathrm{d}\Gamma_C \tag{5}$$

From the above expression it is seen that the spring stiffnesses k_n and k_t play the role of penalty coefficients. If their values are high, the bond is ensured. If certain interfacial condition is violated, the values are dropped according to the prescription of the *Problem A* and the last terms in (5) represent the peak (strength) energies for both normal and tangential directions.

4 Homogenization

Localization and homogenization in general are concisely described by Suquet, [7]. Recall some basic assumptions that we use later on.

First, we denote quantities used in what follows. Two different scales will naturally be introduced. The macroscopic scale, where the homogeneous law (involving the overall material properties) is sought, will be described in the coordinate system $0x_1x_2$, points are identified by $\mathbf{x} \equiv (x_1, x_2)$, and the microscopic scale – heterogeneous – is characterized by the system of coordinates $0y_1y_2$, i.e. the points are identified as $\mathbf{y} \equiv (y_1, y_2)$. Material properties of the composite are generally randomly distributed, but locally – in the microscopic scale – is assumed they are periodic, so that a representative volume element (RVE) may be cut out from the structure, and the appropriate boundary conditions can be introduced in this element. In what follows we rather use a unit cell, or in example even its one quarter (because of symmetry mostly used for simplicity), see Fig. 1.

Let us distinguish the considered mechanical quantities depending on the macroscopic or microscopic scale, in the following manner:
Displacements in the macroscopic level will be denoted as $\mathbf{U} \equiv (U_1, U_2)$ while the same in the microscopic level as $\mathbf{u} \equiv (u_1, u_2)$. Moreover, in macroscopic level strains are denoted as $\mathbf{E} \equiv E_{ij}, i, j = 1,2$, and stresses as $\mathbf{S} \equiv S_{ij}, i, j = 1,2$. In microscopic level stresses are denoted as $\boldsymbol{\sigma} \equiv \sigma_{ij}, i, j, = 1,2$, and strains as $\boldsymbol{\varepsilon} \equiv \varepsilon_{ij}, i, j, = 1,2$. Assuming first the linear behavior of the composite aggregate, let us define the corresponding quantities by

$$S_{ij} < \sigma_{ij} >= \frac{1}{\text{meas}\Omega} \int_{\Omega} \sigma_{ij} \, \mathrm{d}\Omega(\boldsymbol{y}), \quad E_{ij} < \varepsilon_{ij} >= \frac{1}{\text{meas}\Omega} \int_{\Omega} \varepsilon_{ij} \, \mathrm{d}\Omega(\boldsymbol{y}), \tag{6}$$

where <.> stands for the average, meas. Ω is the measure of the domain $\Omega = \Omega^f \cup \Omega^m$, i.e. Ω describes the entire volume of the unit cell.

The real microscopic strain field $\varepsilon_{ij}(\boldsymbol{y})$ will be split into two parts as:

$$\varepsilon_{ij}(\boldsymbol{y}) = \boldsymbol{E}_{ij} + \varepsilon_{ij}(\boldsymbol{u}'(\boldsymbol{y})) = E_{ij} + \varepsilon'_{ij}(\boldsymbol{y}) \tag{7}$$

where $\varepsilon'_{ij}(\boldsymbol{y})$ is the fluctuating term.

Since the locally periodic assumption is accepted, the following properties can be adopted on the boundary:
1) the fluctuating term $\boldsymbol{u}'(\boldsymbol{y})$ possesses the same values on the opposite sides of the unit cell,
2) on the opposite sides of the boundary of the unit cell, the real tractions $p_i(\boldsymbol{y})$ have the opposite directions and the same values.

To start with, no debond is assumed. Then, using Green's theorem yields the average strain as:

$$E_{ij} =< \varepsilon_{ij}(\boldsymbol{u}) >= \frac{1}{\text{meas } \Omega} \int_{\Omega} \varepsilon_{ij}(\boldsymbol{u}) \, \mathrm{d}\Omega = \frac{1}{2 \text{ meas } \Omega} \int_{\partial\Omega} (u_i n_j + u_j n_i) \, \mathrm{d}\partial\Omega \tag{8}$$

Since both the matrix and the fibers are assumed to behave physically linearly and are bonded, the localization problem leads to:

$$\sigma_{ij}(\boldsymbol{y}) = \boldsymbol{L}_{ijkl}(\boldsymbol{y}) \varepsilon_{kl}(\boldsymbol{u}(\boldsymbol{y})), \qquad \frac{\partial \sigma_{ij}}{\partial y_j} = 0, \qquad E_{ij} =< \varepsilon_{ij}(\boldsymbol{u}) >, \tag{9}$$

and (4) is to be taken in the sense of distributions.

The loading is now split into unit impulses of overall strain components E_{ij}. In this way we get "the mechanical concentration factor tensors" A_{ijkl} defined as:

$$\varepsilon_{ij}(\boldsymbol{u}(\boldsymbol{y})) = A_{ijkl}(\boldsymbol{y})\boldsymbol{E}_{\boldsymbol{kl}} \tag{10}$$

The homogenization immediately follows from the above described localization, see, e.g. Suquet [7].

The opening (flaw, gap) is denoted by Ω^* and its boundary as s Γ^*, see Fig. 1. The domain Ω now consists of the disjoint sum $\Omega^f \cup \Omega^m \cup \Omega^*$. For the sake of simplicity, let $\Omega^M = \Omega^f \cup \Omega^m = \Omega - \Omega^*$.

In the case of stress, the modification is immediate: since the stress tensor transforms to traction vector on the boundary, it vanishes on the boundary of Ω^* and we can extend continuously the stress tensor by zero in the interior of the defect. We denote by $\bar{\sigma}$ this extension to obtain that:

$$S_{ij} = \frac{1}{\text{meas}\,\Omega} \int_{\Omega} \bar{\sigma}_{ij}\, \mathrm{d}\Omega = \frac{1}{\text{meas}\,\Omega} \int_{\Omega-\Omega^*} \sigma_{ij}\, \mathrm{d}\Omega \tag{11}$$

More attention should be paid to the treatment with strains, since the microscopic displacements do not vanish on the boundary of Ω^*. We will admit that it can in the regular manner be continuously extended to the inside of Ω^*. Denoting this extension by $\bar{\boldsymbol{u}}$, and using Green's theorem, we obtain:

$$E_{ij} = \frac{1}{\text{meas}\,\Omega} \int_{\Omega} \varepsilon_{ij}(\bar{\boldsymbol{u}})\, \mathrm{d}\Omega = \frac{1}{\text{meas}\,\Omega} \left\{ \int_{\Omega^M} \varepsilon_{ij}(\bar{\boldsymbol{u}})\, \mathrm{d}\Omega + \frac{1}{2} \int_{\Gamma^*} (u_i n_j + u_j n_i)\, \mathrm{d}\Gamma^* \right\}. \tag{12}$$

In order to fulfill the condition of successive application of components of the average strain E_{ij}, Green's theorem can be applied to the surface integral, and instead of (12) we write:

$$E_{ij} = \frac{1}{2\,\text{meas}\,\Omega} \int_{\partial\Omega} (u_i n_j + u_j n_i)\, \mathrm{d}\partial\Omega \tag{13}$$

Generally, the procedure for the localization and homogenization is the same for debonding media as that in the elastic case. The overall stresses are computed only over Ω^M, so that as naturally expected the overall stiffness decreases.

5 Boundary element solution

From the above paragraph we can conclude that the problem is linear in each domain Ω^f and Ω^m. The linearity "fails" along the contact surface Γ_C where the condition of continuity of displacements is no more necessarily valid on entire Γ_C and only the balance condition holds.

The solution of this problem after discretization of the boundaries of both domains into the element-wise linear distribution of both displacements and tractions leads to the solution of two linear algebraic systems on particular phases given by square matrices $\boldsymbol{K}^f$ and $\boldsymbol{K}^m$:

$$\begin{bmatrix} \boldsymbol{K}_{11}^f & \boldsymbol{K}_{12}^f \\ \boldsymbol{K}_{21}^f & \boldsymbol{K}_{22}^f \end{bmatrix} \begin{pmatrix} \boldsymbol{u}_f^{\text{on}} \\ \boldsymbol{u}_f^{\text{out}} \end{pmatrix} = \begin{pmatrix} \boldsymbol{p}_f^{\text{on}} \\ \boldsymbol{p}_f^{\text{out}} \end{pmatrix}, \qquad \begin{bmatrix} \boldsymbol{K}_{11}^m & \boldsymbol{K}_{12}^m \\ \boldsymbol{K}_{21}^m & \boldsymbol{K}_{22}^m \end{bmatrix} \begin{pmatrix} \boldsymbol{u}_m^{\text{on}} \\ \boldsymbol{u}_m^{\text{out}} \end{pmatrix} = \begin{pmatrix} \boldsymbol{p}_m^{\text{on}} \\ \boldsymbol{p}_m^{\text{out}} \end{pmatrix}, \tag{14}$$

where superscript on denotes quantities on the contact and the superscript out means out of contact boundary. Denoting diagonal matrix $\mathbf{k}$ = diag$\{ k_n^1, k_t^1, k_n^2, \ldots, k_t^s \}$, where s is the number of nodal points on the contact boundary. Then using (4) the equations (14) can be rewritten as:

$$\begin{bmatrix} \boldsymbol{K}_{22}^f & \boldsymbol{0} & \boldsymbol{K}_{21}^f & \boldsymbol{0} \\ \boldsymbol{0} & \boldsymbol{K}_{22}^m & \boldsymbol{0} & \boldsymbol{K}_{21}^m \\ \boldsymbol{K}_{12}^f & \boldsymbol{0} & \boldsymbol{K}_{11}^f + \boldsymbol{k} & -\boldsymbol{k} \\ \boldsymbol{0} & \boldsymbol{K}_{12}^m & -\boldsymbol{k} & \boldsymbol{K}_{11}^m + \boldsymbol{k} \end{bmatrix} \begin{pmatrix} \boldsymbol{u}_f^{\text{out}} \\ \boldsymbol{u}_m^{\text{out}} \\ \boldsymbol{u}_f^{\text{on}} \\ \boldsymbol{u}_m^{\text{on}} \end{pmatrix} = \begin{pmatrix} \boldsymbol{p}_f^{\text{out}} \\ \boldsymbol{p}_m^{\text{out}} \\ \boldsymbol{0} \\ \boldsymbol{0} \end{pmatrix} \tag{15}$$

where for given k **the** upper simultaneous system is uniquely solvable. The starting iteration step begins with high stiffnesses $\boldsymbol{k}$ (it is not recommended to select very high numbers as then the system tends to instable system. Fulfillment of conditions in *Problem A* has to be checked in every iteration step and according to the result the spring stiffnesses have to be improved. New diagonal matrix k ***should*** be substituted in (15) and the loop is repeated till almost no change of unknowns is observed (in the sense of certain error prescribed). Then the results are substituted into the approach described in section 4.

6 Numerical results

Several examples were tested by the BEM. In all examples the fiber possess the following material properties E^f = 772 GPa, ν^f = 0.25 while the epoxy matrix is considered, for which E^m = 96.5 GPa, ν^m = 0.3. On the contact the coefficient of Coulomb friction $\tan\varphi$ = 0.22 and the shear bond strength τ_b = 1.5 MPa. The unit cell is formed as a square 1 mm x 1 mm, volume fraction of the fibers is 0.4. The unit cell is loaded by horizontal displacement applied on the vertical boundary at the right hand side, while the left vertical boundary together with both horizontal boundaries are free of horizontal movement and the vertical displacement is not permitted.

The displacement u_x of magnitude 0.01 mm is applied in the x direction in the first case, in the second case the displacement u_x = 0.03 mm is imposed. In the BEM the linear distribution of both displacements and tractions on the boundary involving Γ_C are used.

In Fig. 2, the distribution of contact forces is depicted, and in Fig. 3, the course of relative displacements on the contact is shown. The dashed line describes the first case, the thin line belongs to the second case and the thick line denotes the solution involving temperature influence (eigenstrain tensor with the only non-zero component in normal x-direction expressing the change of temperature by 30^0 C).

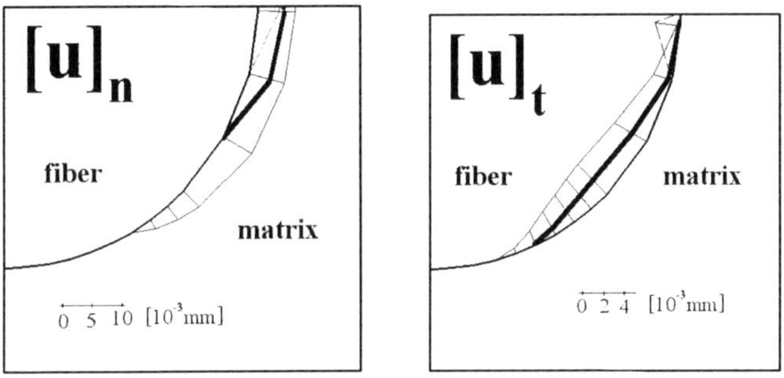

Figure 2: Distribution of relative displacements on the contact Γ_C.

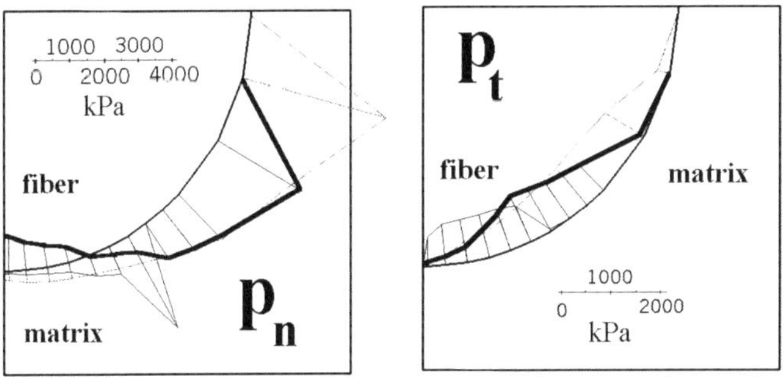

Figure 3: Distribution of contact tractions on Γ_C.

7 Conclusions

In this paper penalty formulation of behavior of debonding fibers out of matrix is described. Boundary element method is used as the numerical tool for solving homogenization on a unit cell. From the examples involved in the study it follows that the penalty formulation seems to be more appropriate then the classical Uzawa's algorithm used by the authors before. The reason consists in the fact that the iterations solving the problem are more stable in the case of improvement the situation on the interfacial boundary by penalty than in that case of solving reactions on interface, which is the case of Uzawa. Three examples are considered: two different displacements at one outer boundary are applied and a change of temperature by given number is considered. A comparison with certain papers provides very good agreement in results, but the algorithm seems to be much more stable and faster in comparison with the other methods.

Acknowledgement

Financial support of this research was provided by Grant agency of the Czech Republic, No. 103/04/1178.

References

[1] Cartié, D.D.R., Cox, B.N. & Fleck, N.A. Mechanisms of crack bridging by composite and metallic rods. Composites Part A: Applied Science and Manufacturing. Volume 35, Issue 11, 1325-1336

[2] Achenbach, J.D. & Zhu, H. Effect of interfacial zone on mechanical behaviour and failure of fiber-reinforced composites. *J. Mech. Phys. Solids* **37**, 1989, 381–393

[3] Karihaloo, B.L. & Vishvanathan, K. Elastic field of a partially debonded elliptical inhomogeneity in an elastic matrix. *ASME J. Appl. Mech.*, 1985, **52**, 835–840.

[4] Tvergaard, V. Model studies of fibre breakage and debonding in a metal reinforced by short fibers. *J. Mech. Phys. Solids* **41**, 1993, 1309–1326

[5] Procházka, P. Homogenization of linear and of debonding composites using the BEM. Engineering Analysis with Boundary Elements Volume 25, Issue 9, 2001, 753-769

[6] Procházka, P. & Sejnoha, M. Development of debond region of lag model. Computers & Structures, Volume 55, Issue 2, 1995, 249-260

[7] Suquet, P. M. Homogenization Techniques for Composite Media. Lecture Notes in Physics 272, eds. E. Sanchez-Palencia and A. Zaoui, 1985, Springer-Verlag Berlin, 194-278.

Author Index

WIT PRESS

Trefftz and Collocation Methods

A. H-D. CHENG, University of Mississippi, USA, Z-C. LI, National Sun Yat-sen University, Taiwan, T-T. LU, National Center for Theoretical Science, Taiwan, H-Y. HU, National Tsing Hua University, Taiwan

This book covers a class of numerical methods that are generally referred to as "Collocation Methods". Different from the Finite Element and the Finite Difference Method, the discretization and approximation of the collocation method is based on a set of unstructured points in space. This "meshless" feature is attractive because it eliminates the bookkeeping requirements of the "element" based methods. This text discusses several types of collocation methods including the radial basis function method, the Trefftz method, the Schwartz alternating method, and the coupled collocation and finite element method. Governing equations investigated include Laplace, Poisson, Helmholtz and bi-harmonic equations. Regular boundary value problems, boundary value problems with singularity, and eigenvalue problems are also examined. Rigorous mathematical proofs are contained in these chapters, and many numerical experiments are also provided to support the algorithms and to verify the theory. A tutorial on the applications of these methods is also provided.

ISBN: 1-84564-153-1 2006 apx 500pp
apx £170.00/US$295.00/€255.00

Computer Aided Design of Wire Structure in the Frequency and Time Domain

D. POLJAK, University of Split, Croatia

As an introduction to the integral equation analysis of wire structures, this book and enclosed software packages contain the user friendly version of the boundary element software for modelling the straight thin wire arrays in both frequency and time domain. This package is designed as a step by step guide for postgraduate students, researchers and also practising engineers to learn CAD of wire antennas immersed in inhomogeneous media. Some electromagnetic compatibility (EMC) applications can be also handled using this package. The package contains detailed description of antenna theory, integral equation modelling and full manuals for software packages.

Series: Advances in Electrical and Electronic Engineering Vol 7

ISBN: 1-85312-884-8 2006

apx 250pp+CD-ROM
apx £119.00/US$215.00/€178.50

WIT eLibrary

Home of the Transactions of the Wessex Institute, the WIT electronic-library provides the international scientific community with immediate and permanent access to individual papers presented at WIT conferences. Visitors to the WIT eLibrary can freely browse and search abstracts of all papers in the collection before progressing to download their full text.

Visit the WIT eLibrary at
http://library.witpress.com

WIT PRESS

Viscous Incompressible Flow

For Low Reynolds Numbers

M. KOHR and I. POP, Babes-Bolyai University, Cluj-Napoca, Romania

This book presents the fundamental mathematical theory of, and reviews state-of-the-art advances in, low Reynolds number viscous incompressible flow. The authors devote much of the text to the development of boundary integral methods for slow viscous flow pointing out new and important results. Problems are proposed throughout, while every chapter contains a large list of references.

A valuable contribution to the field, the book is designed for research mathematicians in pure and applied mathematics and graduate students in viscous fluid mechanics.

Contents: Introduction; Fundamentals of Low Reynolds Number Viscous Incompressible Flow; The Singularity Method for Low Reynolds Number Viscous Incompressible Flows; The Theory of Hydrodynamic Potentials with Application to Low Reynolds Number Viscous Incompressible Flows; Boundary Integral Methods for Steady and Unsteady Stokes Flows; Boundary Integral Formulations for Linearized Viscous Flows in the Presence of Interfaces; List of Symbols; Index.

Series: Advances in Boundary Elements, Vol 16

ISBN: 1-85312-991-7 2004 448pp
£148.00/US$237.00/€222.00

Advanced Boundary Elements for Heat Transfer

M.-T. IBANEZ, Wessex Institute of Technology, UK and H. POWER, University of Nottingham, UK

In this book the authors present an efficient Boundary Element Method scheme for the numerical solution of two-dimensional heat transfer problems.

Partial Contents: Integral Representation Formula for Heat Transfer; Non-History-Dependent Convolution Scheme; Recursive Reinitialization Scheme.

Series: Topics in Engineering, Vol 42

ISBN: 1-85312-898-8 2002 144pp
£65.00/US$99.00/€97.50

Boundary Element Methods for Heat Conduction

With Applications in Non-Homogeneous Media

E. DIVO and A.J. KASSAB, University of Central Florida, USA

A novel contribution to integral equation methods, this book provides the first formulation of a boundary-only integral equation for field problems governed by variable coefficient partial differential equations. The authors concentrate on the heat conduction equation, but the method they propose is general and applicable to a variety of engineering field problems.

Series: Topics in Engineering, Vol 44

ISBN: 1-85312-771-X 2003
264pp + CD-ROM
£89.00/US$138.00/€133.50

WIT PRESS

Boundary Elements XXVI

Editor: C.A. BREBBIA, Wessex Institute of Technology, UK

Featuring the results of state-of-the-art research from many countries, this book contains papers from the Twenty-Sixth World Conference on Boundary Elements and Other Mesh Reduction Methods.
Over 40 contributions are included and these cover specific topics within areas such as: Advanced Formulations; Advances in DRM and Radial Basis Functions; Inverse Problems; Advances in Structural Analysis; Fracture and Damage Mechanics; Electrical and Electromagnetic Problems; Fluid and Heat Transfer Problems; and Wave Propagation.
WIT Transactions on Modelling and Simulation Volume 37

ISBN: 1-85312-708-6 2004 488pp
£172.00/US$275.00/€258.00

Boundary Elements XXV

Editors: C.A.BREBBIA, Wessex Institute of Technology, UK, and D. POLJAK and V. ROJE, University of Split, Croatia

An invaluable aid to understanding the BEM and an excellent source of recent ideas and applications, this book includes most of the papers presented at the Twenty-Fifth International Conference on Boundary Element Methods.
WIT Transactions on Modelling and Simulation Volume 35

ISBN: 1-85312-980-1 2003 368pp
£119.00/US$189.00/€178.50

Find us at
http://www.witpress.com

Save 10% when you order from our encrypted ordering service on the web using your credit card.

Boundary Elements XXIV

Incorporating Meshless Solutions

Editors: C.A.BREBBIA, Wessex Institute of Technology, UK, A. TADEU, University of Coimbra, Portugal and V. POPOV, Wessex Institute of Technology, UK

Contributions from the twenty-fourth conference on this topic.
WIT Transactions on Modelling and Simulation Volume 32

ISBN: 1-85312-914-3 2002 776pp
£239.00/US$369.00/€358.50

Transformation of Domain Effects to the Boundary

Editors: Y.F. RASHED and C.A. BREBBIA, Wessex Institute of Technology, UK

The transformation of domain integrals to the boundary is one of the most challenging and important parts of boundary element research.
This book presents existing methods and new developments.
Partial Contents: On the Treatment of Domain Integrals in BEM; The Multiple-Reciprocity Method for Elastic Problems with Arbitrary Body Force; Generalized Body Forces in Multi-Field Problems with Material Anisotropy; On the Convergence of the Dual Reciprocity Method for Poisson's Equation.
Series: Advances in Boundary Elements, Vol 14

ISBN: 1-85312-896-1 2003 264pp
£85.00/US$136.00/€127.50

WIT PRESS

Singular Integrals in Boundary Element Methods

Editors: V. SLADEK and J. SLADEK, Slovak Academy of Sciences, Slovak Republic

"...a good source...for...researchers investigating advanced BIE formulations."
AIAA JOURNAL

Series: Advances in Boundary Elements, Vol 3

ISBN: 1-85312-533-4 1998 448pp
£125.00/US$195.00/€187.50

Underlying Principles of the Boundary Element Method

D. CARTWRIGHT, Bucknell University, USA

"...very well written...should be purchased by teachers, undergraduate and graduate students, researchers who would like to start working in the field, and certainly by libraries."
APPLIED MECHANICS REVIEWS

Providing a unified introduction to the underlying ideas of the Boundary Element Method (BEM), this book places emphasis on the principles of the method rather than its numerical implementation.
The author includes many worked examples to reinforce understanding while, to aid practice, trial problems are also given in each chapter.

ISBN: 1-85312-839-2 2001 296pp
£99.00/US$158.00/€148.50

Boundary Element Methods for Damage Tolerance Design of Aircraft Structures

N. SALGADO, Embraer S. A., Brazil

"...clearly written...very useful in promoting further research."
APPLIED MECHANICS REVIEWS

Series: Topics in Engineering, Vol 33

ISBN: 1-85312-532-6 1998 192pp
£69.00/US$110.00/€103.50

Boundary Integral Methods

Numerical and Mathematical Aspects

Editor: M.A. GOLBERG, Las Vegas, Nevada, USA

Covers some significant recent mathematical and computational developments in the BEM.
Series: Computational Engineering, Vol 1

ISBN: 1-85312-529-6 1998 392pp
£112.00/US$179.00/€168.00

All prices correct at time of going to press but subject to change.
WIT Press books are available through your bookseller or direct from the publisher.

WIT Press
Ashurst Lodge, Ashurst, Southampton, SO40 7AA, UK.
Tel: 44 (0) 238 029 3223
Fax: 44 (0) 238 029 2853
E-Mail: marketing@witpress.com